D1228561

Introduction to Wine Laboratory Practices and Procedures

Jean L. Jacobson

Introduction to Wine Laboratory Practices and Procedures

With 50 Figures, Including 2 Color Plates

 Springer

Library of Congress Control Number: 2005928372

ISBN 978-0-387-24377-1   ISBN 978-0-387-25120-2   Printed on acid-free paper.

The author has made every effort to ensure the accuracy of the information herein. However, appropriate information sources should be consulted, especially for new or unfamiliar procedures. It is the responsibility of every practitioner to evaluate the appropriateness of a particular opinion in the context of actual situations and with due considerations to new developments.

(SPI/SBA)

9 8 7 6 5 4 3 2 1

springer.com

To my sons, Christopher and Matthew, with profound love.
To my twin sister, Joyce, my guardian angel and soul mate.
To Kaarlo, for believing in me.
To sisters Julie and Jeralyn, for their unconditional love.
To Daddy, my hero.
To Mother, who never gave up.
To all of you who knew I could.
ILYS

Preface

In the beginning, for me, winemaking was a romanticized notion of putting grape juice into a barrel and allowing time to perform its magic as you sat on the veranda watching the sunset on a Tuscan landscape. For some small wineries, this notion might still ring true, but for the majority of wineries commercially producing quality wines, the reality of winemaking is far more complex.

The persistent evolution of the wine industry demands continual advancements in technology and education to sustain and promote quality winemaking. The sciences of viticulture, enology, and wine chemistry are becoming more intricate and sophisticated each year. Wine laboratories have become an integral part of the winemaking process, necessitating a knowledgeable staff possessing a multitude of skills. Science incorporates the tools that new-age winemakers are utilizing to produce some of the best wines ever made in this multibillion dollar trade.

A novice to enology and wine chemistry can find these subjects daunting and intimidating. Whether you are a home winemaker, a new winemaker, an enology student, or a beginning-to-intermediate laboratory technician, putting all the pieces together can take time. As a winemaker friend once told me, "winemaking is a moving target."

Introduction to Wine Laboratory Practices and Procedures was written for the multitude of people entering the wine industry and those that wish to learn about wine chemistry and enology. It is a guide to understanding basic enology, wine chemistry, safety, quality control, wine history, winemaking processes, bottling, and analytical procedures. Areas of study are applicable to the majority of wine laboratories in the world. The reader will gain a firm basic working knowledge of the wine laboratory and enology upon which to build a greater understanding of more advanced theory and technical information.

Make wine with honor.

Healdsburg, California   Jean L. Jacobson

Acknowledgments

With great appreciation, I would like to thank the people who laboriously read chapters and supplied feedback: Marcia Manix, Sharon Dougherty, Cathy Jacobson, Dr. Les Nouget, Joyce Nouget, Dr. Will Brogdon, Christine Brogdon, Jeralyn Heath, Joanne Foote-Lynch, and Galen McCorkle.

Thank you to Kendall-Jackson Wine Estates for your cooperation in this project. I would also like to thank the manufacturers, governmental agencies, and wineries that provided photographs and information used in this book.

Contents

Color Plate 1

HEALTH HAZARD

4 – DEADLY

3 – Extreme Danger

2 – Hazardous

1 – Slightly Hazardous

0 – Normal Material

FIRE HAZARD

Flash Points

4 - Below 73° F

3 – Below 100° F

2 – Above 100° F, Not Exceeding 200° F

1 – Above 200° F

0 – Will not burn

SPECIFIC HAZARD

OX – Oxidizer

ACID – Acid

ALK – Alkali

CORR – Corrosive

W̶ – Use NO WATER

△△△ - Radioactive

REACTIVITY

4 – May Detonate

3 – Shock and heat may detonate

2 – Violent chemical change

1 – Unstable if heated

0 - Stable

FIGURE 2.2. NFPA Hazard Classification (NFPA, 2001). This warning system is intended to be interpreted and applied only by properly trained individuals to identify fire, health, and stability hazards of chemicals. The user is referred to the recommended classifications of certain chemicals in the NFPA Guide to Hazardous Materials, which should be used as a guideline only. Whether the chemicals are classified by NFPA or not, those using the 704 system to classify chemicals does so at their own risk. (Reprinted with permission from NFPA 704-2001, *System of Identification of Hazardous Materials for Emergency Response*, Copyright © 2001, National Fire Protection Association, Quincy, MA.).

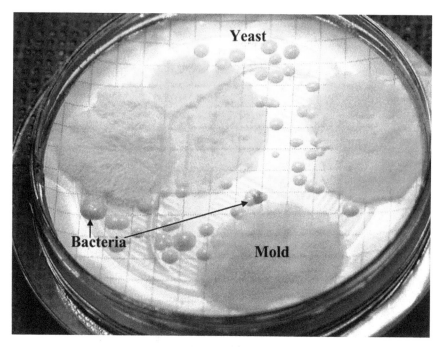

FIGURE 8.2. Contaminated plating results from postbottled wine. Mold colonies are spreading, irregular in shape, off-white growths that appear fuzzy; yeast form small, smooth-surfaced, uniform colonies, milky in color; bacterial colonies are irregular in shape, smooth-surfaced, and are various colors.

1

Chemistry to Remember

1.1 Introduction

Wine chemistry is a complex and interesting subject full of mysteries yet to be solved, a bit of a "new frontier." The first steps toward gaining an understanding of wine chemistry will require a basic comprehension of chemistry fundamentals. This chapter is not a magical shortcut to get around a traditional organic and inorganic chemistry education but as a helpful reminder to the readers who have had some past chemistry education and a learning incentive to those readers who have not.

1.2 Metric System

The metric system is fundamental to all sciences, and wine chemistry is no exception. Serious but common errors made in the laboratory are simply incorrect scientific notations. Misplaced decimal points, incorrect weights, measures, and volumes, and miscalculations contribute to the problem. Correct notation and detailed documentation are the cornerstones of science.

Table 1.1 lists the metric prefixes as they relate to the base units of liter, gram, and meter. Table 1.2 list the metric conversions most frequently used in the wine laboratory and wine cellar. Understanding the metric system and committing it to memory will be a tremendous help to you as you proceed toward your winemaking goal. The Internet has several on-line conversion websites that are very helpful.

1.3 Density and Specific Gravity

Density and specific gravity are physical properties of a substance. Density is defined as the mass per volume of a substance and specific gravity is the *ratio* of the density of a substance measured at a particular temperature to the density of a reference material, often water, also measured at a particular

TABLE 1.1. Metric prefixes.

---

mega = 1,000,000 = (10^6) = (number ÷ 1,000,000) = decimal to the **left 6 places**
   0.002345 megameter (Mm), megaliter (ML), or megagram (Mg)
kilo = 1000 = (10^3) = (number ÷ 1000) = decimal to the **left 3 places**
   2.345 kilometer (km), kiloliter (kL), or kilogram (kg)
hecto = 100 = (10^2) = (number ÷ 100) = decimal to the **left 2 places**
   23.45 hectometer (hm), hectoliter (hL), or hectogram (hg)
deka = 10 = (10) = (number ÷ 10) = decimal to the **left 1 place**
   234.5 dekameter (dam), dekaliter (daL), or dekagram (dag)
deci = 0.1 = (10^{-1}) = (10 × number) = decimal to the **right 1 place**
   23,450 decimeter (dm), deciliter (dL), or decigram (dg)
centi = 0.01 = (10^{-2}) = (100 × number) = decimal to the **right 2 places**
   234,500 centimeter (cm), centiliter (cL), or centigram (cg)
milli = 0.001 = (10^{-3}) = (1000 × number) = decimal to the **right 3 places**
   2,345,000 millimeter (mm), milliliter (mL), or milligram (mg)
micro = 0.000001 = (10^{-6}) = (1,000,000 × number) = decimal to the **right 6 places**
   2,345,000,000 micrometer (μm), microliter (μL), or microgram (μg)
nano = 0.000000001 = (10^{-9}) = (1,000,000,000 × number) = decimal to the **right 9 places**
   2,345,000,000,000 nanometer (nm), nanoliter (nL), or nanogram (ng)

Note: Prefixes are combined with basic units (meter, liter, gram). EXAMPLE: 2345 meter, liter, or gram expressed in respective prefixes.

temperature (Hägglund, 2004). For the purpose of this text, water will be the reference material when discussing specific gravity.

Density is expressed as grams per cubic centimeter (g/cc). The density of gold is 19.3 g/cc, the density of sulfur is 2.0 g/cc, the density of cork is 0.22 g/cc, the density of salt water is 1.025 g/cc, and the density of water is 1.0 g/cc (Hess, 1955). The difference in density allows cork, which is less dense than water, to float and sulfur, which is denser than water, to sink. The density of water at 20°C is 0.998204 g/cc.

According to Hägglund (2004), specific gravity is the *ratio* of the density of a substance measured at a particular temperature (t_1) to the density of water also measured at a particular temperature (t_2). Temperatures t_1 and t_2 are often the same, but in certain cases, they might be different. To simplify, references to specific gravity in this text will consider t_1 and t_2 to be the same (20°C); therefore, specific gravity compares the density of one substance to the density of water at the same temperature. The specific gravity of water at 20°C is the ratio of 20/20 or the density of water at 20°C (0.998204 g/cc) divided by the density of water at 20°C (0.998204 g/cc), which equals 1.000. Specific gravity has no units.

The specific gravity of wine is less than water. The more alcohol present in the wine, the lower the density and specific gravity, because alcohol has a lower density than water. The more solid and insoluble materials present in wine or juice; the higher the density and specific gravity, because these substances make the wine or juice denser than water.

TABLE 1.2. Metric conversions.

Metric to metric	Metric to english	English to metric	
	Capacity		
1 liter (L)			
= 1000 mL	= 1.0567 fluid qt	1 fluid quart (qt)	= 0.9463 liter
= 100 cL	= 0.2642 fluid gal	1 fluid gallon (gal)	= 3.7854 liter
= 0.1 daL	= 4.23 cup		
= 1×10^6 μL	= 33.814 fluid oz		
1 milliliter (mL)			
= 0.001 liter	= 0.0338 fluid oz	1 fluid ounce (oz)	= 29.573 mL
		1 teaspoon (t)	= 4.93 mL
		1 tablespoon (T)	= 14.79 mL
		1 cup (c)	= 236.588 mL
		1 gallon (gal)	= 3785.41 mL
1 microliter (μL)			
= 1×10^{-6} liter			
= 0.001 mL			
1 hectoliter (hL)	= 26.418 gal		
	Weight		
1 gram (g)			
= 10 dg	=15.43 gr	1 grain (gr)	= 0.0648 g
= 0.001 kg			= 64.80 mg
=1000 mg	= 0.0353 oz	1 ounce (oz)	= 28.35 g
= 1×10^6 μg	= 0.002205 lb	1 pound (lb)	= 453.59 g
1 milligram (mg)			
= 0.001 g	= 3.215×10^{-5} oz		
1 microgram (μg)			
= 1×10^{-6} g			
= 0.001 mg			
1 kilogram (kg)			
= 1000 g	= 2.2046 lb	1 pound (lb)	= 0.4536 kg
	= 0.0011 ton		= 0.0005 ton
	Linear Measure		
1 millimeter (mm)	= 0.03937 (in.)	1 inch (in.)	= 25.40 mm
1 centimeter (cm)	= 0.3937 in.	1 inch (in.)	= 2.240 cm
1 meter (m)	= 3.281 ft	1 foot (ft)	= 0.3048 m
	= 39.37 in.		= 30.48 cm
1 hectare	= 2.471 acre		

1.4 Liquid, Solids, and Gases

The physical properties of gases, liquids, and solids are important to understand when working with chemicals and performing analysis. This section will only touch on a few outstanding points.

1.4.1   Pressure and Temperature

Pressure is the force exerted on a unit surface, and gas pressure is measured via a barometer. A barometer measures the displacement of a column of mercury by air pressure. Standard pressure is 760 mm of mercury or 1 atmosphere (atm), specifically; the air pressure at sea level supports a column of mercury 760 mm high. Today's technology uses this basic measurement to calibrate an array of fine instruments that can measure the slightest change in barometric pressure and gas pressure electronically.

Temperature is measured by two different scales: degree Fahrenheit (°F) and degree Centigrade (°C). A thermometer is a column of mercury in a vacuum that expands and contracts depending on the thermometric activity of the substance surrounding it. These thermometric scales have been established using water at sea level. The Fahrenheit scale was developed in 1753 and has 180 degree points; the centigrade scale developed in 1801 uses 100 degree points (Latin for 100 is centi). The freezing point of water is 0°C or 32°F. The boiling point of water at sea level is 100°C or 212°F.

The centigrade scale is used in all sciences and throughout most of the world. Because both temperature scales are used in some form universally, it is important to know the conversion formula: °F = (9/5)°C + 32. Absolute zero (°A) is the complete absence of heat, the lowest possible theoretical temperature equivalent to −273.15°C or −459.67°F.

The contamination problems associated with mercury have led to improved temperature measurement methods. Mercury thermometers and barometers have been taken off the market for general use but are available for laboratory use. Chapter 2 will cover the handling of mercury and other hazardous material.

1.4.2   Liquids

A gas that is cooled or put under pressure, or both, will condense into a vapor as it moves into a liquid state. At room temperature, a liquid has the propensity to vaporize, or evaporate back into the atmosphere. The degree of evaporation is related to the vapor pressure of the liquid. Vapor pressure is the increase in pressure created by a liquid's vapor moving into the atmosphere directly above the liquid. The higher the vapor pressure of a liquid, the greater amount of vapor that will move into the atmosphere, or evaporate.

The boiling point of a liquid is the temperature at which the liquid's vapor pressure equals the pressure of the atmosphere about it. The freezing point is the temperature at which a liquid moves into a solid state.

The gravitational attraction of molecules to one another in a liquid (cohesion) exerts a force in all directions. At the surface of a liquid, the molecules are not surrounded, creating an imbalance of attractive force. The surface molecules are pulled back into the liquid as the molecules below exert a greater attractive force, creating an encasing film on the surface; this is known as surface tension.

Adhesion is the attraction between liquid molecules and the molecules of the liquid's container. Capillary action is the rise or fall of a liquid's surface when a small-diameter tube penetrates the liquid's surface. The degree of capillary action depends on the adhesion of liquid and tube.

1.4.3 Gases

Boyle's Law, Charles's Law, and Dalton's Law of Partial Pressure explain the physical properties and propensities of gases:

Boyle's Law: The volume occupied by a gas is inversely proportional to the pressure at a given temperature: $P_1 V_1 = P_2 V_2$.
Charles' Law: The volume occupied by a gas is directly proportional to the absolute temperature of the gas at a given pressure: $V_1/T_1 = V_2/T_2$.
Combined Gas Law: $P_1 V_1 \div T_1 = P_2 V_2 \div T_2$.
Dalton's Law of Partial Pressure: The total pressure of a gas mixture is the sum of the partial pressure of each gas: $P_{total} = P_1 + P_2 + P_3 + \dots P_n$.

Wineries utilize gases such as carbon dioxide (CO_2) and nitrogen (N_2) to displace oxygen in storage containers. Sulfur dioxide (SO_2) gas is used as a microbial agent and an antioxidant. A variety of wine laboratory analyses might require the use of compressed gas or liquid gases. Understanding the gas laws is important when working with any gas.

1.5 Chemistry Fundamentals

We have looked at some of the physical properties of substances; now we move on to chemical properties. Chemical properties describe the ability of one substance to form a new substance via chemical reaction and the circumstances of that reaction.

1.5.1 Matter

Matter is made up of mixtures and substances. Mixtures are made up of substances held together by physical means that retain their individual properties and can be separated back into those individual substances by physical change. Substances are comprised of compounds or pure elements. Compounds are chemically combined elements that form unique substances. Compounds can only be decomposed into their individual elements by chemical change. Elements are the basis of all matter—the simplest form.

Combination, replacement, double displacement, and decomposition are the principal types of chemical reaction. Two or more simple substances or compounds that join to form a more complex compound are called a combination reaction. Replacement reaction involves compounds that replace one element for another. Two compounds reacting and exchanging substances is a double-displacement reaction and a decomposition reaction is the breakdown of a complex compound into a simpler compound or its elements.

Some reactions require a catalyst to change the speed of the reaction. Reactions will have a change in energy by absorbing or emitting energy in a variety of forms. You will remember the famous Einstein equation $E=mc^2$, where E represents energy, m equals mass, and c is the velocity of light. Energy cannot be created or destroyed.

1.5.2   Structure

The atom is the smallest particle of an element that maintains the properties of the element. The atom is made up of electrically charged particles: protons, neutrons, and electrons. Protons possess a positive charge and neutrons have no electrical charge. Protons and neutrons are contained within the nucleus of the atom and exert a positive charge. Negatively charged electrons orbit the nucleus at different levels called shells, held in orbit by their attraction to the positive nucleus. There are never more than eight electrons in the outermost shell. Atoms are electrically neutral and have an equal number of electrons orbiting the nucleus as there are protons in the nucleus.

The atomic number of an element is the number of electrons orbiting the nucleus. Atomic weight is the *sum* of the relative weights of protons and neutrons. There are a few elements that contain atoms with different atomic weights; these are termed *isotopes*. The atomic weight of an element represents the average weight of the isotopes of the element.

Chemical reaction involves electrons in the outermost shell and, occasionally, electrons in the second outermost shell. Elements with a full outer shell (eight electrons) are termed *inert elements*; they have no chemistry and they combine with nothing.

1.5.3   Periodic Table

Studying the structure of elements will allow the reader to understand the formation of compounds, write formulas, solve equations, and anticipate the outcome of reactions. The periodic table (Fig. 1.1) is an arrangement of elements according to their atomic number. Figure 1.1 includes the number of electrons in the element's outermost shell, valence numbers, and number of shells.

1.5.4   Compounds

An element's valence number is the number of electrons involved in forming a compound through a shift of electronic charge. A valence number of an element can be positive or negative, but the total of all element valences in a compound must equal zero.

Compounds add, subtract, and share electrons to achieve a neutral state. Electrovalence is the transferring of electrons to form a compound and the creation of ions that are electrically charged particles with properties totally different from the atom from which they came. Covalence is sharing pairs of electrons via a single, double, or triple bond.

Radicals are clusters of elements held together by covalent bonds that behave as if they were a single element. There exists a surplus or deficit of electrons; thus, the radical is a complex ion and will combine as a unit with other ions to form electrovalent compounds. Table 1.3 lists the common radicals.

Molecular weight is the sum of all the atomic weights of the elements present in a compound. A mole is a quantity of the compound equal in weight to its molecular weight.

The ratio of the number of atoms of each element present in a compound is a chemical formula, and a chemical equation is an account of the chemical change taking place. Equations must balance with an equal number of each type of atom on either side of the equation. Coefficients are used to balance the number of atoms in an equation (Fig. 1.2).

A pure compound always contains the same elements in the same proportions by mass; this is the Law of Definite Proportions. Based on the Law of Conservation of Matter and Definite Proportions, we are able to mathematically formulate a chemical change and solve problems. Table 1.4 lists several useful equations to express concentration, composition, and proportions.

1.5.5 Solutions

Solutions are mixtures of the solute (the substance dissolved) and the solvent (the dissolving medium). The solute is dispersed into molecules or ions consistently throughout the solution. Any given unit volume of solution will have an equal amount of solute molecules or ions. The amount of solute per unit volume of solvent is the concentration of the solution and can be expressed as molarity, normality, molality, or as a percentage (Table 1.4).

TABLE 1.3. List of radicals.

Valence +1	Valence −1	Valence −2	Valence −3
Ammonium NH_4	Acetate $C_2H_3O_2$	Carbonate CO_3	Phosphate PO_4
	Bicarbonate HCO_3	Chromate CrO_4	
	Chlorate ClO_3	Dichromate Cr_2O_7	
	Hydroxide OH	Sulfite SO_3	
	Cyanide CN	Sulfate SO_4	
	Nitrite NO_2		
	Nitrate NO_3		
	Permanganate MnO_4		

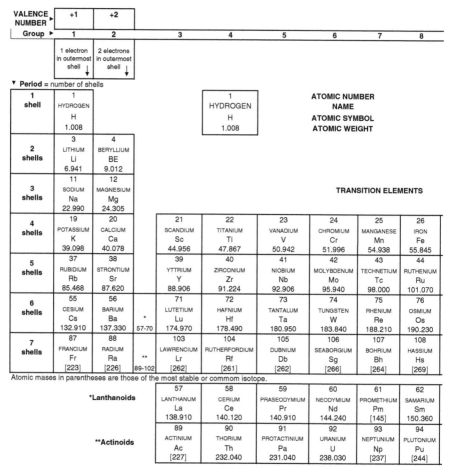

FIGURE 1.1. Periodic table depicting groups, periods, valence numbers, and the number of electrons in the outermost shell. Elements in italics are the more recently discovered elements.

Molarity is the number of moles of solute per liter of solution. Normality is the number of equivalents of solute per liter of solution. Molality is the number of moles of solute per 1000 g of solvent. The percentage composition is expressed by weight when solids are dissolved in liquids or by volume when liquids are dissolved in liquids or gases are dissolved in gases.

A standard solution is a solution with a precisely known concentration. Standard solutions can be diluted and the new concentration calculated using the equation in Table 1.4.

Solubility is the maximum amount of solute that can be dissolved in a given volume of solvent. Factors such as temperature and pressure will affect

				+3	± 4	−3	−2	−1	INERT
9	10	11	12	13	14	15	16	17	18
				3 electrons in outermost shell ↓	4 electrons in outermost shell ↓	5 electrons in outermost shell ↓	6 electrons in outermost shell ↓	7 electrons in outermost shell ↓	8 electrons in outermost shell ↓

PERIODIC TABLE

									2 HELIUM H 4.003
				5 BORON B 10.811	6 CARBON C 12.011	7 NITROGEN N 14.007	8 OXYGEN O 15.999	9 FLUORINE F 18.998	10 NEON Ne 20.180
				13 ALUMINUM Al 26.982	14 SILICON Si 28.086	15 PHOSPHORUS P 30.974	16 SULFUR S 32.065	17 CHLORINE Cl 35.453	18 ARGON Ar 39.948
27 COBALT Co 58.933	28 NICKEL Ni 58.693	29 COPPER Cu 63.546	30 ZINC Zn 65.390	31 GALLIUM Ga 69.723	32 GERMANIUM Ge 72.610	33 ARSENIC As 74.992	34 SELENIUM Se 78.960	35 BROMINE Br 79.904	36 KRYPTON Kr 83.800
45 RHODIUM Rh 102.910	46 PALLADIUM Pd 106.420	47 SILVER Ag 107.870	48 CADMIUM Cd 112.410	49 INDIUM In 114.820	50 TIN Sn 118.710	51 ANTIMONY Sb 121.760	52 TELLURIUM Te 127.600	53 IODINE I 126.900	54 XENON Xe 131.290
77 IRIDIUM Ir 192.220	78 PLATINUM Pt 195.080	79 GOLD Au 196.970	80 MERCURY Hg 200.590	81 THALLIUM Tl 204.380	82 LEAD Pb 207.200	83 BISMUTH Bi 208.980	84 POLONIUM Po [209]	85 ASTATINE At [210]	86 RADON Rn [222]
109 MEITNERIUM Mt [268]	110 DARMSTADTIUM Ds [271]	111 UNUNUNIUM Uuu [272]	112 UNUNBIUM Uub [277]	113	114 Uuq [296]	115	116 Uuh [298]	117	118 Uuo [?]

63 EUROPIUM Eu 151.960	64 GADOLINIUM Gd 157.250	65 TERBIUM Tb 158.930	66 DYSPROSIUM Dy 162.500	67 HOLMIUM Ho 164.930	68 ERBIUM Er 167.260	69 THULIUM Tm 168.930	70 YTTERBIUM Yb 173.040
95 AMERICIUM Am [243]	96 CURIUM Cm [247]	97 BERKELIUM Bk [247]	98 CALIFORNIUM Cf [251]	99 EINSTEINIUM Es [252]	100 FERMIUM Fm [257]	101 MENDELEVIUM Md [258]	102 NOBELIUM No [259]

FIGURE 1.1. (*Continued*)

$$Cu + O_2 = CuO$$
1 atom copper + 2 atoms oxygen = copper oxide
To balance the equation the addition of a coefficient is necessary
$$2\,Cu + O_2 = 2CuO$$
Reactants = Product
2 atoms copper + 2 atoms oxygen = 2 atoms copper + 2 atoms oxygen
Expressed as moles:
2 moles copper + 2 moles oxygen = 2 moles copper oxide

FIGURE 1.2. Balancing an equation.

TABLE 1.4. Useful calculations.

Percentage by weight of elements in a compound

$$= \frac{\text{Total wt. of element present}}{\text{Molecular wt. of compound}} = \% \text{ of element}$$

Weight of one substance in a compound

$$= \frac{\text{Actual wt. of one substance}}{\text{Its equation wt.}} = \frac{\text{Unknown actual wt.}}{\text{Its equation wt.}}$$

Equivalent weight of a compound

$$= \frac{\text{Molecular wt.}}{\text{Net positive valence}}$$

Number of equivalents of a compound

$$= \frac{\text{Actual wt. of compound}}{\text{Equivalent wt. of compound}}$$

Normality

$$= \frac{\text{No. of equivalents of solute}}{\text{Volume of solution in liters}}$$

Number of moles

$$= \frac{\text{Actual wt. of compound}}{\text{Atomic wt. of elements in compound (molecular wt.)}}$$

Molarity

$$= \frac{\text{Actual wt./Molecular wt.}}{\text{Volume (in liters)}}$$

Molality

$$= \frac{\text{Actual wt. of solute/1000 g of solvent}}{\text{Molecular wt. of solute}}$$

Dilution of a more concentrated solution

$$C_1 V_1 = C_2 V_2$$

where C_1 = concentration of initial solution, V_1 = volume of initial solution, C_2 = concentration of final solution, and V_2 = volume of final solution

---

Note: Weight (wt.) is expressed in grams.

the solubility of a substance. The degree that a substance will dissolve into a solvent is dependent on the polarity (amount of electrical activity) and structure of the molecules involved. Water is very polar and is used more as a solvent than any other compound. The more similar the substances, the more likely they will dissolve one another. When a solution can hold no more solute, it is *saturated* and, conversely, a solution that has the ability to hold more solute in solution is termed *unsaturated*.

A solute that dissolves by absorbing moisture from the air is a *desiccant*.

1.5.6 *Electrolyte Solutions*

An electrolyte solution is a solution that conducts an electric current. The electrolyte is the solute, which when dissolved in water, creates the electrically conductive solution. Both electrolyte and nonelectrolyte solutions can change the properties of solvents, but the electrolyte solutions do so to a greater extent. In solution, electrolytes separate into positive and negative ions and these ions are responsible for the conductivity of the solution (Fig. 1.3).

Electrolytes are broken down into three categories:

Acids produce hydrogen ions (H^+).
Bases produce hydroxide ions also known as hydroxyl ions (HO^-).
Salts produce other ions.

Strong electrolytes ionize 100%, whereas weak electrolytes ionize only to some extent. Strong and weak refer *only* to the ionization of electrolytes and do not refer to the concentration of a solution (water is a weak electrolyte):

Strong acids   Hydrochloric acid (HCl)
                Nitric acid (HNO_3)
                Sulfuric acid (H_2SO_4)
Strong bases   Potassium hydroxide (KOH)
                Sodium hydroxide (NaOH)
                Hydroxides of groups 1 and 2 on the periodic table
                (Fig. 1.1)
Ions in solution usually react with each other in three ways:

Forming a weak electrolyte
Forming an insoluble substance
Oxidation (reduction) of molecules

IONIZATION
Sodium Chloride
NaCl dissociates in water = Na^+ and Cl^- ions
Acetic Acid
$HC_2H_3O_2$ dissociates in water = H^+ and $C_2H_3O_2^-$ ions
NEUTRALIZATION
Hydrochloric Acid and Sodium Hydroxide in water:
HCl ionizes = H^+ and Cl^-
NaOH ionizes = Na^+ and OH^-
The ionic formula:
$(Na^+ OH^-) + (H^+ Cl^-) = H_2O + NaCl$

FIGURE 1.3. Electrolytes are solutes that when dissolved in water dissociate into ions creating a solution that conducts electricity. Neutralization is the reaction of an acid and base combined in the presence of water to form water and a salt.

An acid or base mixed with water will disassociate into their constituent ions. Mixing these acid and base solutions creates a neutralizing reaction forming water, heat, and a salt (Fig. 1.3). The degree of neutralization is dependent on the ionization ability of the acid and base (strong or weak) and the concentration used. The salts produced can be of a strong acid and base, a strong acid and a weak base, a weak acid and a strong base, or a weak acid and base. These salts might have ions that can react with the water the salt is dissolved in to form weak electrolytes. This reaction is called *hydrolysis*.

1.5.7   pH

Hydrogen ions in a water solution are responsible for the acid properties of that solution and hydroxide ions in a water solution are responsible for the base properties of that solution.

Water ionizes slightly, 0.00001% (Hess, 1955). The molar concentration of free H^+ in water is very small (1×10^{-7}). In dilute acid solutions, the amount of H^+ is also very small. The pH scale was developed as a tool to indicate very small quantities of free H^+ in a solution containing water.

The pH is defined as the logarithm of the *reciprocal* of the molar concentration of the hydrogen ion, [$pH = \log(1/H^+)$; more simply, $pH = -\log(H^+)$]. Hydroxide is expressed as $pOH = \log[1/(OH^-)]$ or, more simply, $pOH = -\log(OH^-)$.

In pure water, $(OH^-) = (H^+)$. Using the equations for pH and pOH, you will find that $pH + pOH = 14$ (Law of Mass Action). This tells us the pOH is 7 and the pH of pure water is 7. The calculation is

$$
\begin{aligned}
pH &= \log(1/1 \times 10^{-7}) \\
&= \log 1 \times 10^7 / \log 1 \\
&= \log 10^7 - \log 1 \\
&= 7 - 0 \\
&= 7.0
\end{aligned}
$$

Solutions with a pH less than 7.0 have *more* H^+ than water and exhibit acidic properties. Conversely, solutions with a pH greater than 7.0 have *less* H^+ than water and exhibit basic properties.

The pH scale has a range of 0.0–14.0. A slight change in a solution's pH indicates a much greater change in the concentration of free H^+. The pH of juice and wine are normally in the acidic range pH 3.0–4.0.

1.5.8   Oxidation: Reduction and Electrolysis

Oxidation–reduction reactions are commonly referred to as "redox" reactions. The reaction is one of gaining and losing electrons (i.e., a change in positive valence). The oxidation side of a reaction is the process of losing the electrons and increasing the valence number. The reduction side of the reaction is the process of gaining the electrons and reducing the valence number. The number of electrons lost and gained in the reaction must be equal.

Oxidizing agents gain electrons via reduction and the reducing agents lose electrons via oxidation. It is just a bit confusing, but remember that we are looking at the flow of electrons, or electric current, from one substance to another. Passing an electric current through an electrolyte solution will create a redox type reaction *at the electrodes*. This breakdown of the electrolyte by an external electric current is called *electrolysis*.

1.5.9   Halogens

Elements that tend to form salts are called *halogens*. The four primary non-metal halogens are fluorine, chlorine, bromine, and iodine. Fluorine is the most active halogen and forms very stable compounds. Going down the list, iodine is the least active and forms less stable compounds. All halogens are poisonous.

Hydrogen halides are halogens combined covalently with hydrogen. Hydrogen halides are very soluble in water and disassociate into ions similar to electrolytes. They will fume in moist air as they combine with the moisture and then condense, forming drops of acid solution. When hydrogen halides are in a water solution, they are known as hydrohalic acids. The acids produced are hydrofluoric acid, hydrobromic acid, hydrochloric acid, and hydriodic acid.

1.5.10   Sulfur

Sulfur is used throughout wineries and vineyards as an antimicrobial agent and antioxidant. Sulfur compounds are used in laboratories as acids and oxidizing agents in chemical assays. The significant roll of sulfur in winemaking will be discussed in upcoming chapters.

Sulfur combined with heat and a metal forms a *sulfide*. Sulfides react with HCl to form hydrogen sulfide (H_2S).

Sulfur dioxide dissolved in water forms sulfurous acid (H_2SO_3, hydrogen *sulfite*). Salts of sulfurous acid are sulfites. Sodium sulfite (Na_2SO_3) plus sulfur equals sodium thiosulfate ($Na_2S_2O_3$), a common reducing agent used in the wine laboratory.

Sulfuric acid (H_2SO_4, hydrogen *sulfate*) when mixed with water liberates a tremendous amount of heat, creating a dangerous scenario of boiling water and spattering. Dilution of concentrated sulfuric acid requires small amounts of the acid added to the water while stirring constantly to dissipate the heat (Chapter 2).

1.6   Organic Chemistry

Organic chemistry is the study of compounds containing hydrogen and carbon bonds (hydrocarbons). There are approximately 8 million compounds known to date, with 300,000 new compounds found each year (Petrik Laboratories Inc., 2003).

All living things fall under the realm of organic chemistry; we are truly "carbon-based units." Carbon has a valence of 4 and forms covalent bonds with atoms of hydrogen, nitrogen, oxygen, sulfur, halogens, and additional carbon atoms.

Carbon and hydrogen are readily available in our world, thus allowing simple compounds to build on each other to create long carbon chains or rings of carbon atoms with a variety of single, double, and triple bonds. Single bonds are the least active and triple bonds are more volatile. Some compounds contain literally thousand of carbon atoms.

Compounds are divided into two classes: aromatic and aliphatic. Aromatic compounds contain rings of carbon atoms with alternating double bonds (benzene ring) and aliphatic compounds have an array of carbon chains in different configurations. The longest unbroken chain of carbon atoms in a compound is designated with a root name to assist in identification (Table 1.5).

Organic compounds are identified by six different methods. I will use ethanoic acid as an example:

1. **Name** is based on accepted nomenclature.
    Example: CH_3CO_2H is called ethanoic acid.
2. **Molecular formula** is the actual number of atoms of each element.
    Example: $C_2H_4O_2$
3. **Empirical formula** is the simplest whole-number ratio of atoms of each element.
    Example: CH_2O
4. **Structural formula** shows how the atoms are arranged.
    Example: CH_3CO_2

TABLE 1.5. Root name of a compound.

Carbon atoms	Root term
1	meth-
2	eth-
3	prop-
4	but-
5	pent-
6	hex-
7	hept-
8	oct-
9	non-
10	dec-
11	undec-
12	dodec-

5. **Graphical formula** demonstrates how the atoms are spaced and the type of covalent bond.
    Example:

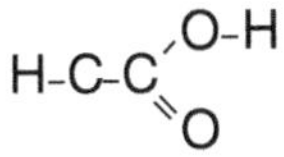

$$H\text{-}C\text{-}C\overset{O\text{-}H}{\underset{O}{}}$$

6. **Skeletal formula** is an abbreviated form of the graphical formula with an understanding that the lines have a carbon atom at each end.
    Example:

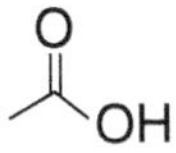

$$\overset{O}{\underset{OH}{}}$$

Aromatic hydrocarbons contain benzene rings that bond to each other or other functional groups. The names of compounds containing benzene rings give the position on the benzene ring containing the functional group.
    Example:

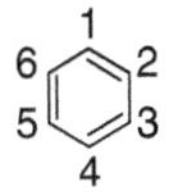

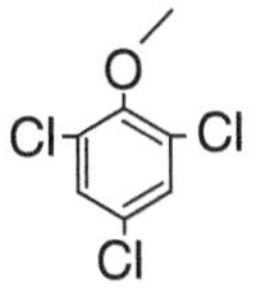

2,4,6-Trichloroanisole

Trillions of compounds have the same molecular formula but entirely different structural formulas; they are called *isomers*. Figure 1.4 illustrates the different molecular and graphical formulas of simple hydrocarbon compounds

ETHANOL C_2H_6O
    Single bonds

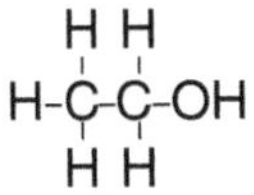

$$H\text{-}\underset{H}{\overset{H}{C}}\text{-}\underset{H}{\overset{H}{C}}\text{-}OH$$

DIMETHYL ETHER C_2H_6O
    NOTE: The molecular formula is identical to ETHANOL but the structural formula is different this is called an isomer

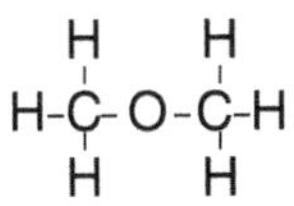

$$H\text{-}\underset{H}{\overset{H}{C}}\text{-}O\text{-}\underset{H}{\overset{H}{C}}\text{-}H$$

FIGURE 1.4. Structural and molecular formulas of selected organic compounds.
(*continued*)

BENZENE C_6H_6
   Single and double bonds in a ring formation

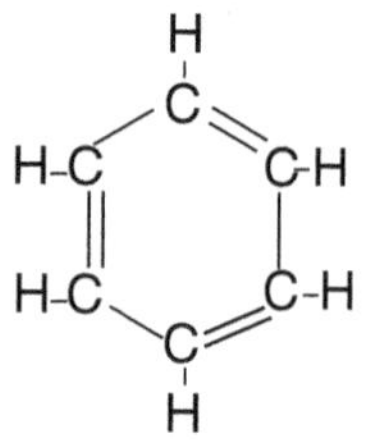

PHENOL C_6H_6O

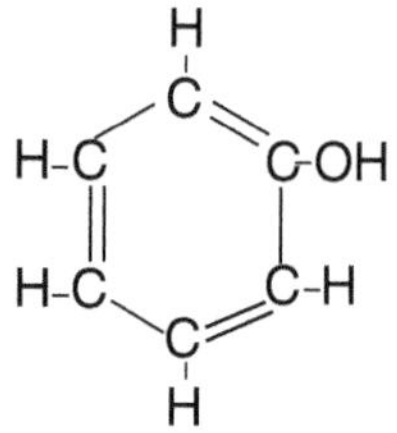

ACETIC ACID $C_2H_4O_2$

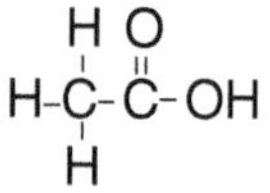

FIGURE 1.4. (*Continued*)

A functional group is made up of an atom or group of atoms that together have their own characteristic properties. Table 1.6 lists the various functional groups of organic compounds. In the wine laboratory, we are very interested in the aromatic hydrocarbons, carbonyl compounds, alcohols, organic acids, amines, and esters. Further study of these groups will be beneficial in understanding wine chemistry.

TABLE 1.6. Homologous series of common organic compounds with their corresponding functional group suffix and typical structure and bonds.

Name	Suffix	Structure
Alkanes (alkyl)	-ane	Single bond
Alkenes (alkyl)	-ene	Carbon to carbon double
Alkynes (alkyl)	-yne	Carbon to carbon triple
Esters	-oate	
Carboxylic acids (organic acids)	-oic acid	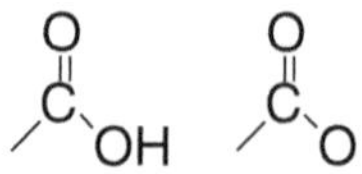

TABLE 1.6. (*Continued*)

Name	Suffix	Structure
Acid halides	-oyl halide	$$\underset{\diagup \overset{\displaystyle C}{}\diagdown HALOGEN}{\overset{O}{\shortparallel}}$$ involved
Amides	-amide	$$\underset{\diagup \overset{\displaystyle C}{}\diagdown NH_2}{\overset{O}{\shortparallel}}$$
Nitriles	-nitrile	Carbon to nitrogen triple
Aldehydes (carbonyl compounds)	-al	$$\underset{\diagup \overset{\displaystyle C}{}\diagdown H}{\overset{O}{\shortparallel}}$$
Ketones	-one	$$\underset{\diagup \overset{\displaystyle C}{}\diagdown}{\overset{O}{\shortparallel}}$$
Alcohols	-ol	$-OH$
Amines	-amine	$-NH_2$

Note: Open-ended lines indicate that a carbon atom or carbon chain is attached.

2

Safety First

2.1 Introduction

The laboratory is a hazardous environment, period. Not taking this fact to heart is a foolish endeavor, a real danger to co-workers and community and detrimental to the environment. Wine laboratory personnel not only face the dangers often found in the laboratory but also the dangers lurking in the cellar and on the bottling lines. Safety must always be in the forefront of your mind, a conscience everyday practice. Know what hazards exist in your workplace, how to avoid danger, and how to respond to an accident. Making safety first can save lives.

Safety and the preservation of a healthy workplace is the moral obligation of every laboratory worker, chemist, supervisor, department head, winery president or CEO, local government, state government, and the federal government. Laws and regulations have been developed to ensure a safe workplace, and these laws require mandatory safety rules and programs making safety a lawful obligation.

The economic impact of accidents in the workplace directly affects all employees. Businesses lose millions of dollars each year due to on-the-job accidents, directly affecting the profitability of a company. These losses add up to higher insurance rates for employees and reduces the ability of companies to grow and compensate employees. Everyone loses.

Despite the abundance of regulations, information, and safety procedures available to employers and employees, a high number of accidents still occur. Unfortunately, most accidents are due to the "can't happen to me" attitude, "skipping" procedural steps, sloppy technique, and just not thinking.

Ultimately, being safe is everyone's responsibility. This chapter is an introduction to basic safety in the wine laboratory and cellar.

2.2 Regulatory Agencies and Acts

Every aspect of the work environment is addressed and regulated by a specific organization via legislative guidelines focused on human health and protection of the environment.

The Occupational Safety and Health Act (OSHA citation 29, USC 651 et seq.) is enforced by the Occupational Safety and Health Administration, the primary national regulatory body instituting and monitoring set regulations and guidelines for safety in the workplace. OSHA outlines the recognition and evaluation of potential hazards, prevention or control of hazards, and training of employees to understand and know how to protect themselves and others from hazards.

The Environmental Protection Agency (EPA) enforces several pieces of legislation dealing with environmental issues:

- The Resource Conservation and Recovery Act (RCRA citation 42 USC 6901 et seq.) sets guidelines for chemical waste and the protection of groundwater and soil contamination.
- The Superfund Amendments and Reauthorization Act (SARA citation 42 USC 9601 et seq. and 11000 et seq.) sets regulations for emergency planning and reporting of hazardous materials.
- The Toxic Substances Control Act (TSCA citation 15 USC 2601 et seq.) is concerned with the testing and restrictions of certain chemical substances.
- The Clean Air Act (CAA citation 42 USC 7401 et seq. and 7409 et seq.) protects air quality by controlling emissions of ozone and air pollutants.
- The Federal Water Pollution Control Act (FWPCA citation 33 USC 1251 et seq.) protects water quality with the regulation of substances discharged into public waters.

The US Department of Transportation (DOT) is the regulatory body for the Hazardous Materials Transportation Act (HMTA citation 49 USC 1801 et seq.), which controls the movement of hazardous materials by regulating packing, labeling, transportation, and education of transport personnel.

Every company and facility is required to have a safety program geared for the winery size, number of employees, and degree of hazard. The safety program should include a chemical hygiene plan as outlined in the OSHA's laboratory standard (29 CRF 1910.1450).

Violation of the OSHA safety laws and regulations has grave consequences. Serious violations where severe injury or death could occur can carry fines in a range from $100 to $7000 per violation. Willful violations by an employer can carry fines up to $70,000 per violation. If a violation results in the death of an employee, fines for a first offense can reach $500,000 for a company and $250,000 for an individual or a jail sentence of 6 months—and in some cases, both.

Workplace safety information and guidelines to prepare and initiate a safety program and chemical hygiene plan are readily available and can be obtained from government and local organizations by contacting them directly or through the Internet.

2.3   Laboratory and Winery Hazards

Common laboratory and winery hazards are listed in Table 2.1. Chemical hazards are dangers that are inherent to a chemical and the potential risks associated by handling, storage, and use of the chemical. Physical hazards are dangers that are inherent to an operation or action posing a potential risk associated by carrying out the operation or action. Be aware of potentially dangerous situations and evaluate the risk potential in each job you perform.

Each laboratory and facility should have an active emergency response system in place to address any emergency that might arise and to reduce and contain losses of property and injury to employees. This emergency plan should include contingencies for evacuation, fires, chemical spills, earthquakes, accidents, and injury. Training all employees in fast-response emergency action is required. OSHA's 29 CFR 1910.38 standard delineates the elements required for an emergency response plan.

TABLE 2.1. Common winery hazards.

Laboratory	Winery Cellar
Chemical hazards	Chemical hazards
Carcinogens	Corrosive substances
Toxic compounds	Allergens
Irritants	Irritants
Corrosive substances	Flammable substances
Allergens	Asphyxiates
Reproductive toxins	
Developmental toxins	Physical Hazards
Flammable substances	
Reactive substances	Compressed gases
Explosive substances	Cryogens
Asphyxiates	Microwaves
	Electrical
Physical Hazards	Cuts
	Slips
Compressed gases	Trips
Cryogens	Falls
Vacuum	Burns
Electrical	Truck traffic
Explosion	Equipment failure
Cuts	Forklifts
Slips	Explosion
Trips	Drowning
Falls	Entanglement
Burns	Fire
Strains	Crushing
Punctures	Punctures
Fire	Strains
	Abrasions

2.4   Chemical Hazards

Before touching any chemical, it is essential to educate yourself on the handling and use of the chemical. All chemical companies and distributors are required to provide information sheets on each chemical shipped to protect the user from possible hazards; they are called Material Safety Data Sheets (MSDS). MSDS contain information about the chemical's properties, health hazards, National Fire Protection Agency (NFPA) ratings, storage, disposal, transportation, neutralizing agents, personal protection equipment, extinguishing media, ecological information, and emergency first aid procedures. Few companies supply complete MSDS with all the necessary information you require. Some sections might have the statement "I/A/W (in accordance with) ALL FEDERAL, STATE, and LOCAL REGULATIONS," which means that you will need to find out what those regulations are on your own. It might take looking at several MSDS from different companies to compile all the information you require, or going directly to the regulatory agencies for clarity. MSDS for chemicals used in the laboratory are required to be readily accessible to all employees.

Another source of information is the Laboratory Chemical Safety Summaries (LCSS). The LCSS is directed specifically toward the laboratory worker providing similar information found in the MSDS but with additional information that directly affects the handling of the chemicals in the laboratory setting (see the Appendix).

2.4.1   Toxic Chemicals

It is a good rule of thumb to assume that no chemical is safe. The risks associated with a chemical are based on the toxicity of the chemical itself, or when combined with other chemicals, and the degree of exposure. The degree of exposure is dependent on the quantity of chemical contact (dose), for how long (duration), how often (frequency), and route of exposure.

Toxic chemicals produce local effects, systemic effects, or both, and are classified by the toxic response in the body: irritant, neurotoxin, corrosive, allergen, asphyxiant, carcinogen, or reproductive and developmental.

2.4.1.1   Dose, Duration, Frequency, and Routes of Exposure

Dose response is the relationship of the quantity of chemical exposure and the effect it produces. As an example, a 2-mg dose of a substance might have no effect and a 2-g dose of the substance might create vomiting; however, a 25-g dose of the substance causes death. It does not matter what the substance is, if enough of the substance is introduced into the body, there will be toxic results.

Lethal concentration 50 (LC_{50}) or lethal dose 50 (LD_{50}) is an indicator of the level of chemical toxicity. The LC_{50} is determined in a controlled environment using test animals. A group of animals are exposed to various dosages of a chemical via inhalation, direct contact, ingestion, or injection. The dose of chemical that is given that kills 50% of the animals is noted as the LC_{50} and is expressed in grams per kilogram of body weight. Toxicity ratings are generally accepted as follows (National Research Council, 1995):

- Extremely toxic = LD_{50} <5 mg/kg
- Highly toxic = LD_{50} <50 mg/kg
- Moderately toxic = LD_{50}=50–500 mg/kg
- Somewhat toxic = LD_{50}=500 mg/kg to 5 g/kg
- Virtually nontoxic = LD_{50} >5 g/kg

Volatile chemicals are reported as LD_{50} and expressed in parts per million (ppm), milligrams per liter (mg/L), or milligrams per cubic meter (mg/m^3).

Chemical toxicity can occur immediately upon contact with a chemical or upon repeated or chronic exposures. Chronic exposure to low-dose toxins is particularly dangerous due to their cumulative effects in the body that might not be apparent for many years. The duration and limits of exposure to toxic chemicals have been established by the American Conference of Governmental Industrial Hygienists (ACGIH) and OSHA.

Inhalation is the number one mode of chemical exposure in most laboratories. Inhalation of toxic materials poisons the body by absorption through the mucous membranes via the nose, mouth, throat, and lungs or by passing into the capillaries of the lungs into the bloodstream.

Threshold limit value (TLV) (ACGIH) and permissible exposure limit (PEL) (OSHA) are analogous terms identifying the concentration of chemical in the air that produces no adverse effects.

The ACGIH has set TLV using two different time frames (National Research Council, 1995):

- TLV–TWA = threshold limit value – time-weighted average; delineates the limit of safe chemical exposure in an 8-h workday for a 40-h workweek.
- TLV–STEL = threshold limit value – short-term exposure limit; used for higher concentrations of chemical indicating the amount of chemical exposure that is safe for a 15-min period.

Workers can be exposed to a maximum of four STEL periods per 8-h shift, with at least 60 min between exposure periods. It is recommended that a fume hood be utilized when using a chemical with a PEL or TLV less than 50 ppm.

Direct skin or eye contact with a toxic chemical is a frequent route of exposure in the laboratory. Skin irritations and allergic reactions are more frequent with direct exposure to a variety of chemicals. Corrosive chemicals might cause severe burns at the point of contact. Chemicals can enter the body through the skin's glands, follicles, cuts, or abrasions, causing systemic toxicity.

Ingestion of toxic chemicals happens. Food eaten in the laboratory can be contaminated by chemicals found on work gloves or eating food that has been in contact with laboratory benches. The chemical might cause contact tissue damage throughout the gastrointestinal (GI) tract, be absorbed through the GI tract into the bloodstream causing systemic damage, or both.

Direct injection of toxic chemical into the bloodstream is the fastest way to introduce a toxic chemical into the body. Most often this occurs with contaminated broken glass cutting into the skin. Some types of laboratory equipment with sharp parts and needle assemblies can also create an injection injury.

2.4.1.2   Classification of Toxins

Chemicals producing a reversible inflammation at the contact site are called *irritants*. Direct contact with the chemicals in this category should be avoided.

Corrosive substances are the most common toxic chemicals found in a wine laboratory and cellar. These substances can come in the form of a liquid, solid, or gas. Corrosive substances react chemically at the contact site, destroying living tissue. Corrosive substances include strong acids, strong bases, and strong dehydrating agents. Examples of common corrosives are hydrochloric acid, phosphoric acid, sulfuric acid, sodium hydroxide, concentrated hydrogen peroxide, nitrogen dioxide, and sulfur dioxide.

Allergic responses can occur with exposure to a toxic chemical after initial sensitization of the immune system. The reaction can range from itchy skin to anaphylactic shock, depending on an individual's level of sensitivity. Some people with high sensitivity can react adversely to the smallest dose, whereas others will have no reaction to a large dose of the same chemical. Reactions can be immediate or delayed for hours or days.

The winery cellar uses a variety of gases to prevent oxidation in wine and as an antimicrobial agent for wine and juice. Nitrogen, argon, sulfur dioxide, and carbon dioxide are the gases most frequently used due to their ability to displace oxygen, thus producing an anoxic environment in the tanks or barrels. Carbon dioxide gas is also naturally produced during juice fermentation. Tanks and cellars that have a lack of breathable oxygen can render a person unconscious and lead to death by asphyxiation. Asphyxiation is the most common cause of death in a winery.

Carcinogens are cancer-producing substances and are highly toxic. Most of these substances affect the body after repeated or extended exposure, with no indication of effects for many years. Chronic exposure can stimulate the uncontrolled growth of cells in an organ and some chemicals can directly affect the cell DNA.

Reproductive and developmental chemical toxins adversely affect the entire reproductive process. There are developmental toxins called *teratogens* that cause developmental malformations or death in a developing embryo or fetus and *mutagens* that cause chromosomal damage with an increase in the frequency of cellular mutations in an embryo or fetus. Fetus growth can be retarded by some developmental toxins as well as creating postnatal functional deficiencies. Reproductive toxins can render the user, male or female, sterile. Cycloheximide (Actidione) is a developmental toxin used by wine laboratories as a standard bactericidal ingredient in selective microbiological plating media.

Neurotoxins adversely affect the central nervous system (CNS) of the body by altering its structure or function. A large number of neurotoxins are chronically toxic due to their long latency periods and most neurotoxins can be most difficult to diagnose. Further investigation of neurotoxins is being conducted to improve early detection.

2.4.2   Reactive Chemical Hazards

Chemicals pose a serious risk due to their reactive natures. They can be flammable and explosive or create intense reactions.

Flammable chemicals can be a gas, liquid, or solid that can ignite and burn in the presence of oxygen. Highly flammable denatured alcohol is used frequently in the laboratory as a fuel source in open-flame burners. The alcohol burns readily by itself but it is more dangerous due to its rapid vaporization. If there is a buildup of vapor due to poor ventilation, the vapor can ignite and enflame the source.

Volatility indicates the ease with which a liquid or solid will pass into the vapor stage and is measured by the chemical's boiling point. The flash point of a chemical is the lowest temperature at which a liquid can form enough vapor pressure to ignite in the presence of air when exposed to a flame. Chemicals with flash points of 38°C or below are the most dangerous. The ignition temperature is the point at which a substance reaches self-sustained combustion without a source of heat or spark; the chemical will simply autoignite.

Information concerning the flammability, flash-point temperatures, and ignition temperatures of chemicals dictate how a chemical should be handled and stored, what precautions to take while working with the chemical, and what safety equipment is required in case of an accident.

The rate of vaporization, the ability to form combustible mixtures with air, and the ease of ignition dictate the level of fire danger associated with a chemical. The NFPA classifies flammable chemicals according to their degree of hazard and has developed a system that incorporates this information in conjunction with health hazards, reactivity levels, and other specific hazards associated with chemicals. The NFPA classifications will be discussed later in this chapter. Table 2.2 lists the NFPA fire hazard rating, flash point, boiling point, and ignition temperature of a few chemicals that could be found in the wine laboratory.

TABLE 2.2. Chemical flammability.

	NFPA rating	Flash point	Boiling point	Ignition temperature
Acetic acid (glacial)	2	39°C	118°C	463°C
Acetone	3	−18°C	56.7°C	465°C
Ethyl alcohol	3	12.8°C	78.3°C	365°C
Hydrogen	4	—	−252°C	500°C
Isopropyl alcohol	3	11.7°C	82.8°C	398°C

Note: 0 = will not burn; 1 = must be preheated to burn; 2 = ignites when moderately heated; 3 = ignites at normal temperature; 4 = extremely flammable.
Source: Reprinted with permission from NFPA 10-2002, *Portable Fire Extinguishers*, Copyright © 2002, National Fire Protection, Quincy, MA. This reprinted material is not the complete and official position of the NFPA on the referenced subject, which is represented only by the standard in its entirety.

The combination of certain chemicals by accident or design can result in potentially dangerous situations due to the intensity of the reaction. Low-intensity reactions might produce a slight increase in the temperature of the chemical mixture, whereas a high-intensity reaction might create an explosion.

It is important to know the classes of chemicals, chemical compatibility, and what *not* to combine in a noncontrolled reaction. The major risk lies in the storage of chemicals where accidental spills of two or more chemicals in large quantities can have devastating consequences. Disasters such as earthquakes, fires, floods, and tornadoes could lead to the damage of chemical storage containers, creating a potential reactive hazard. Table 2.3 is a partial listing of chemicals commonly found in the wine laboratory and the chemicals that are incompatible with them. An excellent reference with additional listings of incompatible chemicals can be found in L. Bretherick's 1986 *Hazards in the Chemical Laboratory*, 4th edition, Royal Society of Chemistry, London, England.

Water can react violently with groups of chemicals, creating tremendous heat that could lead to an explosion or fire. Pyrophorics are substances that when combined with the moisture in the air or oxygen react so fast that the mixture can ignite. Extremely reactive chemicals can begin a rapid reaction by a spark, movement, heat, catalyst, or detonation that will create a violent expansion of gases, heat, and noise. Peroxides, oxidizing agents combined with reducing agents, and auto-oxidative compounds could have explosive tendencies. Explosive chemicals need very special handling, storage, and safety precautions.

2.4.3   Chemical Handling and Storage

Because of to hazards handling chemicals, it is important to institute strict guidelines for every aspect of chemical contact. These guidelines must be accessible to all receiving and warehouse personnel, cellar workers, and laboratory personnel. Key elements to safe handling of chemicals include the following:

TABLE 2.3. Incompatible chemicals.

Chemical	Incompatible chemicals
Acetic acid (glacial)	Alcohols, hydrogen peroxide, ethanolamine, alkali hydroxides, chlorates, chromic acid, carbonates, peroxides, permanganates, metals, aldehydes, anhydrides, ethylene glycol
Acetone	Concentrated nitric and sulfuric acid mixtures, hydrogen peroxide
Calcium carbonate	Acids, fluorine gas, ammonium salts
Carbon (charcoal) activated	Calcium hypochlorite, other oxidants
Carbon dioxide	Alkali metals
Chromic acid	Acetic acid, naphthalene, camphor, alcohol, turpentine, glycerol, other flammable liquids
Copper(II) sulfate	Hydrogen peroxide, acetylene, hydroxylamine, strong reducing agents
Cycloheximide	Strong peroxides, hydrogen peroxide, oxidizing agents; strong acid chlorides, acid anhydrides, alkali, bases
Ethyl alcohol	Peroxides, acids, acid chlorides, acid anhydrides, alkali metals, ammonia, moisture, strong oxidizing agents
Ethylene glycol	Strong bases, perchloric acid, strong acids, chlorosulfonic acid, acetic acid, strong oxidizing agents
Hydrogen peroxide 3%	Copper, rust, brass, zinc, nickel, iron, most metals or their salts, alcohols, acetone, strong oxidizers
Hydrochloric acid (concentrated)	Metals, amines, metal oxides, vinyl acetate, formaldehyde, alkalis, carbonates, strong bases, sulfuric acid
Hydroxides (alkali and alkaline earth metals)	Acids, water, carbon dioxide, hydrogen peroxide, nitro compounds, metals, oxidizing agents, ammonium compounds, chlorinated hydrocarbons
Iodine	Acetylene, ammonia (anhydrous or aqueous), ammonium hydroxide, strong reducing agents, some metals
Malic acid	Bases, strong oxidizing agents, reducing agents, alkali metals
Mercury	Acetylene, nitric acid/ethanol mixtures, ammonia, boron, metals, oxygen, ethylene oxide, oxidants
Oxygen	Oils, grease, hydrogen, flammable liquids, solids, gases
Ozone	Reducing agents, combustible materials
Sodium thiosulfate	Mercury, iodine, strong acids and oxidizing agents
Sulfuric acid (concentrated)	Water, metals, combustible materials, bases, perchlorates, permanganates, chlorates, hydrogen peroxide, oxidizing agents, strong reducing agents
Sulfur dioxide	Zinc, aluminum, iron oxide, sodium, strong reducing agents, chlorates, hydrogen peroxide, strong oxidizing agents,
Tartaric acid	Bases, silver and silver salts, oxidizing agents, reducing agents
Triethanolamine	Acids, peroxides, oxidizing agents

Note: This is a partial listing.

- Ordering and receiving
- Proper labeling
- Chemical compatibility
- Proper storage

2.4.3.1   Ordering and Receiving

Ordering chemicals requires a complete understanding of the types of analysis performed in the laboratory, the quantity of chemicals used in a given time frame, the location and amount of storage available, storage risks, and economic factors.

You never want to run out of chemical for any given assay, but you also do not want a large supply on hand that might reach its expiration date before it has been used. If a facility or company has several laboratories, check with the other locations before you order. They might have excess stock that can be transferred. Keeping a 1-month supply on hand seems to work well. Most chemicals have approximately a 1–2-week lead time, which is the time it takes to order and receive the product. Suppliers can let you know how far in advance to order and the availability of the chemicals you use. This method prevents old chemicals from accumulating, it requires less storage space, and smaller quantities equal less safety risks, less chance of contamination, and little or no waste. Large quantities are less expensive, but if they expire or are unused, the cost of disposing of those chemicals could be very high. Chemicals with long shelf lives that must be ordered in larger quantities should be stored safely with strict attention to expiration dates.

Request the chemical supplier to include the MSDS with each shipment. An MSDS binder with all MSDS listings should be kept up-to-date and located in the receiving departments, cellar, and, of course, the laboratory. Keeping up-to-date MSDS information is critical, and when included with each shipment, it assists receiving personnel in handling a leaking package quickly.

Upon delivery of chemicals in the receiving department, the appropriate person or laboratory should be notified immediately. Some chemicals will require special storage such as refrigeration and need attention as soon as they arrive. Advising receiving of expected chemical shipments will expedite the receiving process. After proper notification, the chemicals should be set aside in a safe area away from other materials until they are picked up or delivered.

Small packages should be carried with both hands; larger or heavy packages should be placed on a cart for transport. If the chemical is moderately to highly toxic or has a glass container, place the package in an unbreakable secondary containment vessel such as a chemical-resistant bucket or pan large enough to contain the entire volume of chemical plus the container and transport via cart. It is against most companies' policies to transport chemicals in personal vehicles. If transporting chemicals in a vehicle, secondary containment must be provided in the event of a spill, especially with a liquid or chemicals in glass containers. The containment vessels should be secured to the vehicle to prevent movement during transportation.

Receiving gas cylinders in the cellar, laboratory, or receiving area requires cautious handling. Cylinders must be secured, with caps on, and transported on a cylinder cart, never dragged or rolled (see Fig. 2.1). Some liquid gas cylinders, such as a Dewar, can be mounted on heavy-duty wheels for ease of transportation. Placing the order for gas cylinders should include a request for safe transport via cylinder carts. If a supplier fails to conform to the request, refuse the order and find another supplier. Use a gas supplier that will not only deliver but pick up the empty cylinders.

2.4.3.2 Proper Labeling

A chemical inventory sheet or database listing the chemical's proper name, Chemical Abstract Service (CAS) registry number, hazard information, stor-

FIGURE 2.1. Gas cylinder and transportation cart.

age location, date received, date of expiration, and supplier is helpful for the laboratory staff, is a quick reference in locating chemicals, and, in some communities, is required. Check with local authorities for their requirements. Keeping a copy of the inventory sheet at all storage locations is helpful as a quick guide in the event of a spill.

Receiving chemicals into the laboratory for storage or immediate use requires time to properly complete paperwork, properly label the containers, and, in some cases, properly place warning signs. Chemical labeling and storage can prevent dangerous situations, so be thorough. This is not a place to save time.

Before opening the package, compare the order to the shipping information. Compare the chemical name, registry number, catalog number, and concentration of the chemical. If correct, look for any damage or leaks and open shipping packages carefully.

Check the labels of the containers and replace torn or damaged labels. Do not accept containers that have no labeling. Contact the chemical supplier immediately upon receipt of damaged or unlabeled chemicals. Each container should have the name of the chemical, molecular formula, CAS registry number, concentration, NFPA hazard classification, and expiration date.

It is good safety practice to affix an NFPA label to chemical containers being stored and chemical containers used in the laboratory. Labels can be made or purchased from a safety supply company. Figure 2.2 (see Color Plate 1) outlines the NFPA hazard classification method used to identify chemicals.

2.4.3.3   Proper Storage

Before storing chemicals, pay strict attention to the NFPA hazard rating, the chemical compatibility, stability, and storage information located on the MSDS, and general OSHA guidelines (OSHA, 29CFR 1910.1450). Avoid storing chemicals on benchtops, on tops of cabinets, ventilation hoods, in drawers, in cleanup/wash areas, in high-traffic areas, and in office areas. When possible, store chemicals in unbreakable containers and use unbreakable containers for benchtop work. Refrigerated chemicals should be stored in a labeled "chemicals only" refrigerator where they will not be in contact with food items. Odiferous chemicals should be stored in a ventilated cabinet, but if such a cabinet is not available, store them in a sealed container. Chemicals should be stored away from extreme temperatures and sunlight.

Store the chemicals in compatible groups. They can occupy the same chemical cabinet but different shelves. Secondary containment is recommended for storage cabinets and benchtop working containers. A few examples of compatible groups are as follows:

- Acids
- Bases
- Flammable liquids
- Alcohols, glycols
- Hydrogen peroxide, hypochlorites
- Hydroxides, carbon, carbonates
- Sulfur, phosphorus
- Sulfates, sulfites, thiosulfates, phosphates
- Chromates, permanganates

It is good common practice to rotate your chemical stock (or any stock item) by dating each container with the date received. Place the newly received chemical containers behind the current stock, making it easier to use the oldest chemicals first. To avoid hazards and problems, check chemical expiration dates frequently and properly dispose of expired chemicals.

Before using a chemical, refilling lab containers, or refilling benchtop working containers, always double check the container for the name and concentration of the chemical. The use, or refilling of a container, with the wrong chemical or chemical concentration is a very common accident in the laboratory and all laboratory personnel must make checking and rechecking second nature. Color coding the working chemical containers and stock containers can make replenishing faster, but it perpetuates the bad habit of not reading and verifying the chemical name and concentration.

Check gas cylinders for proper labeling; never depend on color coding. Flammable, toxic, or corrosive gases are coded with a yellow background and black letters. Inert gases are coded with a green background and black letters. Cylinder storage in earthquake zones requires securing them to a wall with two straps or chains. A single chain or strap is sufficient for other zones. It is best to check your local OSHA and NFPA regulations. Dewars or liquid gas cylinders with wheels must also be secured to a wall to prevent movement in an earthquake. Gas cylinder storage follows the same rules as other chemicals: separate according to compatibility; store separate from other chemicals; when not in use, remove regulators and replace the cap; and segregate empty from full cylinders. It is wise to avoid transport of full cylinders in public elevators.

Cylinders of flammable gases should always be grounded to prevent static electricity buildup. Do not empty a compressed gas cylinder below 25 psi. At this pressure, ambient air could migrate into the cylinder if the valve is left open, requiring notification of the supplier. It is also good practice to remove a cylinder from a closed system while the gas pressure in the cylinder is still higher than the system pressure.

Flammable liquids have flash points less than 100°F (38°C) and must be stored away from heat and any ignition source. Combustible liquids have flash points above 100°F (38°C). Underwriters Laboratory (UL) in conjunction with the NFPA Code No. 30 regulates the manufacturing of flammable and combustible chemical storage cabinets. Flammable storage cabinets or metal safety cans are required for all flammable chemicals. All flammable and combustible chemicals contained in a cabinet must be compatible. Do not store paper, cardboard, or other flammable materials in the same cabinet. Follow the NFPA guidelines for maximum allowable volumes of flammable liquids in the laboratory and in the facility.

The NFPA classifications are as follows:

- Class I Flammable—flash point below 38°C
- Class II Combustible—flash point of 38–60°C
- Class IIIA Combustible—flash point of 60–93°C
- Class IIIB Combustible—flash point above 93°C

Highly toxic chemicals can have extreme storage requirements such as explosion-proof cabinets or cabinets ventilated to a scrubber that will contain toxic fumes.

2.4.4   Chemical Waste

Solid waste is defined by the EPA as discarded materials including solids, semi-solids, liquids, and contained gaseous materials; garbage; or sludge. It is prudent practice to evaluate all chemical assays performed in the laboratory, the amount of waste generated from those assays, and understand the chemical characteristics and possible hazards. It is the responsibility of waste generators to determine the type of hazardous waste produced and to properly dispose of it. Consult the EPA's interpretation and lists of hazardous waste (RCRA 40 CFR 261) as well as the local and state regulations to determine if the waste is regulated as hazardous, and if so, in what hazard classification. Most states operate their own hazardous waste programs and may have more stringent regulations and hazardous waste identification criteria than federal regulations.

2.4.4.1   Classification

Classification of EPA listed hazardous waste is broken into five categories (generalized):

1. F (RCRA 40 CFR 261.31)—nonspecific waste (e.g., reaction waste, processing waste)
2. K (RCRA 40 CFR 261.32)—waste from particular industries
3. U (RCRA 40 CFR 261.33)—unused pure or commercial-grade formulations of chemical
4. P (RCRA 40 CFR 261.33)—unused pure or commercial-grade formulations of acutely toxic chemical (highly toxic having an LD_{50} less than 50 mg/kg)
5. D (RCRA 40 CFR 261, Subpart C)—waste not listed above but exhibits hazardous waste characteristics

Hazardous waste characteristics are described by the EPA as follows:

- Ignitable (I) (RCRA 40 CFR 261.21)—flash point of less than 140°F (60°C) or other characteristics that could potentially cause a fire
- Corrosive (C) (RCRA 40 CFR 261.22)—liquids that have a pH of 2 or less (acid) or a pH of 12.5 or greater (base).
- Reactive (R) (RCRA 40 CFR 261.23)—substances that are unstable, react violently with water, could possible detonate, or are a sulfide or cyanide-bearing waste.
- Toxic (T) (RCRA 40 CFR 261.24)—materials that do not pass the Toxicity Characteristic Leaching Procedure (TCLP), which measures the amount of toxic material leached from waste materials where the leachate from the waste is likely to release toxins into the groundwater

The wine laboratory normally generates very little hazardous waste (much less than 100 kg of hazardous waste in 1 month) and is considered exempt from formal management program requirements. If a facility produces more than 100 kg of hazardous waste, consult RCRA 40 CFR 262. Typical hazardous waste found in the wine laboratory include acids, bases, mercury, ethanol, iodine, and denatured alcohol.

2.4.4.2   Waste Disposal

When waste has been classified, the method of disposal can be determined. The method chosen is dependent on the type of chemical or material hazard, amount of generated waste, and economics. Neutralization of certain chemicals is a method used to reduce the hazards of a chemical prior to disposal. This allows stronger chemicals to be disposed of easily and safely. Information and guidelines concerning methods of neutralization can be found in *Prudent Practices in the Laboratory, Handling and Disposal of Chemicals* (National Academy of Science, Washington, DC, 1995).

Incineration of lab waste is carried out by an RCRA-approved treatment, storage, and disposal facility (TSDF). This is one of the most widely used and expensive methods for the destruction of hazardous waste. Contact your TSDF for instructions on waste preparation for this method of disposal.

The laboratory trash receptacles can accept a large number of laboratory wastes. Larger quantities of chemicals should be in unbreakable containers, there should be no free-flowing liquid or broken glassware, and powders should be in containers. Contact the local municipal agency for any restrictions.

Release of hazardous vapors from volatile chemicals into the atmosphere through an exhaust system is allowed if the exhaust is fitted with the proper traps.

The sewer system is one of the most common wine laboratory methods of waste disposal. Substances that can go down the drain must be water soluble and approved by the local sewer facility and public-owned treatment works (POTW): biodegradable aqueous solutions; inorganic solutions; and corrosive and ignitable wastes that are 1% or less of the total wastewater, or 1 ppm of total wastewater generated by the facility. The waste should flow into the sewer system—never into storm drains or septic systems—and be compatible with the piping material. Flushing the system with water after disposal is required and it is suggested that a 100 : 1 ratio of water to chemical be used to keep the chemical levels low in the sewer system. Some chemicals might require predilution before disposal and flushing.

Hazardous waste that requires off-site disposal via an RCRA-approved TSDF should be collected and contained in a proper chemical-resistant package. The package label should contain the name of the substance, CAS number, concentration, physical description (solid or liquid), and any hazard information. If there is a mixture of substances, list all of the above plus note the percentages of all chemicals (contact your specific TSDF for instructions). Labeled waste packages should be placed in clearly labeled chemical-resistant storage containers containing compatible waste (e.g., flammable, corrosive) and stored in an area where they will not be disturbed. A listing of all waste contained within a storage container should be attached to the container and kept current. Laboratories are allowed to keep up to 55 gallons of hazardous waste (or 1 qt acutely hazardous waste) indefinitely, but it is recommended that waste not be held from more than 1 year.

Table 2.4 lists common laboratory chemicals, their classification, and common recommended waste disposal according to various MSDS information. Check MSDS and always confirm with all regulatory agencies to assure compliance. Chemical containers ready for discard that retain a small amount of chemical and the residue from chemical spills might have more rigid disposal requirements, so check with RCRA 40 CFR 261.33.

TABLE 2.4. Chemical transportation and waste disposal.

Laboratory chemical (small quantities)	DOT hazard classification NFPA rating[a]	Common disposal methods[b]
Acetic acid (glacial) CAS# 64-19-7	DOT: Class 8 Corrosive Class 3 Flammable NFPA: H3, F2, R0, S-CORR	**Undiluted**: Contact approved disposal facility **Diluted**: Small quantities dilute to <1 %, neutralize, flush with copious amount of water to sewer
Calcium carbonate CAS# 471-34-1	DOT: Not regulated NFPA: H1, F0, R0	**Undiluted**: Contact approved disposal facility for landfill burial **Diluted**: Small quantities dilute to <1%, flush with copious amount of water to sewer
Chromic acid CAS# 1333-82-0	DOT: Class 8 Corrosive NFPA: Estimate H4	**Undiluted and diluted**: Hazardous waste ID-D007, contact approved disposal facility for landfill burial, reduce/neutralize

TABLE 2.4. (*Continued*)

Laboratory chemical (Small Quantities)	DOT hazard classification NFPA rating[a]	Common disposal methods[b]
Copper(II) sulfate CAS#7758-99-8	DOT: Class 9 Miscellaneous Hazard NFPA: H2, F0, R0	**Undiluted**: Dilute to <1% **Diluted**: Dilute to <1%, flush with copious amount of water to sewer
Cycloheximide CAS# 66-81-9	DOT: Class 6.1 Poison NFPA: Unavailable	**Undiluted and diluted**: Contact approved disposal facility for incineration
Ethyl alcohol CAS# 64-17-5	DOT: Class 3 Flammable NFPA: H2, F3, R0 Estimated	**Undiluted**: Contact approved disposal facility for incineration **Diluted**: Small quantities dilute to <1%, flush with copious amount of water to sewer
Ethylene glycol CAS#107-21-1	DOT: Class 3 Flammable Class 6.1 Poison NFPA: H2, F1, R0	**Undiluted**: Contact approved disposal facility for incineration **Diluted**: Small quantities dilute to <1%, flush with copious amount of water to sewer
Hydrochloric acid (concentrated) CAS# 7647-01-0	DOT: Class 8 Corrosive NFPA: H3, F0, R0, S-CORR	**Undiluted**: Contact approved disposal facility for disposal **Diluted:** Small quantities dilute to <1%, neutralize, flush with copious amount of water to sewer
Hydrogen peroxide 3% CAS# 7722-84-1	DOT: Not regulated NFPA: H1, F0, R1	**Undiluted**: Dilute to <1% **Diluted**: Dilute to <1%, flush with copious amount of water to sewer
Iodine CAS# 7553-56-2	DOT: Class 5.1 Oxidizer Class 6.1 Poison NFPA: H3, R0, R1, S-OX	**Undiluted**: Contact approved disposal facility for disposal **Diluted:** Small quantities dilute to <1%, flush with copious amount of water to sewer
Malic acid CAS#617-48-1	DOT: Not regulated NFPA: H1, F1, R0	**Undiluted**: Contact approved disposal facility for incineration **Diluted**: Small quantities dilute to <1%, flush with copious amount of water to sewer
Mercury CAS# 7439-97-6	DOT: Class 8 Corrosive Class 6.1 Poison NFPA: H3, F0, R0	**Undiluted**: Hazardous waste ID-U151, contact approved disposal reclamation center
Phosphoric acid (concentrated) CAS# 7664-38-2	DOT: Class 8 Corrosive NFPA: H3, F0, R0	**Undiluted**: Contact approved disposal facility for disposal **Diluted**: Small amounts dilute to <1%, neutralize, flush with copious amount of water to sewer

(*continued*)

TABLE 2.4. (*Continued*)

Laboratory chemical (Small Quantities)	DOT hazard classification NFPA rating[a]	Common disposal methods[b]
Sodium hydroxide CAS# 1310-73-2	DOT: Class 8 Corrosive NFPA: H3, F0, R1, S-CORR	**Undiluted:** Contact approved disposal facility for disposal **Diluted:** Small amounts neutralize, dilute to <1%, neutralize, flush with copious amount of water to sewer
Sodium thiosulfate CAS#10102-17-7	DOT: Not regulated NFPA: H1, F0, R0	**Undiluted:** Neutralize, dilute to <1% **Diluted:** Dilute to <1%, flush with copious amount of water to sewer
Sulfur dioxide CAS# 7782-99-2	DOT: Class 2.3 Poison Gas NFPA: H3, F0, R0 S-CORR	**Dry:** Contact approved disposal facility for incineration, oxidize to inert sulfate salt **Gas:** Bleed gas into water mixture of 15% NaOH to neutralize, flush with copious amount of water to sewer
Sulfuric acid (concentrated) CAS# 7664-93-9	DOT: Class 8 Corrosive NFPA: H3, F0, R2 S-W	**Undiluted:** Contact approved disposal facility for disposal **Diluted:** Small quantities dilute to <1%, neutralize, flush with copious amount of water to sewer
Tartaric acid CAS# 133-37-9	DOT: Not regulated NFPA: H2, F1, R0	**Undiluted:** Contact approved disposal facility for incineration **Diluted:** Small quantities dilute to <1%, flush with copious amount of water to sewer
Triethanolamine CAS# 102-71-6	DOT: Not regulated NFPA: H2, F1, R1	**Undiluted:** Contact approved disposal facility for incineration **Diluted:** Small quantities dilute to <1%, flush with copious amount of water to sewer

[a]H=health; F=fire; R=reactivity; S=specific hazard.

[b]Contact federal, state, and local authorities for disposal guidelines in your area to assure compliance.

2.4.5 Personal Protection Equipment

Knowledge of chemicals is the primary tool for successful personal protection. Proper personal protection equipment (PPE) and chemical protective clothing (CPC) for every type of possible chemical exposure in each individual laboratory must offer the maximum level of protection. The amount of protection required will depend on the degree of chemical toxicity, exposure scenario, and risk potential. OSHA requires PPE and CPC be provided to each employee and the equipment be used and maintained in clean and reliable condition. PPE includes eye, ear, face, head, and extremities protective clothing, respiratory devices, and adequate barriers. Each laboratory should have its own PPE and CPC equipment stored in that laboratory. Consult OSHA 29CFR 1910.133 for details.

For routine working conditions where there is no handling of severely toxic chemicals, the following PPE should be worn:

- A cotton laboratory coat to provide first-line defense protection against chemical splashes, vapor, and particulate matter. Laboratory coats should be free of holes, tears, and frayed edges. The coat should fit well and allow free movement. Sleeve length should cover the arm to the wrist. Rolling sleeves to the wrist creates bulk that could get entangled or caught on equipment. Laboratory coats should not be worn outside the laboratory.
- Splash-proof safety glasses with side shields to provide protection to the eyes. To date, OSHA does not require safety glasses but strongly recommends their use. OSHA does require laboratory staff that wear corrective lenses be supplied with prescription safety glasses or safety goggles that fit over nonsafety glasses.
- Disposable gloves to protect hands and forearms from splashes.
- Shoes that cover the entire foot and have nonslip soles.

What not to wear under laboratory coats are shorts, midriff tops, tube tops, halter tops, cropped pants, open-toed shoes, or sandals. Keep your skin covered and protected. Long hair should be pulled back to prevent obstruction of vision, dipping into chemicals, catching fire, and entanglement. Dangling jewelry should be removed to avoid entanglement in glassware or equipment and dipping into chemicals. Torn or ragged clothing and clothing that is too large should be avoided to prevent entanglement, dipping into chemicals, and hindering mobility.

Before handling severely toxic chemicals, read the MSDS for the manufacturer's suggestions on the proper PPE to wear. Health and safety professionals can provide additional guidelines. Added PPE might include the following:

- Disposable chemical-resistant sleeves
- Splash-proof safety glasses with side shields or goggles
- Face shield
- Chemical-resistant apron
- Chemical-resistant gloves
- Dust mask or respirator
- Chemical resistant pants or coveralls

Figure 2.3 illustrates proper PPE when working with hazardous and volatile chemicals.

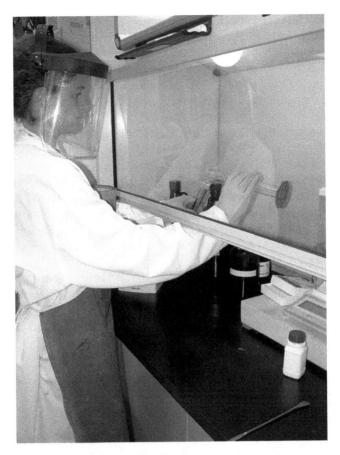

FIGURE 2.3. Example of personal protection equipment.

2.4.6   Chemical Accidents and Emergencies

Working in a hazardous environment dictates that all laboratory staff must be trained to handle chemical spills, accidents, and emergencies. Knowing the procedures for potential problems will make responding to accidents or emergencies faster and more effective. Quick response time and appropriate action can save lives and control damage.

2.4.6.1   Spills

When an accidental spill occurs, immediately notify laboratory personnel in the area; if alone, contact any person in the area, and let them know what has happened. Evacuation might be necessary.

Help any injured person first and notify the emergency contact in your facility for assistance. For persons who have been injured or contaminated with small amounts of chemical, use the following procedures:

Immediately flush contact area with water for no less than 15 min. If the eyes are the contact point, get to an eyewash station, hold the eyelids open and flush with potable water, remove contact lenses quickly, and seek medical attention.

For contamination of the face, hands, and arms, use the sink faucet to flush the site; remove jewelry and contaminated clothing while continuing to flush the site. When there are larger areas of contact with skin and clothing, use the safety shower. Quickly remove all contaminated clothing, shoes, and jewelry while in the safety shower (see Fig. 2.4). DO NOT BE MODEST; every second counts. Be very careful in removing clothing to avoid additional skin contact or spreading the chemical into the eyes. Clothing might best be removed by cutting it off rather than pulling the clothing over an injured person's head.

After flushing, check the contact site. If there is no visible burn, wash the area with soap and water. Seek medical attention for the slightest burn. Do not put anything on the site—no creams, lotions, or salves. Take a copy of the MSDS to the hospital or physician's office to assist the medical staff in treat-

FIGURE 2.4. Eyewash and shower station.

ing the injured person quickly. Exposure to certain chemicals can lead to other physical problems and the doctor and patient will benefit from the MSDS information.

Attempt to contain the spill and avoid putting anyone in jeopardy. Locate the MSDS information as soon as possible and determine the best absorbing material to contain the spill. Use absorbent spill pillows for liquid chemicals such as acids, bases, and solvents. Other absorbents such as clay, sand, vermiculite, kitty litter (for nonoxidizing substances) or the use of a specific spill kit containing the proper material can be used. The MSDS most often contain a list of neutralizing agents that can be applied to a spill or excess chemicals. Sodium bicarbonate is often used to neutralize acids and sodium bisulfate to neutralize bases. Spill pillows, absorbing materials, and, in some cases, neutralized waste are considered solid waste and require proper handling and disposal. Consult the MSDS for disposal instructions.

Turn off hot plates, stirrers, and equipment in the area that could possible ignite the chemical. Set up appropriate warning signs in clear view to protect other personnel. Clean the spill using the appropriate PPE and procedures. Do not attempt to clean a significant spill alone. Use situation-specific tools to clean up the waste and place the waste in a proper receptacle. Wash the area two to three times with a sponge or cloth and a mild detergent and water and then dry with rags or paper towels. Wipe down all containers or equipment that might have been contaminated with rags or paper towels. Place sponges, cloths, rags, and paper towels in the appropriate receptacle. Hazardous material cleanup requires extreme care and all cleaned up materials, tools, cloths, rags, and paper towels must be handled as hazardous waste.

Mercury spills should be isolated before collecting the droplets (wearing the appropriate PPE, of course). Large droplets can be consolidated into a pool using a scraper or piece of cardboard and removed via a pump or similar equipment in the case of large spills. Do not use a regular vacuum cleaner. For smaller spills, use a wet paper towel or a piece of adhesive tape to pick up the droplets. It is prudent to keep a mercury spill cleanup kit at the ready. These kits contain instructions and the supplies to handle a small-scale spill. All waste should be placed in a high-density polyethylene container and sent to an approved reclamation center. The spill area will require decontamination by using a mercury decontamination kit.

2.4.6.2 Accidental Inhalation or Ingestion

Inhalation of a chemical can lead to delayed symptoms. The victim should leave the area immediately and be encouraged to take deep breaths of fresh air. If the inhaled chemical is not an irritant, corrosive, or very toxic, contact a physician if symptoms such as coughing or shortness of breath occur. If the substance inhaled is an irritant, corrosive, or toxin, contact a physician immediately. Do not wait for symptoms to develop. Transport the victim to a hospital.

Chemicals that pose an inhalation hazard or reactions that can generate inhalation hazards should only be worked with and conducted inside an exhaust fume hood. The fume hood should be vented to the outdoors with the appropriate scrubbers/traps installed. The doors of the hood should be closed as much as possible while working and completely closed when not in use. The fume hood should achieve an air velocity at the face of the unit of 100 ± 20 feet per minute at the opening to assure adequate airflow. Chemicals should never be stored in a fume hood.

In the event of ingestion of an irritant, corrosive, or volatile substance, do not induce vomiting. Have the victim drink one or two glasses of water to dilute the chemical. Very toxic chemicals might require the victim to drink activated charcoal slurry to absorb the chemical. Contact the hospital or poison control center immediately.

When a victim has ingested a highly toxic substance call the hospital or poison control center immediately. If the substance has a fatal dose to humans of 1 teaspoon or less, it might be wise to take the risk and induce vomiting by giving the victim salt water or Ipecac syrup.

Nonirritant, corrosive, or volatile substances that have a low toxicity require dilution of the chemical. Have the victim drink several glasses of water and call the hospital or poison control.

Concentrated acid ingestion needs immediate dilution to prevent severe tissue damage. Because the water addition will react with the acid and create heat, give the victim several glasses of very cold water and call the hospital or poison control center immediately. With ingestion of a dilute acid, give the victim cold water and an antacid such as Maalox® or Milk of Magnesia® to neutralize the acid and then call the hospital or poison control center.

If the victim has ingested a concentrated or dilute base, give them cold water to dilute the substance and immediately call the hospital or poison control center.

2.4.6.3   Fire

The use of flammable chemicals such as ethanol, denatured alcohol, hydrogen gas, acetic acid, and acetone is a daily event in the wine laboratory. Denatured alcohol burners are used in a variety of assays; ethanol is used in chemical standards and for sanitizing in the lab, cellar, and bottling lines. Hydrogen is used as a carrier gas for gas spectrometry. Ignition sources such as stirring plates, equipment motors, and static are everywhere and great care must be taken to avoid the chance of a fire.

Proper training in the use and the appropriate choice of fire extinguishers is imperative, and it is the responsibility of the laboratory worker to know the location, operation, and limitations of the fire extinguishers in the work area.

There are four classes of fires and specific fire extinguishers for each class (NFPA):

- Class A: Ordinary combustible materials such as wood, paper, cloth, some plastics, and rubber can be extinguished using either water or a dry-chemical extinguisher.
- Class B: Flammable liquids, oils, greases, tars, oil-based paints, lacquers, and flammable gases can be extinguished using a carbon dioxide or dry-chemical extinguisher.
- Class C: Electrical fire energized by electrical equipment can be extinguished using a carbon dioxide or dry-chemical extinguisher.
- Class D: Combustible metals such as titanium, sodium, magnesium, and potassium can be extinguished using a Met-L-X® extinguisher.

Carbon dioxide extinguishers are recommended for fires involving computer equipment, delicate instruments, and optical systems because the carbon dioxide will not damage the equipment. It is recommended that laboratories be equipped with carbon dioxide and dry-chemical fire extinguishers.

Fires in small containers can most often be extinguished by simply covering the container and suffocating the fire. Containers of burning materials should never be picked up and moved to a sink or outdoors. Small fires can be put out with the appropriate extinguisher. Do not attempt to put out a fire with an extinguisher unless trained and confident in its use. The situation can be worsened by spreading or blowing flammable material to other sections using pressurized extinguishers.

For more serious fires, evacuate the laboratory and activate the nearest fire alarm. Do not attempt to put out the fire. If possible, shut off the gas lines and take the chemical inventory list with you as you leave the site. Be prepared to inform the fire department of the hazardous substances in the laboratory as well as the location of compressed gas cylinders.

In the event that a person's clothing catches fire, have them drop to the floor and roll until the flames are extinguished, or if the safety shower is close, use it to extinguish the fire. Do not let them run or try to beat the flames. Quickly relocate the victim to a safe location, remove the affected clothing immediately, and douse the victim with water. Apply clean, wet, cold cloths on the burned areas until medical help arrives.

2.4.6.4   Gas Cylinder Leaks

Check the cylinder valve and make sure that it is closed. A suspected leak in a cylinder can be confirmed by applying soapy water or a 50% glycerin–water mixture to the suspicious area and looking for bubbles. Laboratory staff should never try to force a leaking valve closed. If the leak at the cylinder valve cannot be closed, reinstall the packing nut or valve gland and notify the gas cylinder provider. If the leak continues, immediately contact the appropriate gas cylinder provider to assist with the problem. Gas cylinders that develop leaks should be removed from the laboratory to a well-ventilated area using the appropriate PPE and transport equipment.

Warnings should be posted to keep the area clear and describe the type of hazard.

Flammable, inert, or oxidizing gas in cylinders should be handled with great care to prevent ignition. A leak in a cylinder of corrosive gas might increase in size as the gas releases, requiring immediate relocation. Toxic gases should be removed from the laboratory immediately by personnel properly protected including self-contained breathing apparatus.

For minor leaks of corrosive or toxic gas cylinders evacuate the area; place a plastic bag over the top of the cylinder, tape it shut, and then transport it to a safe area. Larger leaks might require evacuation of laboratory staff and EPA notification. Handling larger leaks in corrosive or toxic gas cylinders will require proper PPE including self-contained breathing apparatus.

After isolation, toxic and corrosive gas cylinders can be equipped with gas lines incorporating antisiphon valves or traps. The gas is then bubbled through an appropriate neutralizing liquid.

2.5   Physical Hazards

The physical hazards we face in the wine laboratory, cellar, and bottling lines are numerous (Table 2.1). Wine laboratory responsibilities require staff members to traverse into the grape crushing areas, the cellar, bottling line, tasting rooms, receiving, and shipping. We obtain samples, monitor fermentations, inspect bottles, fix equipment, perform analyses, monitor quality, move cases of wine, clean glassware, pick up supplies, ship packages, prepare paperwork, put in computer time, change compressed gas cylinders, and so much more that the risk of an accident or injury can be high if unprepared.

2.5.1   Physical Hazards Associated with Laboratory Work

Common hazards found in most wine laboratories involve equipment, vacuum systems, glassware, cryogenic material, compressed gas cylinders, ergonomics, and the general working environment. Know the risks associated with daily laboratory function, as it is often these small daily events that create the greatest hazards for laboratory staff. The adage "better to be safe than sorry" certainly applies here.

2.5.1.1   Equipment Hazards

It is always prudent to read the instruction manuals for all equipment *before* you begin using a piece of equipment. Keeping equipment in good working order and performing the recommended scheduled maintenance not only keeps the equipment safe but prolongs the life of the equipment. Laboratory equipment is expensive and, for some companies, difficult to justify, so keep it in the best working condition. Working with laboratory equipment poses several risk factors:

- Shock
- Electrocution
- Spark generation
- Fire
- Punctures
- Burns
- Eye damage
- Explosions
- Cuts
- Entanglement

Electrocution, shock, and accidental ignition of flammables can be prevented by using common sense and simple safety rules and practices:

- Use nonsparking induction motors without variable autotransformers in flammable environments.
- Inspect and replace any frayed or damaged cords; check cords monthly.
- Use electrically insulated equipment only.
- Do not use household appliances with series-wound motors in flammable areas.
- Ground all equipment and use with ground-fault circuit interrupter (GFCI).
- Place equipment in a safe location away from spills, flammable vapors, or condensation unless the equipment is designed for that purpose.
- Unplug equipment prior to any repair or maintenance.
- Use only equipment that meets the NFPA National Electrical Code.
- Equip or use equipment with thermal shutoffs.
- Do not attempt repair on equipment unless you are qualified.
- Know the location of all power main switches and circuit breakers in the area.
- Keep work areas clear and clean spills immediately.
- Dust equipment regularly.
- Use caution when working with moving parts.

Heating devices such as hot plates, heating mantles, aquarium heaters, distillation stills, heat baths, and autoclaves should be equipped with variable autotransformers to control the heating level and with temperature-sensing devices to shut off power in the event of overheating, which can cause fires, explosions, and destruction of equipment.

Cooling systems for condensers such as water recirculation units or water flowing from a tap can create unsafe flooding conditions in a short time. Inspect all connections on a regular basis and replace any connection or tubing showing signs of wear. Use proper hose clips and secure the drain tube to the sink or drain to prevent the tube from moving. Quick connects with check valves should be used as much as possible, but if cost is a factor, plastic locking disconnects can be used.

Centrifuges should be equipped with shut-off switches that turn off the power when the lid is raised. Keep your hand out of a moving centrifuge, as serious injury can occur. Balance the sample load in a centrifuge to prevent spilling and ensure smooth operation.

Specialized instrumentation that requires the use of hydrogen or other flammable gases should be well grounded and placed in a room that has static-free grounded flooring or grounded floor mats to prevent any static electric spark.

2.5.1.2   Compressed and Liquid Gas Hazards

Compressed gas cylinders used in the laboratory require extreme caution. First and foremost, any container of gas under pressure should be regarded as a ballistic missile.

Due to the danger of compressed gases, it is wise to order the smallest volume needed to avoid unwanted larger cylinders. The larger the cylinder, the more hazards, and it is not only the cylinder itself but also the handling, storage, and movement of the cylinder that creates additional hazards. Storage, movement, and leak control of cylinders have been addressed earlier in the chapter (see Sections 2.4.3.1, 2.4.3.3, and 2.4.6.4).

Cylinder valve connections are standardized by the Compressed Gas Association (CGA) to prevent mixing of incompatible gases. Cylinder valve connections are straight threaded nipples and are specific to a type of gas. The connection threads will vary in diameter, be right or left handed, and might be inside or outside threads (female/male). The seal is formed by a metal-to-metal interface, eliminating the use of Teflon tape, which can weaken the connection. Only CGA standard combinations of valves and fitting should be used. Adapters should never be used and do not cross-thread a fitting.

A compressed gas cylinder should be equipped with a pressure-compensated regulator that will break down, or reduce, the higher pressure to a manageable working pressure. This break down of pressure is called a stage. Most often for laboratory use, a "double-stage" regulator is used, but the number of stages will depend on the degree of gas control required for the job. A pressure-compensated regulator has more accurate gauge pressure readings because the line pressure does not affect the reading. Unlike other gauges, a higher line pressure can show a lower pressure reading because of back-pressure buildup in the regulator valve. Corrosive gases require regulators made with corrosion-resistant materials and carbon dioxide requires a specially designed regulator.

Cylinders and regulators must be equipped with pressure-relief systems to avoid excess pressure buildup, resulting in the weakening of the cylinder leading to possible explosion. The majority of regulators manufactured today have built-in pressure-relief valves. It is always helpful to know the pressure-relief limit, or pop-off, of the regulator used. Cylinders are never filled to maximum capacity allowing for gas expansion and increased pressure.

Never use a screwdriver or pliers when removing caps and opening valves. Metal to metal can create sparks. When cleaning connections or attaching regulators, never use oil or grease that could react with some gases, (e.g., oxygen). Depending on the gas, it is wise to crack the tank valve slightly, letting out a small burst of gas. This will blow out any debris at the connection that might damage the regulator.

Liquefied gases such as liquid nitrogen are used in the wine laboratory as a coolant in certain assays. Hazards associated with liquid gases include frostbite, degradation of structural materials, asphyxiation, fire, and explosion. In addition, use of a liquid gas cylinder in the laboratory generates an icy buildup along the supply lines that will melt and create a slip hazard. Insulation of the pipes leaving the cylinder can help reduce the amount of water condensation.

Containers for the small volumes of liquid gases that are commonly used in the laboratory are called Dewar flasks. Cylinders built for liquid gases in volumes of 100 to 200 liters are especially built for high pressure, are insulated, and have multiple built-in pressure-relief valves.

Liquid gas cylinders should not be filled to more than 80% of their capacity to allow for thermal expansion. These cylinders should be located in a well-ventilated room to prevent the chance of oxygen displacement by the vaporized gas.

Personnel handling a cylinder exchange should wear appropriate PPE that includes dry gloves and a face shield. First-time users should ask for training from the liquid gas supplier or an experienced co-worker.

2.5.1.3   Vacuum Hazards

Laboratory vacuum systems or vacuum pumps are used on a daily basis in the wine laboratory to evacuate gases from samples or to aid in filtration of substances. Implosion, flying glass, and chemical spattering are the most common hazards associated with vacuum use.

Glass vessels should be inspected for any sign of weakness (star cracks, scratches, and etching marks) before use. Never use a glass vessel for high-vacuum work; glass is used only in low-vacuum operations. Do not evacuate repaired glassware, thin-walled flasks, or Erlenmeyer flasks larger than 1 liter; they just will not stand up to the negative pressure.

Vacuum systems should be outfitted with the proper traps and exhaust. Check with the pump manufacturer for guidelines on the proper care of the system and the types of substance that is or is not allowed in the pumping system. Common building vacuum systems that are used by several departments can create hazards if the wrong chemicals or gases are aspirated and mixed.

2.5.1.4   Glassware Hazards

Cuts, cuts, and more cuts! Inspect the glassware for chips and cracks daily and before using it in any type of assay. Glassware that is chipped or cracked should be discarded into a proper glass waste container or set aside for repair.

Wine bottles used for samples should be inspected daily as well. Chipped or cracked bottles should be discarded into glass-recycling bins. Use a carrier to transport sample bottles to the laboratory. Sample bottles with fermenting juice can explode if they are capped tightly. Always leave juice sample bottles loosely capped.

Secure glassware in dishwashers to prevent damage. Drawers and cabinets can be lined with nonslip materials to protect glassware from moving and possible damage. Never carry glassware in wet or soapy hands. Use two hands when transporting larger pieces of glassware to prevent dropping.

Replacing polished rounded-edge glass tubes in stoppers or corks requires extreme caution. It might be prudent to assess the cost–risk factor when replacing these glass tubes. It can be less expensive to buy new tube/stopper setups than risking injury. If replacing the tubes, use the proper PPE with the addition of a towel or rubber barrier around the glass tube in case it breaks. The tubes should be lubricated well before inserting them into the stoppers.

Use the proper PPE when opening glass vials. Snapping off the tops can send glass flying or the vials can break in your hands.

Leaving a distillation still's heating elements on and unattended during use can lead to evaporation of water in the boiling chamber and expose the heating element. This will result in overheating of the glass. The hot glass can render severe burns and can explode, showering the area with glass (see Fig. 2.5). Never leave a distillation still unattended. Never add cold water to a hot dry still, and always turn the heating element off and let the glass cool.

2.5.1.5   Slip, Trip, and Fall Hazards

Accidents due to slips, trips, and falls are high on the accident lists of most laboratories. One of the main causes of these types of accident is bad housekeeping. Floor spills must be cleaned immediately, be they chemical or plain water. Wet floors during rainy or snowy weather can create a big hazard and putting down nonskid door mats is a simple solution. Placement of nonskid floor mats in potentially slippery areas such as in front of the dishwashers, sinks, wet chemistry areas, and liquid gas tank storage areas can eliminate the chance of slipping.

Electrical cords, boxes on the floors and in stairwells, uneven floors, stairs, and computer connections are examples of trip hazards that could result in a fall.

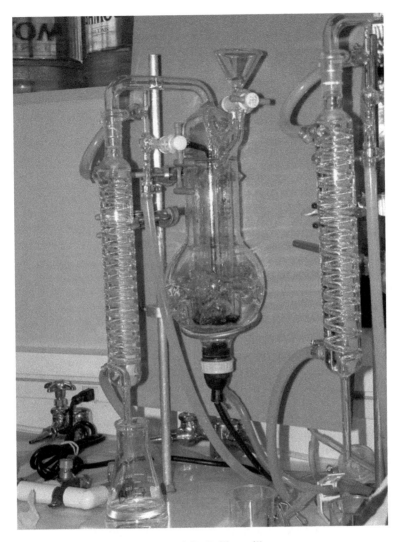

FIGURE 2.5. Boiling still.

2.5.1.6   Ergonomic Hazards

All employees should be trained in the proper ergonomic practices for their particular work area. Injuries to joints, ligaments, muscles, tendons, spine, cumulative trauma disorders, and eye strain are on the increase in the workplace. The National Institute for Occupational Safety and Health (NIOSH) guide U.S. DHHS, 1994, should be consulted for more information.

Starting the day out with a good stretch and taking stretching breaks throughout the day can be helpful to keep muscles loose and promote good circulation. Most people are prone to strains and pulls after they eat because the blood tends to migrate from the muscles to the internal organs. Stretching after a meal will get the blood stirring and back into the musculature. To keep alert, blood circulating, and help prevent injury, take all assigned breaks. Take the time to put your feet up and relax or take a slow relaxing walk.

Ergonomic hazards include the following:

- Sitting for long periods
- Standing for long periods
- Lifting
- Carrying
- Pushing/pulling
- Repetitive motion

Long periods of sitting are most often associated with computer work. It is recommended to use a chair with a backrest that provides constant support of your lower back curve and allows sitting deep in the seat without pinching or rubbing the lower thighs. Sit 18–24 in. away from the computer monitor, with the screen just below eye level. Sit the keyboard on a flat surface or angled down just below elbow level and at a distance that will let the elbows remain at the side. Hands should be in a neutral and relaxed position with no bend. Do not remain in the same position of long periods of time. Get up, walk around, or stretch frequently.

Many laboratory workers stand for long periods of time performing analysis. It is important to have good supportive shoes and well-placed stress-reducing nonskid floor mats. A footrest can be used to vary stance during the day. Try not to slump, and keep the shoulders erect to prevent strain on the upper back. Tilting the pelvis will help to relieve lower back strain, and knee strain can be reduced by keeping the knees flexed slightly. Sit when you can and stretch occasionally.

Repetitive strain injuries occur from continuous physical stress over a long period of time. Computer work, manual pipette work, turning stopcocks, and so forth can create damage of the nerves, ligaments, and tendons of the wrist. Try to keep wrists in a neutral and relaxed position. Frequent shaking and rubbing of the hands and wrists helps increase circulation. When performing a twisting motion, stretch the opposing muscles occasionally. Opening and closing the hand into a fist will help relax it and increase blood supply.

Lifting is the major cause of back strains in a laboratory. Laboratory staff will move boxes of supplies, boxes of wine samples, 5-gal water containers, and equipment. Proper lifting techniques help reduce strains and back injuries. Lifting should always be done from a stable position, leverage advantage, and functional comfort. Straddle the object, maintain spinal curves, lift with the legs, never twist the body, and smoothly rise up into a standing position.

Never reach around and pick up an object when the body is twisted. To pull or push an object, stabilize the elbows against the legs, shift the body weight using the legs, maintain spinal curves, and do not overextend or reach.

2.5.2   Physical Hazards in the Cellar

Most of the safety precautions discussed in Section 2.5.1 applies here, but there are a few hazards indigenous to the cellar that necessitates mentioning.

The winery cellar is wet most of the time from constant wine movements and cleaning. The use of electrical pumps, stirrers, and filtration equipment pose a grave danger for electrocution and great care must be taken to ensure that the equipment is well grounded and in perfect working condition. Slips are very common in this wet environment, so wear shoes with good traction, maintain good housekeeping, and post hazard signs to warn others.

The cellar contains tanks that hold wine and they can vary in capacity from a few hundred gallons to 100,000 gallons or more. Drowning is a very real danger when working around these tanks. Untrained personnel should never be allowed to obtain samples from the tops of tanks. Never lean into a tank or stick your head in a tank.

Hoses used in the winery cellar are large, high pressure, rugged, and very heavy. Strains, pulls, and back injuries are very common. Large hoses should never be moved by a single person; always ask for help. The hoses are everywhere and create a trip hazard at every step. Winery cellar personnel should be diligent in keeping unused hoses put away in hose beds and alleviate as much hose congestion as possible.

A full tank of wine puts a great deal of pressure on hose connections. If a hose is not connected properly or the hardware is weakened, it is not uncommon for hoses to blow off tanks. This serious danger could cause severe injury or death. Double-check the hose connections before the valve is opened. Leaking connections should be reported to the cellar staff immediately. Leaks can be a sign of failure.

Because of the tremendous pressure of a full tank, care should be taken when opening a tank valve for any reason. If a valve is opened too fast, wine can gush out at a tremendous velocity, knocking a person to the ground. Go slow, crack the valve open, and then open it very slowly.

Tanks are usually filled with an inert gas after cleaning to keep the tank sanitary. *Never* enter an empty tank without the proper training and breathing equipment and *never* enter it alone. A person can be rendered unconsciousness after just a few breaths. Sadly, the majority of deaths in wineries are the result of asphyxiation caused by the worker entering a gassed tank.

Around the tops of large tanks there are normally stainless-steel walkways called "catwalks" that allow workers to access the tops of the tanks. These catwalks and the stairs that lead to them can become very slippery. Attention must be paid to where a person steps and both hands should be kept on the handrails at all times. Figure 2.6 shows a typical work area, tanks, catwalks, and forklift traffic.

Wooden barrels in the cellar normally have a capacity of 55 gal and create their own hazards, such as crushing, splinters, cuts, abrasions, muscle strains, and back injury. Wear adequate PPE to prevent splinter, cuts, and abrasions. *Never* attempt to move a full or partially full barrel. Barrels should only be moved on a barrel rack using a forklift. Never climb barrels or stacked barrel racks to obtain samples. Contact a cellar employee with a forklift for assistance.

Forklifts are in constant motion in a cellar—moving barrels, pallets of materials, and cases of wine. When laboratory staff members are working in areas of low visibility, such as in the midst of stacks of barrels, wearing a bright safety vest is advised. It is difficult for forklift drivers to see the workers among the barrels. Barrels can be stacked on racks several barrels high and close together in long rows (see Fig. 2.7). The vest will alert the drivers to the worker's presence. In addition, some wineries require that brightly colored flags be place outside the row of stacked barrels to alert the forklift drivers that someone is working in that row of barrels.

During fermentation of the grape juice carbon dioxide is produced as a by-product from the conversion of sugar to alcohol. Working in the cellar at this time of the year requires extreme caution and implementation of procedures

FIGURE 2.6. Work floor, tanks, and catwalks.

FIGURE 2.7. Stacked barrels in cellar.

to prevent the buildup of carbon dioxide in the cellar areas. Most modern-day wineries have air-exchange units installed that will cycle on when the carbon dioxide reaches a certain level, exhausting the heavier carbon dioxide-laden air near the floor while bringing in fresh outside air. Older wineries keep their doors open and use fans to circulate the air. Regardless of the method, it is prudent to manually check the carbon dioxide levels with a meter several times a day. Staff working amid fermenting juices might require carbon dioxide-metering alarms that will allow them to exit the area if the carbon dioxide level begins to rise. Taking frequent breaks and getting some fresh air is a wise practice to incorporate during this period.

Equipment used to bottle wine is large, noisy, and dangerous. PPE must be worn at all times and should include proper eye protection, proper clothing, and hearing protection.

Bottles smashing to the floor as they fall off the moving conveyor belts, bottle tops shattering as the corker jaws (used to hold the bottles in place while the cork is inserted) lose their adjustment, and bottles exploding from a misaligned filler nipple are all elements of the cacophony of bottling. A noise conservation program is required by OSHA if the noise level exceeds the 8-h time-weight average of 85 decibels. Hearing protection in the form of earplugs or earmuffs is required in areas where the noise levels exceed 90 dB. Comprehensive monitoring, personnel training, and hearing tests for employees must be carried out by a facility. Eye protection is mandatory in this area.

Entanglement is a very real hazard when working around the bottling equipment. There are many moving parts that can catch hair, hair nets, hands, sleeves, and jewelry. Use extreme caution and adhere to safety policies and procedure for emergency shut-off. Shut down equipment prior to repair or maintenance.

The floors are usually wet and in some areas very slippery due to lubricant used on the conveyor belts. Wear shoes with good traction, maintain good housekeeping, and post hazard signs to warn others.

2.5.3   Damage Control

The professional laboratory worker follows a few commonsense rules to ensure everyone's safety. Table 2.5 lists the laboratory safety "rules" that most laboratories follow. Every laboratory should create its own set of rules and guidelines depending on the type of work performed in that laboratory and the current company policies.

TABLE 2.5. List of safety rules for laboratory professionals.

- Do not startle or distract other workers.
- Do not allow practical jokes or horseplay.
- Do not prepare food, eat, or drink in areas using chemicals.
- Do not smoke.
- Do not apply makeup or take medication in areas using chemicals.
- Do not use laboratory refrigerators for food storage.
- Do not taste or smell chemicals.
- Do not siphon or pipette by mouth.
- Do not use the telephone while performing analysis.
- Do not store chemical containers on the floor.
- Do not obstruct access to exits and emergency equipment.
- Do not use stairways or hallways as storage areas.
- Do not leave a chemical reaction unattended.
- Do not work with chemicals when alone.
- Do not run with chemicals or glassware.

TABLE 2.5. List of safety rules for laboratory professionals. (*Continued*)

- Do not wear laboratory coat and gloves to the restroom.
- Always be familiar with the chemicals you are using.
- Always wash your hands thoroughly after using chemicals, gloves or no gloves.
- Always clean spills immediately.
- Always carry glass containers using two hands.
- Always secure gas cylinders.
- Always use an exhaust fume hood when working with inhalation hazards.
- Always properly store chemicals.
- Always keep your work area clean and neat.
- Always label every beaker, flask, or container containing any substance, even if it is temporary.
- Always unplug equipment before repairing.
- Always remove your laboratory coat and gloves when leaving the laboratory.
- Always wear appropriate personal protection equipment.

It is the responsibility and requirement of the employee to understand and act in accordance with the safety requirements established by the laboratory and facility, wear and properly maintain PPE, follow good chemical hygiene practices, participate in all required training programs, read and understand health and safety standard operating procedures, divulge all facts pertaining to accidents, and provide required information to examining physicians.

Hey, be safe out there.

3

What's Your Number?

3.1 Introduction

For ages, a winemaker's mouth and nose were the only detection methods needed (or available) to evaluate wine quality. Frequent tasting and smelling of each barrel and tank required dedication and a great deal of time. Unfortunately, these simple methods were not perfect and winemakers expected to lose a percentage of the year's wine to undetected problems.

As the wine industry continues to change and grow, and the demand for wine in the marketplace increases, more burdens are placed on the winemaker. Time spent with the wine shortens as the marketing, sales, and business needs increase. With rising fruit and production costs, wineries today cannot afford to lose upward of 20% of a year's production to spoilage and other undetected problems.

The evolution of wine science (*enology*), wine chemistry, and the wine laboratory is directly related to the needs of the industry. Understanding and controlling the winemaking process leads to the production of quality wines. The ability to break down the primal sensory methods of taste and smell into a science allows us to identify and quantify components in juice and wine for evaluation and study. The establishment of enology and wine chemistry research programs in colleges and universities around the world continues to add to our base of knowledge.

Today's breed of winemaker is well educated and has the ability to meld sensory evaluation with the analytical test results. The wine laboratory tests and monitors the juice and wine, providing the winemaker with valuable information. The laboratory can quickly alert the winemaker when a problem is detected. Taste and smell are, and always will be, one of winemaking's basic tools for front-line detection of problems and in the winemaking process. Using all available methods of evaluation avoids unnecessary loss of wine and improves quality.

Commercial wine laboratories cater to smaller wineries, where "in-house" basic wine analyses (performed or carried out at the winery) might not be cost-effective. These laboratories are equipped to provide basic analyses in addition to sophisticated wine analyses for wineries that do have a basic in-house laboratory. Most commercial laboratories are certified by a quality assurance (QA) organization such as the Association of Official Analytical Chemist International (AOAC), International Organization for Standardization (ISO), International Electro-technical Commission (IEC), American Association for Laboratory Accreditation (A2LA), or the Alcohol and Tobacco Tax and Trade Bureau (TTB). To ensure the safety of the public, the TTB and most countries require certified laboratory test results for national and international commerce. Certified laboratories perform and document the analysis.

Chemical analysis of wines by a commercial laboratory can be very expensive and the results might be delayed. Wineries should assess their need for fast results, cost of analysis, and labor considerations when evaluating the need for an in-house laboratory.

For winemaking, the wine laboratory can be a godsend or a curse. In the past, the ability of the laboratory to provide accurate results to the winemaker has been a key concern for many wineries. This lack of confidence in laboratory results usually stems from poor staffing and equipment.

Many wineries do not provide the funds for qualified laboratory staff, which results in hiring people off the street or taking them out of the cellar to perform analyses. Typically, the winemaker or enologist is responsible for training these people, but many of them have had limited laboratory experience. As the industry continues to grow and becomes more technologically advanced, the need for quality analyses is, and will be, imperative to quality wine production. It is important to remember that one error in analysis can ruin wine and cost a company thousands of dollars.

Lack of confidence has perpetuated the "do-it-again" mindset. When a winemaker receives laboratory results that might indicate a problem, the laboratory is asked to "do-it-again." If the laboratory staff performs a test over and over, depending on the amount of error in the test, they are bound to produce a result that might be more palatable. Laboratory staff might feel compelled to manipulate the original results, making them more acceptable to the winemaker. The "do-it-again" mindset increases the cost of analyses because of redundancy, it brings down moral, and it creates a stressful atmosphere. The bottom line, wine quality, suffers.

Building confidence in laboratory analyses, improving performance, and increasing efficiency can be accomplished by hiring trained personnel, developing a QA program, and creating a chain of command with accountability. There must be recognition of objectives and goals and a commitment by management and staff for any QA program to be successful. The amount of time and money required to institute a QA program will depend on the winery and the extent of their commitment to wine quality.

3.2 Quality

"Production of high quality analytical data through the use of analytical measurements that is accurate, reliable, and adequate for the intended purpose"* is the principle objective of an analytical laboratory (Garfield, 1984, p1). Quality analytical data are the key to the development of confidence in the wine laboratory. Quality control (QC) involves setting down a working program of systematic procedures that promotes the creation of a quality product. Quality assurance is a planned program that assures the effectiveness of the quality control measures in place.

The National Institute for Occupational Safety and Health (NIOSH) has suggested a number of elements for QA programs (United States Dept. of Health and Human Services Center for Disease Control, 1976); see Table 3.1. The more elements incorporated into a particular QA program, the more effective the program, leading to a higher quality of wine produced.

The Achilles' heel of any good QA program is inferior staffing, old or misused equipment, poor analytical techniques, bad samples, inadvertent errors, incomplete recordkeeping, and lack of support and commitment from management.

3.2.1 Setting up a QA Program

For an effective QA program, it is beneficial to meet with a member of management, the laboratory supervisor, and a winemaker to design a program that best fits the winery and the level of wine quality desired. This group of professionals can set the QA objectives, policy, and implementation time lines, agree on standard operating procedures (SOP); determine the quality of personnel hired, design a QC format for each stage of wine production and acceptance ranges for each analysis, develop a line of authority, and make the commitment to achieve the highest level of quality possible.

Large companies have a corporate or upper-management laboratory position designated as the QA Coordinator or Laboratory Director. This person

TABLE 3.1. NIOSH QA program elements.

Statement of objectives	Preventative maintenance
Policy statement	Reagent and reference standards
Organization	Procurement and control
Quality planning	Sample identification and control
Standard operating procedures	Laboratory analysis and control
Recordkeeping	Interlaboratory and intralaboratory testing
Chain of custody procedures	programs
Corrective action	Handling, storage, and delivery
Quality training	Statistical QC
Document control	Data validation
Calibration	System audits

sets QA policy and guidelines for the winery. The coordinator or director meets with all laboratories to develop, implement, and monitor compliance with the QA program. Smaller companies rely on the laboratory supervisor or lead technician to implement and monitor the QA program.

The responsibility of compliance with the set QA program is defined by the line of authority, or accountability. The winery or laboratory staff member who obtains samples from the cellar is the first level of QC in the wine laboratory. The bench technician, the person performing the test, is directly responsible for the quality of their work and is the second step in QC. From this point, the QC responsibility is passed to the laboratory supervisor; then the laboratory director or QA coordinator, and on to top management.

Interlaboratory and intralaboratory testing assists with evaluating the success of the QA program. These tests might be conducted by the laboratory director, QA coordinator, laboratory supervisor, winemaker, or management. This is an effective way to ensure that all QA measures are in compliance and working effectively. Testing allows the opportunity to evaluate where additional personnel training can be effective or indicate problem areas needing attention. It is important to document all results, problems, and the steps taken to correct those problems.

Issuance of a QA manual is a valuable reference for wine laboratory staff. The QA standard operating procedures (SOP) manual should delineate the elements of the QA program and include the following:

- Organizational chart
- QA objectives and policy
- Performance and system audit requirements
- Corrective action procedures and verification
- Forms
- Description and definition of required QA reports to management
- Distribution list

3.2.2 Laboratory Staff

Planning, organization, staffing, and leading the personnel of a wine laboratory can be a challenge, but it is essential to a good QA program.

Placing a properly trained person in a specific laboratory position is half the battle in fighting poor QA. It is an invitation to disaster to place an unskilled person in a position where they are bound to fail.

The laboratory director manages all personnel and resources within an organization. The director's duties might include setting analysis qualification requirements, developing standard operating procedures, managing and creating budgets, communicating with all departments, recruiting and hiring new personnel, training and educating laboratory staff, dealing with personnel issues, and providing the resources required by the staff to fulfill the commitment to QA.

A laboratory supervisor will support the laboratory director, supervise and direct workflow, provide guidance to staff, perform appraisals, review completed work, implement training and orientation, begin personnel actions, and is the front-line support of the QA program.

The laboratory technicians are placed in key positions determined by their education, experience, and training. This allows each technician to successfully perform a certain group of tasks. A successful laboratory staff member possesses and displays a variety of qualities that ensures professionalism and commitment to quality:

- Good judgment
- Ingenuity
- Inquiring mind
- Independent thought
- Understanding of scientific principles
- Understanding of scientific procedure and notation
- Understanding of analytical methods
- Striving for quality in their personal performance and work
- Awareness of analytical changes
- Recognizing a potential source of error
- Proficiency using equipment and instrumentation
- Understanding the benefits of SOPs

Laboratory technical support staff and harvest temporary staff perform noncomplex analyses, housekeeping, sample preparation, data entry, and monitoring fermentations. They possess the ability to understand the work assignments and importance of those assignments.

3.2.3   Analytical Methods

Analytical methods are procedures or processes for separating a substance into component parts or constituent elements. A valid method provides evidence and identification of the analyte, verification that the analyte can be isolated from interfering substances, determination of the lowest measurement of analyte attainable via the method, assurance that a reasonable degree of precision between laboratories can be achieved, and proven accuracy of the measurement.

The use of standards and controls provides the evidence of identification and isolation of the analyte by the method, verifying that the method is performing satisfactory for its intended use. Validation and verification of the analytical methods used in the wine laboratory give legitimacy to the results produced.

The exactness of determination is directly affected by the analyst, laboratory, substance concentration, analyte stability, interference, contamination, and the limitations of the method. Minimally, each analytical method should have verification of the degree of error, variation, accuracy, ruggedness, precision, limits, range, correlation coefficient, and linear regression.

3.2.3.1   Method Validation

Accuracy is termed as the degree of agreement (how close) an individual measurement, (or mean of a set of measurements) is to the true or known reference value of the property being measured. Basically, accuracy is comparing results of one measurement, or a mean value of measurements, to the measurement of a known reference.

The difference between the largest and the smallest measured value within a small sample set of less than 10 is called a range.

The calculation for range (R) is

$$R = X_L - X_S,$$

where X_L is the largest value and X_S is smallest value. The range mean $\bar{R}$ is the averaged midpoint of several smaller sets of ranges:

$$\bar{R} = \frac{\sum R}{K},$$

where R is the ranges of the sets; and K is the number of sets. The smaller the range number, the closer the results are to the known reference. In a perfect world, we could say that a range or range mean of zero is ideal; (i.e., no difference between the results and the known value). In the real world, we have systematic errors and random errors that merge and create a variable composite quantity with each determination that prevents the range from ever reaching zero.

Random errors vary in magnitude and scatters measurements of replicate samples above and below the calculated range mean, affecting the precision of a set of measurements. Random error is determined by taking a measurement and subtracting the mean result of a larger number of repeated measurements of replicate samples. The results can be positive or negative.

A systematic error, or bias, is a constant error whose magnitude does not vary among determinations made on replicate samples. Systematic error causes the difference between the known value and the analytical result to have predominately the same sign, positive or negative. It will cancel out the majority of random errors that are of the opposite sign. Systematic error is calculated to establish the magnitude of error of the method and allows for it when using the method.

Systematic error and random error create a variance that is simply how much spread there is between measurements, or precision. The precision of a method can be defined as the degree of repeatability, or reproducibility, of individual measurements of a set of replicate samples. Variance (s^2) affects the precision of the method and is the statistical measurement of the errors squared. Standard deviation (s, σ) is the square root of the variance and is the most common statistical measure of precision. The calculation of variance is

$$s^2 = \sum \frac{(X - M)^2}{N - 1},$$

where X is the individual measurement value, M is the mean value of all replicate sample measurements (mean = sum of all replicate measurement values ÷ number of measurements), and N is the total number of measurements. The best standard deviation calculation for small datasets is

$$s = \sqrt{\sum \frac{(X - M)^2}{N - 1}},$$

where X is the individual measurement value, M is the mean value of all sample measurements (mean = sum of all measurement replicate values ÷ number of measurements), and N is the total number of measurements. For larger sets (or populations) the standard deviation uses the root-mean-square deviation from the true mean:

$$\sigma = \sqrt{\sum \frac{(X - \mu)^2}{N - 1}},$$

where X is the the individual value, μ is the true mean value of all measurements in a population, and N is the total number of measurements.

The coefficient of variation (relative standard deviation) is the standard deviation of a set of values, divided by the mean, and multiplied by 100 to give a percentage. In a normal distribution, 68% of the results will fall within one standard deviation of the mean value. Ninety-five percent of the results will fall within two standard deviations of the mean value (Garfield, 1984).

The measurement of the dispersion of values relative to their mean is called the coefficient of variation (CV) and is expressed as a percentage:

$$CV = \left(\frac{s}{M}\right) \times 100,$$

where s is the standard deviation and M is the sample mean.

A variate is the variable quantity under study—the variable measurement. Variate values become more complex as the statistical determinations become more sophisticated. Table 3.2 defines the equations for the variate values.

TABLE 3.2. Variate values.

Mean of variate values	Sum of variate values	Sum of the squares of variate values	Sum of the squares minus the sum of variate values squared divided by the number of measurements
$\overline{X}$	X	X^2	$xx = X^2 - \dfrac{X^2}{n}$
$\overline{Y}$	Y	Y^2	$yy = Y^2 - \dfrac{Y^2}{n}$
		XY	$xy = XY - \dfrac{XY}{n}$

Source: Adapted from Garfield, 1984.

The correlation coefficient (CC) indicates how a known value or values (independent variates X_1, X_2,...) and a measured value or values (dependent variates subject to error Y_1, Y_2,...) vary together. A coefficient of +1 or −1 indicates a very tight linear relationship between the measurements. Coefficients near zero tell you there is no relationship at all between measurements; so back to the drawing board. The calculation for the correlation coefficient is

$$CC = \sum xy \left(\sqrt{\sum xx \sum yy} \right)^{-1}.$$

The correlation coefficient should be taken to a minimum of four decimal points; the more decimal points, the more defined the result.

Confidence is a statistical term used to express the degree of confidence in a measurement. The upper and lower limits for an interval associated with a confidence coefficient of 1, minus the constant (a), is called the confidence interval. A confidence interval is a range of values that has a high probability of containing the substance being studied and is expressed as a percentage. In most cases, a confidence interval of 95% is considered to have a 95% chance of containing the population mean. Calibration curves should have a confidence interval of 95% or higher.

Linear regression is another way to look at the relationship between variation and the determined results. In linear regression, you are to assume that the relationship between all of the result variations is linear and there is one independent variable (i.e., measurement of one substance). Linear regression (Y) is expressed in terms of a constant (b_0) and a slope (b_1 regression coefficient) times the variable (X), basically a one-dimensional surface in a two-variable space:

$$Y = b_0 + (b_1 X) \quad \text{for a sample of variate values}$$

$$Y = \beta_0 + (\beta_1 X) \quad \text{for a population of variate values}$$

where b_0 is a constant (intercept); b_1 the slope, and X is the variable determined results.

Slope is calculated as

$$b_1 = \frac{\sum xy}{\sum xx}.$$

Multiply the slope by 100 if expressed as a percentage. A slope between 97.5% and 102.5% is considered excellent.

The intercept is calculated as

$$b_0 = \overline{Y} - b_1 \overline{X}.$$

Figure 3.1 illustrates linear regression for one variate where the Y is the mean value of Y corresponding to a given value X. When there are several

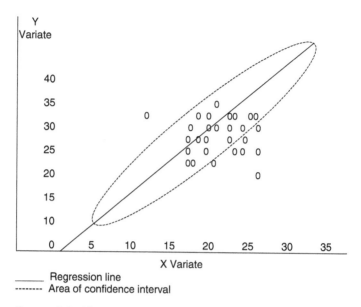

_____ Regression line
-------- Area of confidence interval

FIGURE 3.1. Linear regression and confidence interval illustration.

variables being measured (i.e., acetic acid and pH), you will calculate for multiple regressions, which are the relationships between several independent variate values and a dependent variate.

The coefficient of determination (CC^2) is the square of the correlation coefficient: the explained variation divided by the total variation. It is the percentage of variation that can be explained by the regression equation, or simply how close the points are to the regression line. The points farther away from the line are less likely to be explained. Take the number of points close to or on the regression line and divide by the total variation; this is the explained variation. The points that remain are the unexplained variations. A CC^2 between 0.9750 and 1 has superb precision, whereas a CC^2 between 0.9500 and 0.9750 has good precision (Garfield, 1984).

Figure 3.2 illustrates the relationship between the determined mean of each variate population to its known concentration and how the points sit with regard to the regression line. Note the intercept and slope relationship to the regression line.

Many methods have been developed, validated, and verified. The AOAC maintains a list of these methods. All recognized methods are categorized as follows:

- Official method: required by law or regulation
- Reference method: developed by a college or organization and tested
- Screening method: determines the need for more accurate testing
- Routine method: an official or reference method used on a routine basis
- Automated method: automated equipment used to run a method

Variate Mean to Known Concentration

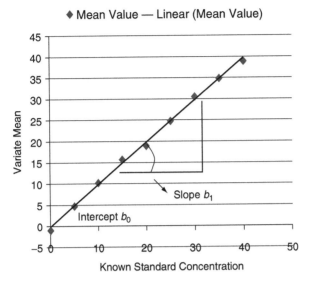

FIGURE 3.2. Relationship of the variate mean to its known concentration with regard to the regression line. Intercept b_0 and slope b_1 relationship to the linear regression line.

- Modified method: taking the official or reference method and changing it to accommodate a different matrix or to remove an interfering substance that was not accounted for in the original method

To validate a method, it is imperative that the laboratory look at similar methods to decide what controls, synthetic formulations, spiked matrices, or standards should be used. Taking the time to design the validation method will be well worth it.

Know the range of measurements encountered with all types and varieties of juice and wine being analyzed, and the upper and lower detection limits. Choose or create replicates of samples (standard solutions) with known concentrations that span the measurement range and use them as variables to validate the method. Using standard solutions will tell you whether the method can accurately measure the analyte, but it does not tell you how the method will respond to the wine and juice matrices.

Research the analyte of concern for the possibility of interferences from substances found in juice and wine, such as high concentration of sugars, alcohol, or acetic acid. If the standard solution verifies the method, the next step is to spike all possible real wine matrices with the known analyte at concentrations that span the range, plus the addition of several ranges of substances that could cause interference.

Most of the literature states that 10–20 times as many measurements (observations) should be made per variable with the outliers (extreme errors) eliminated from your data. The more variables, the more measurements of the known replicate must be made. Variables can include more than one analyst performing the analysis, running the analysis at different times of day or days of the week, or using spiked standards of different concentrations. The validation results will be much more accurate when considering as many variables as possible and accounting for them in your statistical evaluations.

Methods should be evaluated for their practicability and ruggedness for routine laboratory work. Varied changes in the operating conditions or a situation where the method requires specialized equipment or dangerous reagents could be a major deterrent to incorporating the method into the laboratory routine.

Many laboratories purchase statistical calculators or statistical software as an aid in making calculations. Software such as EP Evaluator by David Rhoads is beneficial and helps eliminate mathematical errors, which are the number 1 statistical mistake. If this is not an option or you require additional information, you can purchase references that range from *Statistics for Dummies* by Deborah Rumsey to advanced textbooks. The Internet contains a wealth of information with numerous websites to choose from. In addition to many textbooks and reference manuals on the market, the AOAC publishes several reference books specifically targeting method validation that can be purchased or referenced from their website (AOAC.org).

3.2.3.2   Reagents, Standards, and Controls

Standards, controls, and precisely prepared reagents are essential elements of any analytic method. Reagents and standards are graded on purity:

- Primary: analyzed by lot and certified to a specific percentage of purity
- Analyzed reagent: conforms within specified tolerances also referred to as reagent grade
- USP and NF: chemical reference standards for which the strength and identity have been established and guaranteed, used as a comparative standard (control)
- Pure: chemically pure, uncontaminated
- Practical: preliminary substance
- Technical/commercial: varies widely in purity

The wine laboratory uses a variety of reagents, standards, and controls to perform analyses, correctly calibrate equipment, and ensure proper QC. Reagents, standards, and controls can be purchased or made in-house. Those that are made in-house are made with high-quality chemicals, solvents, and water. The laboratory should maintain a supply of deionized (DI) and pure distilled water. A common error is using poor quality water in making solutions. Deionized water is just that—deionized; it does not have impurities removed. Use purified water that is graded Type 1 by the American Society for Testing and Materials (ASTM).

A standard is a prepared solution with a known concentration used for calibrations, method validation, method troubleshooting, or any analyses that requires precise recovery of an analyte. Primary reference standards are best to use because they have been certified by a qualified laboratory. Many laboratories prefer to use a working standard made in-house. Working standards should be analyzed by the best available method in the laboratory. Maintain all information concerning the making of standards in a log book; include the analyst name, date analyzed, raw data, method of preparation, and standardization calculations. It is recommended that two analysts initial each entry. The container must display the identity, purity, potency, date made, and expiration date of the standard. It is more desirable to use a synthetic formulation. A wine matrix spiked with a known quantity of the analyte under study is not as rigorous as synthetics.

In the wine laboratory, controls are usually stable, routine wine samples that have been analyzed by the method. They are used as a comparison for monitoring the QC of the method. Depending on the method, controls can be analyzed week to week, day to day, or analysis to analysis. Control data should be charted in a log book along with the date, run number, and analyst's initials. It is prudent to analyze several control replicates when a new control of unknown concentration is introduced. This establishes a mean, standard deviation, and acceptable limits. The use of a purchased certified or preanalyzed control eliminates the need to establish the statistical information, it is usually more stable, and the expiration date is established. Unfortunately, these controls can be expensive, but considering the labor costs in troubleshooting method problems, they are well worth the money. The control container should display the date analyzed, the mean measurement, the expiration date, if any, and the analyst's initials. It is imperative that the control be stable and remain pure throughout its use. If there is any chance that the control has been altered, discard it immediately and replace it with a new control.

When the control data are analyzed, a method's measurement accuracy is considered out of control and analysis is stopped when the following occurs:

- ≥ 1 point is outside three standard deviations
- ≥ 4 points are outside one standard deviation

A shift in a control's mean might be due to poor standard preparation, contamination of reagents, analyst error, inaccurate instrument calibration, or equipment failure. If the control's measurement has a tendency to increase, it is most likely the result of the deterioration of the standard or reagent. A decrease in the control's measurements can indicate deterioration of the reagent or concentration of the standard due to evaporation. The analyst might be the source of error when the control's measurements are variable.

Reagents are mixtures or substances that are used in chemical analysis or synthesis. The majority of reagents used in the wine laboratory are working reagents, and a great number of the guidelines that apply to working standards applies here. Most common working reagents are used in large quantities and are made frequently. Most high-volume reagents are identified with the name, concentration, and date made and are verified by standardization or analysis.

3.2.4   Analytical Techniques

The most expensive and time-consuming laboratory errors that directly affect the quality of analyses are errors in making reagents, standards, controls, and measuring sample aliquots. These errors are usually caused by poor analytical techniques. Typical errors can include the following:

- Miscalculations
- Use of the wrong equation
- Conversion errors
- Errors in expressing concentration
- Poor measuring techniques
- Sloppy technique
- Use of dirty glassware and pipettes
- Use of the wrong chemical
- Distilled or deionized water not used

All of these errors are preventable.

Measuring equipment can cause errors but these tend to be infrequent if a good preventative maintenance (PM) program is in effect (equipment PM will be discussed later in this chapter).

3.2.4.1   Calculations

Laboratory work requires understanding calculations for volume and weight and the expressions of those relationships.

Weight per volume (w/v) expresses the weight of a unit of solute in a known volume. For example, 6 g/L malic acid contains 6 g of malic acid for each liter of solution. The calculation for weight per volume is

$$w/v = \frac{\text{Unit weight of solute}}{\text{Unit volume of solution}}.$$

A milligram per liter (mg/L) expresses the number of milligrams of a solute in 1 liter of solution. For example, 13 mg/L sulfur contains 13 mg of sulfur in each liter of solution. The calculation for milligrams per liter is

$$mg/L = \frac{\text{Unit weight of solute (in mg)}}{\text{Volume of solution (in liters)}}.$$

Parts per million (ppm) is a concentration expressed as parts of solute per million equivalent parts of solution. In highly diluted aqueous solutions, mg/L equals ppm and the solution weight is equated to the solution volume. The expression mg/L is preferred over ppm.

Volume percentages (% v/v) express the concentration of 1 unit of 1 solute in 100 units of the same solution. For example, 70% ethanol by volume contains 70 mL of ethanol in 100 mL of solution and is expressed as 70% v/v. The calculation for volume percent is

$$\% \text{ v/v} = \frac{\text{Unit weight of solute}}{\text{Unit weight of solution}} \times 100.$$

Weight percentages (% w/w) express the weight of 1 unit of solute in 100 units of the same weight of solution. For example, 10% tartaric acid by weight contains 10 mg of tartaric acid in 100 mg of solution and is express as 10% w/w. The calculation for weight percent is

$$\% \text{ w/w} = \frac{\text{Unit weight of solute}}{\text{Unit weight of solution}} \times 100.$$

Weight per volume percentages (% w/v) expresses the weight in grams of solute in 100 mL by volume of solution. For example, 1.5% acetic acid by weight and volume contains 1.5 g of acetic acid in 100 mL of solution and is express as 1.5% w/v. The calculation for weight per volume percent is

$$\% \text{ w/v} = \frac{\text{Unit weight of solute}}{\text{Unit volume of solution}} \times 100.$$

The winery laboratory uses many different calculations to make reagents, standards, and controls. The most common calculations determine the following:

- Proportions
- Additions
- Stock additions
- Dilutions

Proportional calculations separate the substance (S) of interest from a compound (C). As an example, from the compound potassium chloride (KCl), 4 g of K^+ is needed. To calculate the weight of KCl needed to obtain the 4 g of K^+, the proportion of the substance K^+ (in g) contained in the compound KCl (in g) is determined. The first step is to find the proportion of the substance (S_p) in the compound (C):

$$S_p = \frac{\text{Atomic mass of S}}{\text{Molar mass of C}}$$

or

$$K^+ = \frac{39.10}{74.55}, \tag{1}$$

$$K^+ = 0.52. \tag{2}$$

Next, find the weight of the compound (C_w) required to give you the weight of the substance (S_w) needed. Take note of the proportion of the substance to compound. When the weight proportion of the compound is larger than the weight proportion of the substance alone, use the following inverted calculation:

$$C_w = S_W = \frac{\text{Molar mass of C}}{\text{Atomic mass of S}} \tag{1}$$

or

$$KCl = 4g \times \frac{74.55}{39.10}, \tag{1}$$

$$KCl = 4 \text{ g} \times 1.91, \tag{2}$$

$$KCl = 7.64 \text{ g}. \tag{3}$$

Additions involve adding a solid compound or substance or a liquid stock solution to a known final volume of liquid. The targeted rate of addition (RA) is the amount or weight of the desired compound or substance per liter of liquid and is expressed in grams per liter. (The mathematical calculations must be in the same weight units and volumes.) If you know the RA and you know the final volume in liters (V), you can calculate the gram weight of the compound (C_w) or substance (S_w) that must be added to that volume to achieve the targeted RA:

$$C_w \text{ or } S_w = RA \times V.$$

Example: Making a K^+ solution with an RA of 4 g/L for a volume of 10 liters, the calculations would be

$$K^+(S_p) = \frac{\text{Atomic mass of } K^+}{\text{Molar mass of KCl}}, \tag{1}$$

$$K^+ = \frac{39.10}{74.55}, $$

$$K^+ = 0.52 \quad \text{(note the weight proportion)}, \tag{2}$$

$$KCl = 4 \text{ g}, \tag{3}$$

$$KCl = 4g \times \left(\frac{74.55}{39.10}\right), \tag{4}$$

$$KCl = 4 \text{ g} \times 1.91, \tag{5}$$

$$KCl = 7.64 \text{ g.} \tag{6}$$

To find the amount of K^+ to add to the 10 liters volume to reach the 4 g/L RA, the calculation would be

$$K^+ = 4 \text{ g/L} \times 10 \text{ liters} \tag{1}$$

$$K^+ = 40 \text{ g;}$$

$$4 \text{ g } K^+ = 7.64 \text{ g of KCl; therefore,}$$

$$\left(\frac{40 \text{g } K^+}{4 \text{g } K^+} \right) \times 7.64 \text{g KCl} \tag{2}$$

= Total amount of KCl to add to 10 liters of liquid to make a 4-g/L K^+ solution or

$$KCl \text{ addition} = 76.4 \text{ g.}$$

Stock solutions usually contain a high known concentration of a compound or substance in a liter of liquid volume. The use of high concentrations of stock solution reduces the volume of stock solution needed for an addition, which makes the addition more accurate. To add a highly concentrated stock solution to a known final volume of liquid (V) where the RA is known, you must first calculate the weight of compound (C_w) or substance (S_w) required to meet the targeted RA from the addition calculation:

$$C_w \text{ or } S_w = \text{RA} \times V.$$

Next, determine the volume of stock solution (V_s) that contains the weight of the compound (C_w) or substance (S_w) required:

$$V_s = \frac{C_w \text{ or } S_w}{\text{SC}},$$

where SC is the concentration of the stock solution (in g/L). Simplified to one calculation,

$$V_s = V \times \left(\frac{\text{RA}}{\text{Sg}} \right)$$

where Sg is the concentration of stock solution in grams.

Dilutions are often made from highly concentrated stock solutions of known concentration to form less concentrated solutions. To find the dilution factor (DF), use the concentration of the stock solution (SC) and the concentration of the diluted solution (SD) as follows:

$$\text{DF} = \frac{\text{SC}}{\text{SD}}.$$

To determine the volume of stock solution (V_s) required for making a specific volume of diluted solution (VSD), we use

$$V_s = \frac{VSD}{DF}.$$

The diluted solution is made up by adding distilled water, deionized water, or a solvent to the determined volume of stock solution V_s. A dilution where the dilution factor is 20 means that the final volume of the diluted solution will be 20 times larger than the stock solution volume. As an example, the dilution factor of 20 can be expressed in several ways:

$$1 \times 20, \qquad 1 \text{ in } 20, \qquad 1: 20, \qquad 1 + 19.$$

It is prudent to check and recheck your formulas and calculations before you start making your measurements. A quick and simple recheck can save money in chemicals and labor, as well as time trying to ascertain and isolate the source of a problem when your method goes awry.

3.2.4.2   Proper Measurement

Methods are designed for the analyses of a precise aliquot of sample; sample measurements are critical in any analysis. Preparation of solutions, reagents, controls, and standards requires excellent measurement techniques in order to deliver the exact weight or volume of liquid, dry chemical, or solution required. Improper measurement will throw off your calculations, setting the method up for failure. Reading a meniscus (level of fluid in a column), proper pipette technique for both manual pipettes and autopipettes, glassware quality and cleanliness, failure to obtain fluid temperature, using the wrong or uncalibrated measuring device, improper burette technique, and poor housekeeping are a few examples of common errors that can wreak havoc with methods.

Use quality Class A volumetric glassware and pipettes that do not require recalibration for all liquid measurements. The majority of graduated glassware, pipettes (auto and manual), and burettes are accurate when the liquid is at 20°C. This is a very important fact to remember. Cold liquids condense; measurements below 20°C will give you an inaccurately high volume when warmed to 20°C. Conversely, warm liquids expand; measurements above 20°C will give you an inaccurately low volume when cooled to 20°C. The magic temperature of 20°C is the standard ambient laboratory temperature used throughout the world. Not only glassware but also many instruments are designed to function optimally at 20°C.

Accurate weighing equipment is a must for making reagents, standards, controls, or any solution. There are many types of balance on the market in ranges from micrograms to kilograms with varying degrees of accuracy (linearity). Weighing micrograms of chemicals requires a balance with an extremely high degree of accuracy. Chemicals weighed in kilograms use a balance with a lesser degree of accuracy. Weighing a bottle of wine for QC will require a balance able to accurately accommodate the weight of a full bottle. Find the proper range of weighing equipment appropriate for the needs of the laboratory. Most wine laboratories use smaller quantities of accurately weighed chemicals, requiring an analytical balance with a range of 0 to 200 g and having a high degree of accuracy within the range of ± 0.2 mg. Analytical balances are enclosed by movable doors that allow chemicals to be placed on the scale and close to keep air movement out of the measuring chamber, which can adversely affect small measurements. If the balance is in an area that experiences vibrations, the use of a vibration-reducing stand might be required. Before making a measurement, read the balance's instructions to ensure proper measurement. Inspect the scale for cleanliness and make sure the balance is level. Most chemicals are weighed in some sort of weighing vessel, such as a beaker, weighing dish (commonly called boats), weighing funnel, flask, or weighing paper whose weight must be accounted for before a measurement is made. To determine the weight of the vessel, place the weighing vessel onto the scale and TARE. TARE means a weight removed, or simply, the weight of the container that is not part of the substance to be measured.

Clean glassware is critical. Glassware is considered clean when it maintains a continuous film of distilled water over the entire inner surface. Glassware can be etched by chemicals, so it is good laboratory practice to methodically rinse all glassware as soon as possible. Use only laboratory-grade nonfilming detergents to clean soiled glassware and rinse thoroughly with distilled or deionized water. Noncritical glassware (not used for measurements) found in the wine laboratory is often cleaned without the addition of detergent agents. A dishwasher connected to a clean water source, preferably distilled water or deionized water, is sufficient to keep most glassware clean for a time, but the glassware will need periodic thorough cleaning. Glassware should always be completely air-dried before returning to storage to prevent any unwanted mold or fungus growth. The use of wet glassware that has not been rinsed with the wine or juice sample might alter your results. Figure 3.3 shows a variety of glassware and measuring equipment commonly used in the laboratory.

Correct use of pipettes for measuring liquids is a skill all laboratory personnel must master. Manual pipettes used in the wine laboratory include bulb or volumetric, Pasteur, and serological (graduated) and are manufactured in a variety of materials, some of which are disposable. Class A glass pipettes are recommended for precise measurements, whereas pipettes made from other materials are useful for nonexact measurements and transfer of liquids. The key steps to developing correct manual pipette techniques are as follows:

1. Choose the appropriate pipette for the type of measurement. Volumetric pipettes are calibrated to deliver one exact volume, such as 5 mL, 10 mL, and so forth. Serological pipettes are graduated and calibrated to deliver a precise volume within the range of the pipette. Serological pipettes vary in the graduated increments and total volume. Always choose a serological pipette that meets the needs of the measurement and is large enough to accommodate the entire needed volume. Always pipette slightly more vol-

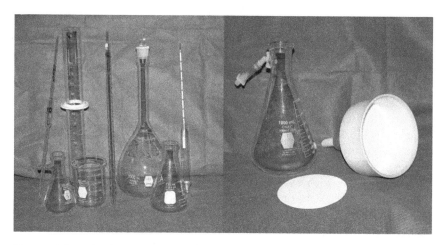

FIGURE 3.3. Laboratory glassware. From left to right: volumetric pipette, vacuum flask, graduated cylinder, Griffin beaker, serological pipette, volumetric flask, Erlenmeyer flask, hydrometer, large vacuum flask, and Buchner funnel and filter.

ume than required to allow for inadvertent loss while making your final measurement adjustment. Using a pipette that is too small will require several measurements to reach the targeted volume, creating a larger margin of error in the measurement.

2. Read the pipette's calibration information printed on the neck of the pipette. The information will tell you what temperature the liquid must be for exact measurement, how to deliver the measured volume, and the degree of accuracy. Pipettes are designed to deliver the measured volume several ways. "To Deliver" (TD) indicates that the pipette be held vertically with the tip against the sidewall of the container, the volume is delivered when the pipette has been allowed to drain completely. "To Deliver/Blow Out" (BO) indicates that the pipette must be allowed to drain as with the TD, but the drop that is left in the tip of the pipette must be blown out and added to the original measured volume. "To Contain" indicates that the pipette holds the exact measured volume and all of the volume must be delivered (see Fig. 3.4).

3. Place an adequate volume of the liquid to be measured in a separate clean container, this makes measuring a bit easier. Use a pipette bulb (or other filling/dispensing device) attached to the top of a clean pipette and draw up a small amount of liquid (*never* use your mouth to pipette chemicals). Quickly remove the bulb and stopper the top with your fingertip. Place the pipette in a horizontal position and slowly rotate it between your fingers. Invert the pipette over the sink, or a waste container, and expel the contents. The simple act of thoroughly rinsing the pipette with the liquid reduces the error of contamination by water, chemicals, or particles. Rinsing the pipette (wetting) also reduces the surface tension between the pipette and

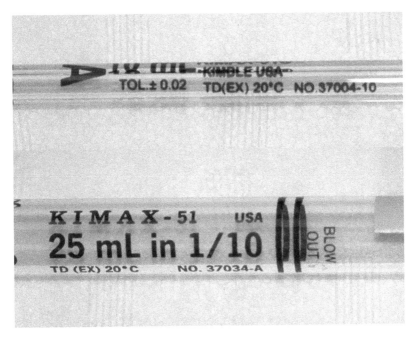

FIGURE 3.4. Glass volumetric pipette delivery designations.

the liquid. Liquid will bead up on a nonwet surface; wetting allows for more accurate measurement. Clean and dab the tip of the pipette with wipes designed for laboratory use and reinsert into the liquid. Draw up more volume than needed and wipe the tip of the pipette to remove any excess liquid. Remove the bulb; stopper the top with your fingertip; hold the pipette vertically with the graduation line at eye level; check the pipette for any bubbles and remove or repipette; slowly rotate the pipette (allows a small amount of liquid to drain); bring the bottom of the meniscus (Fig. 3.5) to the graduation line; then, holding the pipette vertically, place the tip against the receiving container's sidewall (tilting the container) to deliver the volume; and, finally, rinse the pipette thoroughly with deionized or distilled water. Delivering the volume to the sidewall of the receiving container prevents excessive aeration and possible loss of volatiles and eliminates splashing that can result in possible loss of volume and create a safety hazard.

The use of automated pipettes reduces measurement and contamination errors by mechanically or electronically drawing up the desired volume using disposable tips. Wetting the automated pipette tips is still required. Use the automated pipette in the vertical position (90° to liquid surface) to aspirate. Keep the same pipette position but deliver the volume along the sidewall (see Fig. 3.6). Currently, the following automated pipettes are on the market:

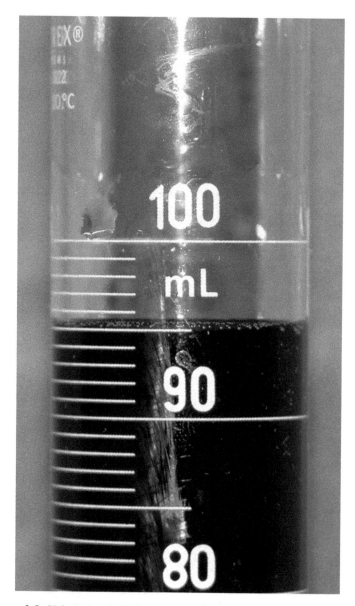

FIGURE 3.5. Volume level of 95 mL read at the bottom of a concave meniscus.

- Single channel: one dispensing point delivering one single accurate volume
- Single channel or multichannel repeat: holds more volume to enable the user to deliver a repeated accurate and precise volume (doses) multiple times without refilling
- Single channel or multichannel repeat adjustable: delivers repeat doses and the volume can be adjusted

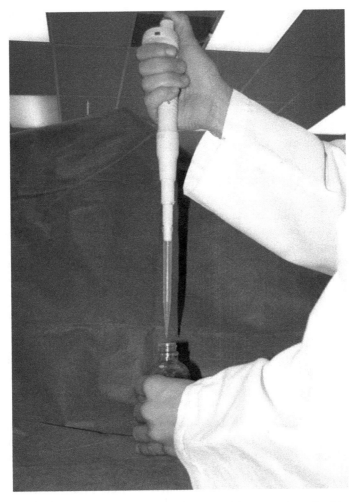

FIGURE 3.6. Correct automatic pipette position to aspirate liquid.

- Electronic (digital) single channel or multichannel: electronically measures a preset volume for more accuracy and precision
- Electronic (digital) single channel or multichannel repeat adjustable: electronically measures an accurate and precise volume repeatedly and can be adjusted

For less accurate delivery of larger volumes, a dispenser is utilized. Dispensers usually sit on top of a bottle filled with a reagent, solvent, or solution. The dispensers will have either a preset or adjustable volume and deliver the volume via a plunger-type device. These are commonly referred to as bottle-top dispensers.

Liquid measurements in graduated cylinders, beakers, and volumetric flasks use similar measurement techniques. They must be clean and dry; of good quality; wetting is not required in these larger vessels; and volume is added or removed with a pipette. Reading the meniscus is the same for all liquid measuring devices.

Burettes are used to deliver an incremented volume of measured liquid from a larger volume of measured liquid. A manual burette is a glass graduated column in a stand with an adjustable stopcock (regulates the flow of liquid). Glass burettes are used as follows:

1. Burettes must be clean, dry, and of good quality.
2. Fill the burette from the top with a clean funnel held slightly above the burette.
3. Adjust the height of the burette and place a waste beaker large enough to contain the burette volume under the tip of the burette. Open the stopcock and pour enough of the liquid to be measured into the funnel to thoroughly rinse and wet the burette. Close the stopcock and fill the burette higher than the desired level.
4. Open the stopcock and let it flow to remove any air bubbles from above and below the stopcock. Bring the liquid to the desired graduation line and discard the waste.
5. To deliver the measured volume, note the beginning measurement, open the stopcock slowly and allow the liquid to drain until the desired volume has been delivered, quickly close the stopcock, and then touch the receiving flask or beaker sidewall to the burette tip to ensure delivery of the entire volume.

If you are using a burette to add a measured amount of liquid solution to another solution, this is generally referred to as a *titration*. When performing a titration, it is best to use a conical flask and slowly swirl the contents of the flask while the measured solution is being added.

Digital burettes are electronic and deliver an electronically measured volume from a reservoir containing the liquid solution or reagent. These devices improve the accuracy of the delivered volume, reducing errors and maintaining consistency between measurements. Several companies produce digital burettes with attached stir plates that are very useful when performing titrations.

The general preparation of standards, controls, and reagents is an additional area of potential problems. Measuring the liquid or solid and mixing it with solvent or distilled water might seem easy enough, but unless proper preparation techniques are employed, all of the best measurements in the world will not deliver an accurate solution.

For preparing a solution from a solid, the steps should include the following:

1. Correctly weigh the solute on a calibrated analytical scale using a beaker, weighing dish, weighing funnel, or weigh paper. The fewer transfers of chemical, the easier it will be to maintain the accuracy of the measured substance as well as good safe laboratory practice.

2. Place the weighed solute into a container that meets the desired end volume (not larger) and add a small amount of the solvent or distilled water to the solute. Mix gently with a glass stirring rod or gently swirl the container to begin the dissolving process. Using the solvent or distilled water, rinse the stir rod over the container to allow any solute on the stir rod to be rinsed back into the solute mixture.
3. Using a clean glass funnel, pour the solute mixture into a volumetric flask large enough to accommodate the final volume (0.5 liter, 1 liter, etc.). Hold the solute container over the funnel and thoroughly rinse with the solvent or distilled water, allowing any of the solute mixture in the beaker to be rinsed into the volumetric flask. Rinse the funnel, allowing the solvent or distilled water to be rinsed into the flask.
4. Fill the flask with the solvent or distilled water by pouring it down the sides of the flask to rinse any undissolved solute into the solution while periodically swirling the flask to help mix the solution. Fill the flask to just below the graduation line.
5. With a dropper or Pasture pipette, add the final volume of solvent or distilled water to bring the volume to the graduation line. This is commonly termed *bringing to volume* or *coming to volume*.
6. Stopper the flask and hold it in place with one hand. Take the base of the flask and hold it next to your body to secure it with your other hand at the base of the neck. Invert the flask and return upright in a back-and-forth motion until the solution is well mixed.

Some solutions might require additional stirring using a stir bar or heat to thoroughly dissolve the solute.

Liquids use the same steps in preparation with the exception of using a pipette to deliver a volume of solute to the volumetric flask rather than the weighed solute (liquids might also be weighed).

3.2.4.3   Sample Quality

The theory behind sampling is to get the most representative sample of a substance to elicit an analysis that will reflect the condition of the entire body of that substance. Sampling in a winery is a daily event, and as with all laboratories, the quality of the sample directly affects the quality of the analysis, the method, and the end product.

Wine samples from barrels should be a composite. A composite is a larger sample consisting of a precise volume of wine or juice from a predetermined percentage of barrels that is mixed together to create the sample. The larger the composite percentage, or barrels, sampled, the more representative the sample will be of the entire group of barrels.

Tank wine samples (nonfermenting) are best if taken from a tank that has recently been mixed. Dormant tanks should be sampled as close to the middle of the tank as possible. Sampling tanks and barrels is discussed further in Chapter 6 (Section 6.4.3.1).

Samples are collected in proper contamination-free containers. Container labels should list every bit of information available to identify the sample. In the wineries, identifying information is usually the date, vintage and blend identification, tank or barrel identification, location, what analysis is to be conducted, initials of the person obtaining the sample, and any information that might be pertinent to the sample or analysis. Miscellaneous information could be tank or barrel temperature, time sampled, sterile sampling, or notation of problems detected while sampling, such as off-odors or film yeasts.

Cross-check the identification of the sample label with the tank or barrel to ensure that the correct wine is being sampled. Use the proper clean or sterile sampling devices.

Collect a small amount of the sample, rinse the sample container thoroughly and discard, fill the container to the brim, and close. The space between the wine and the lid, cap, or cork is called the *headspace*. A headspace containing oxygen reacts adversely with wine and might affect the analysis. If the headspace cannot be avoided, use an inert gas to displace the oxygen and then quickly close the container.

Wine or juice samples should never be subjected to high temperatures. The use of a cooler might be required to transport the sample to the laboratory. Samples received into the laboratory are analyzed as soon as possible. If the samples must be kept for an extended period of time, they might require refrigeration to maintain the integrity of the samples. Use an inert gas to fill the container headspace of wine samples that have been opened, tested, and are being held over for possible reanalysis. Wine samples that have been filtered through a ≤ 0.45-μm filter for analysis will be stable and can be kept refrigerated for up to 1 week.

Juice samples are always transported and maintained in the lab with a loose container closure to allow for escaping gases. Due to the fermentation process, juice chemistry changes rapidly, requiring samples to be transported quickly and analysis conducted immediately. Juice samples are never held for extended periods of time.

3.2.4.4   Preventative Maintenance

Preventative maintenance (PM) is orderly attention given to equipment to maintain accuracy and longevity by keeping them in prime working condition. The facts about equipment are the following:

- They do not last forever.
- They do get dirty.
- They do require calibration.
- They do directly affect QA.
- They do destruct.
- They do break down.
- They are not human.

Keeping these facts in mind, every laboratory needs to implement a solid PM program geared for each piece of equipment in the laboratory. Without a PM program that is strictly adhered to, the quality and confidence of analysis generated will surely deteriorate.

Additional benefits resulting from instituting a PM program include the following:

- Decrease in the variation of test results
- Fewer breakdowns during production hours
- Lower repair costs
- Fewer service calls
- Prolonged life of the instrument
- Saved time troubleshooting problems
- Confidence builder for operator
- Backup for questionable analysis

To set up a PM program, thoroughly read all instruction manuals for equipment used in the laboratory and note the manufacturers' recommended maintenance program.

Attempt to place equipment in protected draft-free areas of the laboratory, keeping the ambient temperature at 20°C. Many electronic instruments are adversely affected by temperature change.

Routine maintenance for most devices should include a wipe down of the exterior and thorough dusting (compressed air) of keyboards, inlet ports, electrical connections, heat fins, and internal wiring (always check with manufacturer's recommendations). As far as electrical instrumentation and some other equipment are concerned, dust is not a friend. Buildup of dust can lead to overheating of electrical components, create blockage of small orifices, and could alter or interfere with some analysis especially those that detect particle size. Many pieces of equipment are air-cooled with filtered ambient air that will require the filters to be cleaned or changed on a routine basis. Keep instruments and equipment covered when not in use.

Always rinse equipment tubing or chambers when the analysis is complete. Wine, juice, reagents, controls, and standards can cause degradation of susceptible parts (O-rings, seals, etc.) in some instruments. Wine and juice can also cause the buildup of sugars or tannins, leading to blockage of small orifices in equipment.

Set up log books for each piece of equipment or instrument to document PM activities, including the following:

- Equipment information

    Purchase date and company
    Serial and model numbers
    Vendor and catalog numbers for supplies and parts
    Service department contact
    List of warranties and service agreements with renewal dates

- Scheduled maintenance

  Calibration
  Routine maintenance
  Major maintenance

- Cleaning requirements
- Corrective action

  Identification of problems
  Corrective action taken
  Resolve of the problem

- Accountability

  Initials of person performing the PM and date
  Maintaining service and repair records
  Frequent inspection of logs by supervisor

Log books should be kept close to their corresponding piece of equipment or instrument. Table 3.3 lists a few very general PM recommendations for a variety of equipment.

3.2.5   Records

Laboratory work is very detail oriented, as this chapter has shown. You can never check and recheck data, computations, conversions, formulas, or notations too many times. This attention to detail generates a tremendous amount of data that substantiates the accuracy and precision of all laboratory methods, reagents, standards, controls, equipment, and instruments. Every detail that leads up to obtaining a test result is documented.

Raw data encompasses analysis results and their computations and verification of accuracy, procedures, methods, actions, and observations. Raw data include the following:

- Notebooks
- Instrument printouts
- Analytical record sheets
- PM logbooks
- Proficiency testing
- Solution preparation logs

Computer records have been a subject of controversy over the years. Many high-quality computerized instruments store their raw data files. It is a simple process to download and copy the data, making the information readily accessible and eliminating storage problems. Unfortunately, these records can be altered, making computer-generated data questionable from a legal standpoint.

Analyses results are transferred to other forms in order to communicate those results to the appropriate requestor. To eliminate transcription errors

TABLE 3.3. General preventative maintenance guidelines for wine laboratories.

Equipment type	PM scheduled	Action taken	Attention
Autopipette	Each use	Cleanliness, volume check	
	Weekly	Lubricate	Disassemble and clean
Autoburette	Each use	Standardize, function check	Remove air bubbles
	Each use	Level, cleanliness	Clean after each use
Analytical balance	Monthly semiyearly	Accuracy precision	
Gas and liquid chromatography	Each use	Septum check, resolution, retention time, sensitivity, noise, reproducibility	Compare to previous results
	Weekly	Glass insert check, oven temp- erature check	Damping check
	Quarterly	Gain, linearity, general maintenance	Service representative
	Yearly	Electronics check, recorder check	
Incubator	Daily	Temperature	
	Each use	Cleanliness	
Laminar flow hood	Daily	Particle count	
	Yearly	Leaks	HEPA filters
pH meters	Daily	Calibration accuracy	4 and 7 buffer
Refractometer	Each use	Accuracy	Distilled H_2O
	Quarterly	Calibrate	20° Brix
Cash stills	Each use	Cleanliness	Tubing connections
	Weekly	Chemically clean, accuracy	
Aeration oxidation	Each use	Cleanliness, standardize	Water bath temperature, glass chips and cracks
	Weekly	Accuracy	
Autoclave	Daily	Clean drain screen	
	Weekly	Thoroughly clean	
	Bimonthly	Time/temperature relationship	
Ebulliometer	Each use	Cleanliness	Barometric pressure
	Daily	Accuracy	
Thermometer	Yearly	Accuracy	Reference thermometer, ice point

and make data available more quickly, computer-based instrumentation has the ability to directly input the data into the winery's database via laboratory information management systems (LIMS).

LIMS are sophisticated software programs used in larger laboratories. These systems use bar codes to track and follow samples through the system, ensuring correct sample identification and completion of required analyses. Analyses results are directly input into the database, allowing the analyst to check for analysis trends on any analytical method. State-of-the-art QA and QC programs can be maintained with these systems. Reports and records can be designed specifically for an individual facility with an abundance of storage available. Smaller laboratories and laboratories that have not reached such a level of sophistication continue to use the standard recordkeeping methods. Regardless of the size or sophistication of a laboratory, there are always records that will need to be manually maintained and stored.

It is generally recommended that records be kept in an active file for 2 years. After 2 years, the records can be sent to storage for an additional 6 years. This should be sufficient time to cover any questions that might arise concerning the analyses and associated data. It is always prudent to verify the record maintenance protocol with a legal professional specializing in the wine industry.

3.2.6   Inadvertent Error

We have looked at variable error, random error, systematic error, explained error, unexplained error, and errors in technique. There is one more category to look at—inadvertent error. Inadvertent errors are human mistakes that happen every day.

Common inadvertent errors as listed in *QA Principles for Analytical Laboratories* by Garfield (1984) are as follows:

- Poor sample identification
- Failure to report observations
- Failure to follow method directions
- Mistakes in reading instrument data or volumetric measurements
- Mistakes recording data
- Errors in calculation of results
- Mistakes in transposing data between records
- Typos when entering data
- Improper reduction of analytical data
- Incorrect reading of retention times or peak heights
- Use of glass cells in an ultraviolet region
- Improper standardization
- Improper interpretation of results
- Nonrepresentative sampling
- Failure to purify or use reagents of suitable quality
- Incorrect reading of injection volumes
- Improper dilutions
- Use of inadequately cleaned glassware

Attention to detail, following SOP, asking questions, avoiding presumptions, reading instruction manuals, avoiding distractions, careful penmanship, and just paying attention to what you are doing can eliminate a large number of these types of error.

4

Berry to Bottle

4.1 Introduction

'Fill ev'ry glass, for wine inspires us, and fires us with courage, love and joy.'

John Gay, 1728.

Praised in poetry and song, used in religious rites, consumed as a dietary staple, and traded throughout the world, wine has been part of the human experience for thousands of years. Wine production and consumption continues to evolve and change as our civilization changes.

This chapter will briefly touch on the history, appellations, cultivation, and production of wines to give the reader a general overview of the "wine world." It is important to understand the evolution of the wine industry to gain an appreciation for wine and the traditions that surround winemaking. Further study is always recommended.

4.2 Wine History

Investigations of prehistoric sites in eastern and central Europe has revealed evidence that wild grapes were gathered to make wine as early as 10,000 years ago. There is additional evidence that cultivation of the wild European wine grapes (*Vitis vinifera*) began in the area between the Black and Caspian seas and in Mesopotamia (near present-day Iran) approximately 6000 years ago. Phoenician ships carried wine for trade from the eastern Mediterranean to the Iberian Peninsula.

From approximately 2700 to 2500 BC, ancient Egyptian royalties, priests, and priestesses drank wine as part of religious ceremonies to honor the gods. The Egyptians developed the first arbors and pruning methods.

The spread of the Greek civilization and their worship of Dionysus (Latin: Bacchus), the god of wine, spread Dionysian cults throughout the Mediterranean areas during the period of 1600 BC to the year 0. Greek physicians, including Hippocrates, used wine for medicinal purposes and readily prescribed it. Greek wines and their varieties were well known and traded throughout the Mediterranean. The Greeks learned to mask spoiled wine by adding herbs and spices. As the demand for wine increased, the Greeks introduced *Vitis vinifera* (*V. vinifera*) into the Greek colonies, southern Italy, and southern Spain. The *V. vinifera* grape thrives in temperate climates near coastal areas with mild winters and warm dry summers. It adapted well and flourished in the northern Mediterranean areas.

Around 1000 BC, Romans began introducing their grape varieties and winemaking methods to their colonies. The Romans influenced and established the foundation of viticulture (cultivation of grapes for wine) in western Europe. The Romans are credited with the following:

- Classifying grape varieties (varietals) and colors
- Charting observations of ripening characteristics
- Identifying diseases
- Recognizing soil-type preferences
- Improving pruning skills
- Increasing production through irrigation and fertilization
- Using wooden cooperage
- Using glass bottles

Exportation of wines from Italy to Spain, Germany, England, and Gaul (France) took place from 300 BC to 500 AD. Wine production was also increasing in these areas during this period, creating direct competition for the Italian wines. Around 100 AD, the Roman Emperor banned importation of French wines to eliminate the competition. In the next several centuries, French wine became dominant in the world market.

The decline of the Roman Empire led to invasions of barbarians that devastated Europe. The church became the stable entity in Europe, and because of the need of sacramental wine, the wine industries methods and tradition were preserved through the Dark Ages. By 1100, some large monasteries in France and Germany had become winemaking centers. Monastic wineries were established in Burgundy, Champagne, and the Rhine Valley.

As time passed, London, Brussels, Amsterdam, and other large cities in northern Europe became major markets for wine. Commercial trade routes sprang up along navigable rivers in the Bordeaux and Rhone regions in France and the Rhine and Mosel regions of Germany. Burgundy was land locked and remained associated with the church.

Britain influenced the European wine regions from around 1152 to 1453 by controlling Bordeaux. After the Hundred Years War in 1453, Britain's only territory in France was Calais. Trade between the two countries was nonexistent. The English set their sites on Portugal, where they established a treaty. The English were instrumental in the increased port production and the development of the sherry industry in southern Spain.

Exploration, conquest, and subsequent establishment of settlements in the New World from 1500 to 1600 introduced European wine culture and traditions to new wine regions where none had previously existed. The main wine-growing areas were in Mexico, Argentina, and South Africa. Unsuccessful attempts were made to grow grapes along the Atlantic, Gulf Coast, and the Mississippi River basin valleys of North American.

Hernando Cortez planted Spanish grape vines in Mexico in 1525. By 1595, the King of Spain banned new plantings and new vineyard development in Mexico, fearful that the settlements would become independent. This ban lasted until 1745.

The 1700s brought significant changes to the well-established major European wine regions. The French Revolution of 1789 resulted in the seizure of church-held vineyards in Burgundy. These vineyards were subdivided among the people of the area, and, today, those land ownerships have been handed down from generation to generation.

In 1769, Father Junipero Serra, a Catholic missionary, planted the first vineyard in California at Mission San Diego. Known as the "Father of California Wine," Father Serra continued up the coast of California to Sonoma, establishing missions and planting vineyards until 1823. The missions trained people in grape growing and winemaking methods. It is probable that the grape vines planted were a Spanish variety (now called Mission) and dominated wine production until around 1880.

Los Angeles, California, and areas in Northern California attracted many European settlers, and by 1850, wine production was a viable industry in California. Jean-Louis Vignes planted the first documented European imported wine vines in Los Angeles in 1833.

The Hungarian considered the Founder of the California Wine Industry was Agoston Harazsthy. Between 1850 and 1860, Harazsthy imported cuttings from 165 of the greatest vineyards in Europe into California. During his career, he introduced 300 different grape varieties and founded the Buena Vista Winery in Northern California, which is still in production today.

The 1800s established several scientific advances for the worldwide wine industry. Louis Pasteur discovered that wine spoilage was caused by micro-organisms. This was the first giant step in controlling wine spoilage. In 1860, Dr. Jules Guyol wrote the first of three essays describing regional traditional viticulture practices, his observations, and arguments on the economy of grape growing. Up to this point in time, viticulture information was apprenticed from generation to generation with no written records or formal instruction.

The year was 1863 and it would change the course of viticulture and wine-making history. The Botanical Gardens in England imported native American vines, a few from the Mississippi River basin. These few vines carried a root louse indigenous to the Mississippi River Valley called *Phylloxera vastatrix*, the larva of a green fly that attacks and feeds on leaves and the vine roots of the plants. The Mississippi River Valley host vines were (and are) resistant to the louse; unfortunately, the European vines were not.

Within 2 years, phylloxera had spread to Provence. In 1865, a shipment of native American vine cuttings arrived at the port of Bordeaux and, unknowing, planted soon after. Three years after the planting, they died. Phylloxera spread rapidly from region to region. By 1885, phylloxera had decimated all the vineyards of Europe.

North American grape varieties such as *Vitis labrusca, Rotundifolia,* v. *riparia,* and *Rotundifolia* v. *rupestris* were found to be resistance to phylloxera due to their thick and tough root bark. Grafting *V. vinifera* vines to North American rootstock saved the European wine industry from extinction. Many native European grape species were not grafted because of their low commercial value and, subsequently, were lost forever.

Phylloxera was not the only import to Europe from North America, the fungi powdery mildew and downy mildew were also introduced around this time. It is thought that downy mildew along with other fungi was transported on the very roots that were sent to save the industry. Many grape and wine producers were ruined and immigrated to other countries, taking their talents and knowledge with them. These emigrants were a huge influence on the development of the worldwide wine industry.

The phylloxera crisis and the replanting of European vineyards created a severe wine shortage for several years. During this time, the American wine industry flourished, and by 1900, it was commercially producing quality wine for the world market.

This worldwide wine shortage and skyrocketing prices led to illegal adulteration of the wines that were produced. To combat these ruinous acts, the French wine growers developed a method to protect wine trade reputations and authenticate the wine to the consumers. This method is known today as Appellation Controlée and has set the precedence for the worldwide development of appellations and wine identification.

At the turn of the century when the American wine industry was enjoying the benefits of the wine boom, disaster was brewing. It began in 1816 with the first prohibition law enacted by Indiana. Alcohol abuse and alcoholism was increasing at an alarming rate during the mid-to-late 1800s. In 1855, 13 of the 31 States had enacted a statewide law prohibiting the manufacture and sale of liquor. By 1880, Kansas had become the first "dry" state, and by World War I, the number of dry states increased to 33. Wartime prohibition was enacted in 1919.

The Eighteenth Amendment to the United States Constitution and the Volstead National Prohibition Act were enacted in 1920 prohibiting the manufacturing, sale, or transportation of intoxicating liquors in the United States. The government did allow each household to make a maximum of 200 gal of nonintoxicating cider and fruit juices. A few wineries were allowed to make wine for medicinal, sacramental, and nonbeverage additives, but 94% of the existing wineries were closed. There were 2500 commercial wineries prior to Prohibition, and by 1933, less than 100 remained. It took 27 years to bring the number of wineries up to 271. California had 713 bonded wineries prior to Prohibition; after Prohibition, it took 53 years (1986) to reach that same number of wineries.

During Prohibition, more police officers were killed than in anytime in American history due to the surge of organized crime, bootlegging, and hard liquor production.

Home winemaking ventures sprung up everywhere, creating a tremendous market for hardy wine grapes that traveled well, for sale across the United States. Vineyard property prices went from $200 per acre in 1918, to $2500 per acre in 1923. Grape varieties such as Alicante, Bouschet, and Carignane were the ideal choice for planting. Existing premium grape vines were pulled out and replaced with these hardy varietals.

Within 5 years, the market was flooded with grapes and land prices dropped to $250 per acre in 1926. The vineyards planted during this boom produced a consistent grape surplus that lasted through 1971.

National repeal came in 1933, too late to save the wine industry from ruin. Premium winemaking was nearly nonexistent. The majority of wines sold until the 1960s were sweet desert wines. During the past 40 years, the wine industry in the United States has made a tremendous comeback. Once again, North American premium wines are excelling in the world market.

4.3 Grape Varieties

There are literally thousands of varieties of grapes used for the production of wine but most wines are made from *V. vinifera* varieties. The following are the most popular wine grape varietals around the world:

- Alicante Bouschet: red grape variety from France
- Cabernet Franc: red grape variety from the Bordeaux region, France
- Cabernet Sauvignon: red grape variety from the Bordeaux region, France
- Carignan: red grape variety from northern regions, Spain
- Chardonnay: white grape from the Burgundy region, France
- Gewürztraminer: white grape from the Baden region, Germany
- Grenache: red grape variety most likely from Spain
- Merlot: red grape variety from the Bordeaux region, France
- Monastrell: red grape variety from the Penedés region, Spain
- Müller-Thurgau: white grape from the Rheinhessen and Rheinpfalz regions, Germany
- Nebbiolo: red grape from the Piedmont region, Italy
- Pinotage: red grape variety South Africa
- Pinot noir: red grape variety from the Burgundy region, France
- Riesling: white grape from Germany
- Sangiovese: red grape from the Tuscany region, Italy
- Sauvignon blanc: white grape from the Bordeaux and Loire regions, France
- Semillon: white grape from the Bordeaux region, France
- Syrah: red grape from the Rhone region, France

- Tempranillo: red grape from the Rioja region, Spain
- Viognier: white grape from the Rhone region, France
- Viura: white grape variety from the Penedés region, Spain
- Zinfandel: red grape variety (juice used for white Zinfandel) from the Balkan peninsula

Table grapes and raisins are made from the *V. vinifera* varieties Thompson Seedless and Flame Seedless. *Vitis labrusca* (Concord grape) is indigenous to North America and used primarily for grape juice and jelly. *Vitis rotundifolia* is indigenous to the southeastern United States and used primarily for table grapes and jelly. Its greatest claim to fame is its immunity to Pierce's disease, a bacterial disease (*Xylellafastidiosa*) fatal to *V. vinifera* grapes that is spread by the glassy-winged sharpshooter (*Homalodisca coagulate*). The range of the glassy-winged sharpshooter is primarily in the warmer climates of the United States, totally eliminating those areas from possible cultivation for wine grapes. The glassy-winged sharpshooter has migrated to the southern California San Juaquin Valley and has been found as far north as Sonoma County. Pierce's disease poses a very real threat to the California wine industry. Several methods to control the disease are being attempted. Traps and insecticides have had positive effects, and there have been proposals to introduce a tiny wasp called *Gonatocerus triguttatus* into the area. This wasp lays its eggs inside the sharpshooter's eggs, thus destroying the egg.

4.4   Appellations

The term "appellation of origin" denotes a geographical name indicating the origin of the grapes used to make the wine (e.g., Bordeaux in France and Napa Valley in the United States). Non-European countries use the appellation of origin in addition to the varietal designation, such as Merlot, Alexander Valley. In many European wines, the appellation of origin identifies the wine itself such as Chablis and Champagne in France.

The term "controlled appellation of origin" denotes the geographical origin of the grapes, the grape variety, how they were grown, and the production standards. Production standards can include, but are not limited to, grape yield, sugar concentration (expressed as °Brix, °Baumé, or Oechslé) and regional typicality. Wine labels that include the controlled appellation's name must legally meet all criteria set for that particular area. The European Union uses controlled appellations for the majority of wine-producing nations.

France has the most complicated appellation system because of the variety of regional requirements. Understanding the French system will ease you into any other countries' appellation system.

4.4.1 France

The French regulatory system has served as the model for the European Union wine laws. The controlled appellation of origin wines are identified as Appellation d'Origine Controlée, or AOC. Approximately 35% of all French wines meet the AOC guidelines. One of those guidelines is the addition of sugar to the grape must prior to or during alcohol fermentation (chaptalization). In northern France, chaptalization is legal; in the southern regions, it is illegal. The majority of French wines produced are non-AOC wines that are designated Appellation d'Originé Vin De Qualité Supérieure (AOVDQS) (higher quality wine of origin), vin de pays (VDP) (regional wine), or vin de table (table wine). There are hundreds of regional wineries across the country. Wine producers do not have to meet AOC guidelines, but if they do, their wines usually will sell for a higher price. The Institut National des Appellation d'Origine (INDA) is the organization that administers the AOC and AOVDQS systems.

The hierarchy of the AOC system contains large areas with lenient requirements down to very small areas with stringent requirements. The more requirements met, the more prestigious the wine and the higher the price. The large wine-producing areas are called regions, such as Burgundy and Bordeaux. Regions are broken down into smaller areas called subregions, followed by the village or commune name, and, in some areas, the single vineyard name or chateau name. If the wine meets all of the requirements for a village appellation, the owner can either use that appellation, the subregional appellation, or the regional appellation depending on their perception of the quality of the wine compared to others in the area. They cannot, however, use the single vineyard or chateau appellation. Figure 4.1 shows examples of French wine labels. France is divided into nine regions: Alsace, Bordeaux, Burgundy, Corsica, Côte du Rhône, Languedoc-Roussillon, Loire Valley, Provence, and the South West.

4.4.1.1 Burgundy

The Côte d'Or (loosely, slope of gold) is at the heart of this region and the wines are 100% varietal (no blending with other varieties). The Burgundy varietals are Chardonnay (the white variety) and Pinot noir (the red variety). Chaptalization is legal in Burgundy because of the cool growing conditions.

Burgundy uses regional and village appellations with two individual vineyard appellations: grand cru (great growth) and premier cru (first growth). Grand cru vineyards are considered the best. There are over 650 appellations in this region:

- Regional AOC appellation (22) contains Bourgogne (Burgundy) or a specific district such as Beaujolais Villages, Brouilly, or Côte de Beaune.
- Village AOC appellation (44) contains the name of the village only such as Pommard or Meursault.

FIGURE 4.1. French wine labels: (**A**) Pomerol, Bordeaux appellation; (**B**) Saint-Estephe, Bordeaux appellation Grand Cru Classé; (**C**) Margaux, Bordeaux appellation Premier Grand Cru Classé; (**D**) Chablis, Premier Cru; (**E**) Puligny-Montrachet, Burgundy appellation.

- Premier cru AOC vineyard appellation (562) contains the village name and the vineyard plot (climates) name, such as Chassagne-Montrachet Premier Cru Les Jolie
- Grand cru AOC vineyard appellation (33) contains the vineyard plot name only, such as Les Jolie Grand cru.

The Burgundy bottle is unique with its sloping shoulders and is used throughout the world to bottle Chardonnay and Pinot noir wines.

The regions of Beaujolais and Chablis are encompassed by Burgundy, but retain their individual regional identity. The Chablis region produces only white wines made from Chardonnay grapes. The appellation system includes 19 municipalities, 40 premier cru, and 7 grand cru vineyard designations, including Valmur, Bougros, and Preuses.

Beaujolais wines are made from Gamay noir grapes. The releasing of the new wine (Beaujolais Nouveau) occurs on the third Thursday in November, close to the end of fermentation. There are 12 appellations, which include Beaujolais Villages, Brouilly, and Saint Amour.

4.4.1.2 Bordeaux

Bordeaux is located in the southwest region of the country near the Atlantic. Chaptalization is legal and often used because of the cool and wet summers. Most red Bordeaux wines are blended (known as Meritage in California) from three main varieties and two lesser used varieties. The three main varieties are Cabernet Sauvignon, Cabernet Franc, and Merlot. Petit Verdot and Malbec are used to a limited extent. White Bordeaux wines are usually a blend of Sauvignon blanc and Semillon. All are referred to as Bordeaux varietals. The squared shouldered Bordeaux bottles are standard in the world wine industry for bottling any Bordeaux varietal.

In 1855, at the Exposition Universelle de Paris, Napoléon III requested a classification system for Bordeaux wines. Sixty-one châteaux in Bordeaux were evaluated and classified by their reputation and wine prices into five categories: first growth through fifth growth. The 1855 Bordeaux appellation system uses the region, two subregions (Médoc and Haut-Médoc), village designations, and, unofficially, the winery or château name. The château is the primary identifier of the wine and will appear in the largest letters on the label, whereas the official AOC appellation is in small print.

First-growth châteaux had the best reputation and commanded the highest prices. These châteaux were designated premier grand cru classé (first great growths class). There were four premier grand cru classé châteaux: Château Haut-Brion, Château Lafite Rothschild, Château Latour, and Château Margaux. Château Mouton Rothschild was elevated from a second-growth to a first-growth vineyard in 1973.

Second-growth through fifth-growth châteaux were simply labeled grand cru classé (great growths class). These designation of the great châteaux remains today much to the chagrin of those that did not make the initial cut and subsequent châteaux. The Médoc and Haut-Médoc subregions contain the majority of the great châteaus producing predominately Cabernet Sauvignon. Merlot is the predominate wine around the village of Pomerol, and Merlot and Cabernet franc are predominate near the village of Saint Emilion.

Sweet white wines were classified into two growths (cru) that included 26 châteaux. The wine of the Grave subregion is a blend of predominantly Sauvignon blanc with some Semillon. The village area of Sauterne produces a blend predominantly of Semillon with some Sauvignon blanc.

Bordeaux appellations and label information are as follows:

- Regional AOC appellation is Bordeaux or a subregion such as Médoc.
- Village AOC appellation contains the name of the village such as Margaux or Pomerol.
- Château or producer's name is unofficial but is predominant on the label such as Château Margaux.
- The 1855 classification (if any) will appear in the second predominant print on the label such as Premier Grand Cru Classé.
- The date of 1855 classification is unofficial but will be listed under the château name.

Further classification has occurred over the years. In 1954, the area of Saint Emilion was classified into Premier Grand Cru Classé containing 13 châteaux and Grand Cru Classé containing 55 châteaux. This subregion is adjusted every 10 years and will be up for revision in 2006. Graves classified 16 châteaux in 1959 into one classification designated as Cru Classé.

The Bordeaux region has been unofficially divided into six geographical and stylistic areas:

- Bordeaux and Bordeaux Supérieur
- Golden sweet whites, including the city of Sauterne
- Elegant dry whites, including the city of Entre-Deaux-Mers
- Côte de Bordeaux, including the area of Côte de Castillon
- Saint Emilion, Pomerol, and Fronsac
- Médoc, Graves, and Pessac-Léognan

4.4.1.3 Rhône

The Rhone contains two regions: Northern Rhone and Southern Rhone. Northern Rhone produces wines primarily made from the red grape Syrah and the white variety Viognier. The Southern Rhone produces mostly blended wine using primarily the red grapes Grenache, Syrah, and Mourvédré. Chaptalization is illegal in the Southern Rhone region.

Rhone appellations are the following:

- Northern Rhone AOC appellations are primarily village names such as Hermitage.

- Southern Rhone AOC appellations are regional and listed as Côtes du Rhône (loosely, slopes or coast of Rhône River) but could also contain village names such as Chateauneuf du Pape.

4.4.1.4   Loire

The Loire Valley is located in the northwestern part of the country. Chaptalization is legal in the Loire Valley. This is primarily a white wine area featuring Sauvignon blanc, Chenin blanc, and Pouilly-Fumé. Cabernet Franc is grown in the central part of the Loire region. The appellation system is based on region, village, and near villages.

Loire appellations are the following:

- Regional AOC appellations would include Le Vignoble du Centre, L'Anjou – Saumur, Le pays Nantais, and La Touraine.
- Village AOC appellations are predominant such as Muscadet de Sévre & Maine, Clermont-Ferrand, Pouilly, Côtes Roannaise, and Sancerre.

4.4.1.5   Champagne

Champagne is the sparkling wine region of France and is located in the northernmost part of the country. Chaptalization is legal in Champagne. There are three varieties approved for this region: Chardonnay, Pinot noir, and the hardy red grape Pinot meunier. Champagne is predominately blended with different varieties, vineyards, and vintages (years). This system of blending allows for consistency year to year. The regional appellation, Champagne, is the only appellation used for sparkling wine. Still wines produced can use a subregional appellation.

4.4.1.6   Languedoc-Roussillon

The Languedoc region is located in the southwestern part of the country where chaptalization is illegal. The primary red varieties of the area are Carignan, Grenache, Cinsault, Aramon, and Syrah. Muscat and Grenache blanc are the primary white wines. Languedoc has three subregions: Hérault, Gard, and Aude. Appellations are regional and subregional and include some villages such as Fitou and La Clape.

Côtes du Roussillon has one sub region, Pyrénées Orientales, and several village appellations, including Maury and Banyuls.

4.4.1.7   Provence

Provence is located in the southernmost part of the country. Chaptalization is illegal. Rosé, red, and white wines are produced. The style of wine is similar to Burgundy and Côtes du Rhône. Cinsault, Grenache, Syrah, and Mourvédré are the primary red varieties used and the white varieties are Clairette and Sémillon.

The regional appellation is Côtes du Provence and within the region, Côte d' Azure (loosely, coast of blue). There are many vin de pays and AOVDQS appellations.

4.4.2   Italy

Italy has over 1000 indigenous grape varieties that have been propagated over the centuries in independent areas. The appellation laws were developed in the 1960s to begin the long processes of controlling the quality of wine produced. The Italian appellation system is the Denominazione di Origine Controllata (DOC) (loosely, name of origin control). The DOC appellation is geographically specific and dictates the grape varieties and methods of production. The wines are to be typical of the area, but there is no consideration of quality. The Denominazione di Origine Controllata e Garantita (DOCG) (loosely, name of origin guaranteed) distinguishes certain wines of quality. Recently, a classification for the lowest category of Italian wines became effective; the Indicazione Geografica Tipica (IGT) (loosely, indication of typical geographic area).

The appellation system is not hierarchical. DOC, DOCG, IGT, and vino da tavola (table wine) are separate categories; there are no layers. If a wine does not qualify for a DOC or DOCG appellation, it will be classified as IGT or da tavola. If a wine does not qualify for a DOCG appellation, it does not automatically fall into the DOC classification, as it would in the hierarchical French system, but is placed in the IGT or da tavola appellation. Prior to 1992, there were no regional appellations, but a new law allows for the gradual development of regional and vineyard appellations.

Italy has 22 regions, including Rome, Puglia, Sicily, Sardina, Val de Aosta, Umbria, Piedmont, Tuscany, Emilia Romagna, and Veneto. The last four will be discussed.

Figure 4.2 illustrates typical Italian wine labels.

4.4.2.1   Piedmont

Piedmont is located in the northwestern section of the country. The red grape Nebbiolo is used to produce some of the most renowned wines in Italy. The area around the two villages of Barolo and Barbaresco are the two DOCG appellations in Piedmont. A DOC red grape variety of the area is Barbera.

A sparkling white wine is made in the area around the village of Asti, a DOCG appellation, from the Moscato (Muscat blanc) grapes. The production methods used for making the sparkling wines are primarily the charmat method, but the méthod champenoise is also used (see Section 4.5.6).

4.4.2.2   Tuscany

Tuscany is a well-known area for wine, history, and art in the north-central part of the country. The cities of Florence and Siena are in the heart of the Tuscan wine region. The primary red wine of the area is Chianti (DOCG).

FIGURE 4.2. Italian and United States wine labels: **(A)** Chianti, Italy DOCG; **(B)** Monferrato, Piedmont, Italy DOC; **(C)** Napa Valley, Napa County, California; **(D)** Alexander Valley, Sonoma County, California; **(E)** Piffero Vineyard, Mendocino County, California.

Prior to 1996, Chianti was a blend of primarily Sangiovese, with the addition of a second red varietal (i.e., Colorino or Canaiolo) and a small amount of two white varietals (Trebbiano and Malvasia).

Chianti Classico is wine produced by a group of traditional winemakers located in the central part of the Chianti region. These winemakers formed an association, and in 1996, they were able to get Chianti Classico designated as an independent appellation. With this designation, the DOCG restrictions changed to include an increase of the minimum amount of Sangiovase base wine from 75% to 80%. Chianti Classico can now be made with 100% Sangiovase, or it can be blended with up to 20% of the typical red varietals, or, according to the new regulations, it can include Cabernet Sauvignon, Merlot, or other red varietals. The addition of white wine can no longer be used beginning with the 2006 vintage.

One association of the Chianti region includes winemakers who do not follow the IGT, DOC, or DOCG regulations and call their wines "Super-Tuscans." These wines are primarily of Sangiovase, but many are blended with varying percentages of European varietals.

4.4.2.3   Emilia Romagna-Veneto

Emilia Romagna is the area directly north of Tuscany. This area is a large producer of a lower-quality grape called Lambrusco. Veneto produces large volumes of both red and white wines. Soave is the most well-known city in this area. These areas produce a lower quality of wine mainly for export.

4.4.3   Spain

The Spanish appellation system follows closely to the Italian system where the traditional winemaking methods and geographic location have been the qualifying factors for appellation determination. The Denominacion de Origen (DO) (loosely, name of origin) is not concerned with the wine quality. Like the Italians, the spanish have developed a second system, the Denominacion de Origen Calificada (DOCa) for wines that show distinction. Spain has incorporated a stamp that appears on the label, giving some authenticity to a region's regulations. The stamps might read Denominacion Origen Consejo Regulador, which loosely translated means "name of origin regulatory council." The stamps will have the name of the geographic origin in the center or will just have the name of the geographic origin.

As in Italy, Spain is attempting to improve their winemaking techniques and encourage good winemaking practices to promote the production of premium wines.

4.4.3.1   Rioja-Penedés-Cava

Spain's most famous premium wine is made in the Rioja region located in the northeastern part of the country. The wine is primarily a blend of Tempranillo with some Grenache and is barrel aged for years.

The DO regulations are very liberal in the Penedés region, which has led to a diversity of wine varieties. The Penedés area is located in the eastern part of the country on the Mediterranean. Some wines are blended; some are 100% varietal. Wines from this area can be labeled with a proprietary name that denotes the type of wine or blend. The primary varieties of this area are the following:

- Monastrell
- Garnache
- Tempranillo
- Cabernet Sauvignon
- Merlot
- Viura (white)
- Parellada (white)
- Xarello (white)
- Chardonnay
- Riesling

Cava (sparkling wine) is made using the méthode champenoise. Although Cava is classified as a DO, the name itself describes a method, not a geographic location. When the Spanish went to the appellation system, a location of origin had to be assigned, so they used Cava. Because Cava was a method, the sparkling wine producers are spread out over much of the same area as Penedés covers. Cava is made using Viura, Parellada, Xarello, Chardonnay, and Pinot noir.

4.4.3.2 Jerez

Jerez (formally Xeres, pronounced sher – Ez) is located in southern Spain. As the pronunciation indicates, this is the sherry producing area of Spain. Subregions include Valencia and Ribeiro.

The white grape Palomino is used primarily for sherry. There are many categories of sherry, the two primary categories are fino and oloroso. Sherry is a fortified wine, which means alcohol has been added to the dry wine to increase the overall alcohol level. Brandy is added to the sherry to increase the alcohol to around 15%. The sherry remains in partially filled barrels that allow air to interface with the wine for 1–3 years. The wine becomes slightly oxidized by the development of flor yeast and is called fino sherry. Wine that does not develop the flor yeast will be fortified up to 18% alcohol and warmed, thus becoming oloroso sherry. Sherry is usually dry. Sweet concentrated juice is sometimes added to the sherry, creating cream sherry.

Unlike most wines, sherry does not use the vintage year as a descriptor, nor is it regulated. Sherry is fractionally blended with many different years, which homogenizes each barrel and prevents year-to-year variations. New barrels of sherry are stacked atop the previous years' barrels. The barrels on the bottom (oldest barrels) are partially emptied for the years bottling, that volume is replaced by the second row of barrels. The second row of barrels atop the oldest barrels contains sherry a year younger. The second row's volume is replaced by the third row's sherry, which is a year younger, and the third row's sherry is replaced by the fourth, and so on. This system of fractional blending is called the solera system.

4.4.4 Portugal

Denominação de Origen Controlada (DOC) is the Portuguese appellation system. There are three primary winemaking areas in Portugal: Douro, Vinhos Verdes, and Maderia. Dao, Setubal, Bucelas, Colares, and Carcavelos are lesser known wine regions Douro, Vinhos Verdes, and Maderia regions each produce very different wines.

4.4.4.1 Douro

Douro is the home of the sweet fortified red port wine. Port is named after the city of Oporto. This area is located in the northeastern portion of the country near the Douro River. Port is fortified with brandy before it has finished fermentation, so it remains sweet but has a higher alcohol. Port, like sherry, is blended with wine from different years, matured in wood for 3–4 years, and bottled as ruby port. Vintage port is wine from one good year; matured for 2 years, and bottled; it is then aged in the bottle (bottle-aged) for several years. Tawny port is aged in wood for many years (refer to Chapter 8 for more details on bottle aging).

4.4.4.2   Vinhos Verdes

Literally Vinhos Verdes means "green wines." The wines have the very high acid and low sugar of an immature or green grape due to the short growing season in the northwestern Atlantic area of the country. These wines can be red or white, are low in alcohol, and have a bit of a fizz.

4.4.4.3   Madeira

Wine produced on the Portuguese island of Madeira is a fortified wine. Historically, this wine traveled the seven seas in the hull of ships subjected to varied temperature changes; these temperature changes added a distinct flavor to the wine. Today, the white wine is artificially heated and cooled to duplicate those original conditions. Madeira can sustain a long life in the barrel and is matured for many, many years. Today, it is possible to purchase a Madeira wine that might be over 100 years old.

4.4.5   Germany

Germany is at the northernmost limit of viticulture. The growing season is short and produces grapes high in acid and typically low in sugar. Chaptalization is allowed for certain quality categories. A dry, trocken, wine might be highly acidic, whereas a partially dry halbtrocken (loosely, half-dry), wine might contain the same high acid but more balance due to a higher sugar content. White grapes such as Riesling and Müller-Thurgau are cold-hardy and grow better than red varieties.

In Germany, the land is considered neutral, so there are no regional appellations. The quality of the wine is the determining factor for appellations and quality categories are based on the amount of sugar in the grapes at the time of harvest.

Germany has 13 defined wine regions which can be broken down into villages and vineyards. These geographical names only indicate where the grapes were grown and do not indicate the nature of the wines as it does in most of the European countries.

Quality wine classification has two categories: Qualitätswein mit Prädikat (QmP) (loosely, quality wine with distinction) and Qualitätswein bestimmter Anbaugebiete (QbA) (loosely, quality wine from a particular region).

The QbA wines are lower quality, having a lower harvest sugar (13° Brix). Chaptalization is allowed in these low-sugar juices. A large percentage of wines in this category are exported.

The QmP wines are the highest-quality category, having a higher harvest sugar level (around 17° Brix). Chaptalization is not allowed in these juices. The QmP category is divided into several subcategories based on increased harvest sugar content (approximate required sugar in °Brix):

- Kabinett: 17° Brix
- Spätlese: 19° Brix
- Auslese: 21° Brix
- Beerenauslese: 28° Brix
- Trockenbeerenauslese: 39° Brix

Figure 4.3 shows an example of a German wine label.

The regions of Rheinhessen and Rheinpfalz, between the Rhine and Mosel rivers, contain approximately 25% of the total vineyard area within Germany. The wines from these areas are typically lower-quality wine made with Müller-Thurgau grapes. The Mosel-Saar-Ruwer region in the westernmost area of Germany produces premium Riesling.

4.4.6   South Africa

South Africa produces a variety of wines that include Cinsaut, Steen (Chenin blanc), Palamino, Cabernet Sauvignon, Merlot, Chardonnay, Sauvignon blanc, and Pinotage. A semigovernmental cooperative organization, KWV, controlled the wine production and marketing from 1918 until 1997. This organization set wine prices, determined surplus, provided technical support, and promoted exportation. In 1997, the cooperative became a group of companies and was divided into the Wine Industry Trust, concerned with administration and regulations, and the commercial side named KWV International.

South Africa's appellation system, Wine of Origin (WO), includes 13 regions. There are a few subregions and individual estate names used. Similar to Germany, the WO appellation indicates the geographic origin of the grapes only and does not indicate the nature of the wine. Winemaking was introduced to the country, so there are no indigenous traditions.

The majority of the viticulture areas are located in the southernmost coastal portion of South Africa. Constantia is the site of the first vineyard planted in the 1600s. Constantia produces fine white wines, and Stellenbosch produces both red and white wines of high quality. Some other regions are Paarl, Caledon, Malmesbury, Piquetberg, Tulbagh, Worcester, Robertson, Olifants River, Swellendam, and Little Karroo.

FIGURE 4.3. German and New Zealand wine labels: **(A)** Appellation Wehlener Sonnenuhr, Wehlen, Mosel, Germany QmP; **(B)** Appellation Ranschbacher Seligmacher, Rheinpfalz, Germany QmP; **(C)** Appellation Marlborough, New Zealand.

4.4.7   Chile

Chile's central regions produce premium wines. The climate is rain-free in the summer with mild and wet winters. The wine-producing regions follow the coastline from Cautin to Atacama. Pisco, a type of brandy and the national beverage, is produced from grapes grown in the warmer northern regions. To date, there is no phylloxera in Chile and a strict quarantine system is in place for all imported vines. The majority of grapes grown are French varietals, particularly Cabernet Sauvignon, Malbec, Petit Verdot, Pinot noir, Semillon, Sauvignon blanc, and Chardonnay. Winemaking was introduced to the country, so there are no indigenous traditions.

The wine regions are called Regiones Vitivinicolas (RV). Grape varieties are not specified, but the grape yields and alcoholic strength are controlled. Chaptalization is not permitted. There are 7 main RVs containing 30 smaller RVs within the best RVs. The best RV regions are Maipo, Rapel, and Maule.

4.4.8   Argentina

Argentina is a large wine producer. Approximately 70% of the vineyard area is located near Mendoza. The majority of wines produced are of lesser quality, but it is a young industry with great potential. Chaptalization is not permitted.

The primary varietals grown are Syrah, Malbec, Cabernet Sauvignon, Merlot, Pinot noir, Barbera, Tannat, Gewürztraminer, Sauvignon blanc, Riojano, and the mission grape Criolla.

Argentina's new geographical appellation system consists of four major regions with subregions. The four major regions are Mendoza, San Juan, Northeastern, and South. Mendoza contains five subregions: The northeastern region contains four subregions; and the south region includes the Rio Negro area. Winemaking was introduced to the country, so there are no indigenous traditions.

Determination of quality is dependent on the approved type of varietal and where it is grown, produced, and bottled. All table wine is under the Indicación de Procedencia (IP) category, indicating a medium-quality product. The Indicación Geográfica (IG) category is for good quality wines. IG regulations require specifically approved grape varietals from a specific region, where the wine is produced and bottled.

The highest quality carries the Denominación de Origen Controlada (DOC). Grapes that go into these wines must be of high quality from approved varietals from a specific region, where the wine is produced, matured, and bottled.

4.4.9   Australia

Australian wines are produced primarily in the southeastern portion of the country near the coast. Classic French varietals are grown, such as Shiraz

(Syrah), Cabernet Sauvignon, Pinot noir, Merlot, Cabernet Franc, Riesling, Semillon, Chardonnay, and Sauvignon blanc. Chardonnay and Semillon are frequently blended as are Shiraz and Cabernet Sauvignon. The wines are labeled by varietal. Winemaking was introduced to the country, so there are no indigenous traditions.

The appellation system geographically indicates the origin of the grapes and does not indicate the nature of the wine or the quality. A few of the regions include Hunter Valley, Coonawara, Barossa Valley, Yarra Valley, and Tasmania.

4.4.10   New Zealand

The grape-growing regions of New Zealand are spread out over the country, but the majority of vineyards are on the north island. Typical varieties produced are Müller-Thurgau, Cabernet Sauvignon, Chardonnay, Gewürztraminer, and Sauvignon blanc. Winemaking was introduced to the country, so there are no indigenous traditions.

There is no appellation system in New Zealand, but there are several regions that produce premium wines such as Poverty Bay, Hawkes Bay, Martinborough, and Marlborough. See Figure 4.3 for an example of a New Zealand wine label.

4.4.11   United States

The United States ranks fourth in the world in wine production (tied with Argentina). The United States produces hundreds of varietal wines through-out 40-states, with the majority (90%) of the production in the state of California. The warmer parts of the country have not fared well in their wine production due to phylloxera and Pierce's disease. The majority of wine production is in the cooler coastal regions.

The United States has several indigenous grapes: *V. labrusca* (Concord grape), which grows in the northeastern part of the country, *V. rotundifolia*, which grows in the southwest, and *V. riparia* and *V. rupestris*, which grow in the southern part of the country. *Vitis labrusca* is used for limited winemaking but most of the others are used for jelly, jams, and juice. In the case of *V. riparia* and *V. rupestris*, their rootstocks are resistant to phylloxera and are used throughout the world in phylloxera-threatened areas.

Other than the limited production of wine made from *V. labrusca*, all other wine varietals and winemaking techniques were introduced to the country, so, basically, the United States has no traditional winemaking practices that could influence the appellation system. The classic French varietals predominate: Chardonnay, Merlot, Cabernet Sauvignon, Sauvignon blanc, and Pinot noir. Riesling, Gewürztraminer, and Zinfandel are widely produced in the western United States.

Federal guidelines and regulations set forth by the Alcohol and Tobacco Tax and Trade Bureau (TTB) pertaining to the wine industry is standard, but allows individual states to set and enforce stricter regulations. The TTB controls all alcoholic beverage production, importation, exportation, and transportation. Production issues include alcohol content and enological material that can and cannot be added to the wine (listed on website www.TTB.gov). An example of federal and state differences is chaptalization, where federal regulations allow chaptalization to 25 °Brix and an individual state such as California bans it.

According to the TTB, wine labels must contain certain information including alcohol content or the words "Table Wine" or "Light Wine," which indicates an alcohol content of 7% to 14% by volume, the appellation of origin, the net contents, the name and address of the bottler or importer, the country of origin for imports, the declaration of sulfites for interstate trade, and health warnings.

The American Viticultural Area (AVA) is the geographical appellation system in the United States established by the TTB to designate distinct growing areas. The TTB does not require viticultural areas on the label, but if they are used, they must follow the set regulations. AVA label designations became mandatory in January 1983. There are well over 145 AVAs in the United States, the majority being in California. An AVA is given to an area that is geographically distinct.

Varietal information is not required on a wine label, but if it is used, that information must also follow set regulations. Table 4.1 breaks down the federal appellation and varietal requirements along with the state exceptions to those regulations.

4.4.11.1   New York

There are three primary wine-growing areas in New York: Long Island, Finger Lakes, and the Hudson River Valley. Phylloxera is indigenous to the area and all *V. vinifera* are grafted to resistent rootstock.

Sparkling wines and still wines are produced. *Vitis labrusca* accounts for approximately two-thirds of the grapes produced, with only a small portion of *V. vinifera* grown. Cold-hardy Riesling and some Chardonnay and Pinot noir are grown. Long Island is somewhat more temperate, allowing additional less cold-hardy varieties to be grown.

4.4.11.2   Washington

The southeastern section of the state constitutes the growing area of Washington. This area is near the Yakima and Colombia rivers. Two-thirds of the grapes produced are *V. labrusca*. Other varieties grown are Chardonnay, Cabernet Sauvignon, and Merlot. Phylloxera has been found in the area, but there is limited use of resistent rootstock.

TABLE 4.1. Alcohol tobacco tax and trade bureau wine label regulations.

Label indication	Federal requirement	State exceptions
Appellation of origin (mandatory for importation)	75% Fruit origin political name (e.g., country, state, county)	None
Viticultural area (AVA)	85% Fruit origin	100% Fruit origin: California; 100% Fruit origin: Oregon
Vineyard designation	95% Fruit origin from named vineyard	None
Estate bottled	100% Fruit origin from named vineyard in same AVA as winery; 100% fruit origin from vineyard controlled by winery; all production, aging, and bottling must be conducted at the winery	None
Geographic regional wine names produced elsewhere	Use of designations such as Chablis, Chianti, or Burgundy must Indicate the true Location of origin (e.g., California Burgundy)	Asti Spumanti, Bordeaux Oregon prohibits the use of European place names
Vintage date	95% of wine produced from grapes harvested in stated year with an appelation of origin smaller than a country	None
Alcohol content (mandatory)	% Volume statement of 7–14% must be within 1.5% or words "Table Wine" used; % volume statement of 14–24% must be within 1%	None
Net contents (mandatory)	Stated in metric	None
Declaration of sulfites (mandatory for interstate commerce)	Sulfur dioxide level greater than 10 ppm	None
Health warning statement (mandatory)	"Government Warning" must appear on label in capital letters and bold type followed by the government warning statement	None
Bottling (mandatory)	Name and address of bottler or importer	None
Produced and bottled by	75% of wine production and bottling must be conducted at the location listed	None
Vinted and bottled by	75% of wine production, maturation, and bottling must be conducted at the location listed, excluding fermentation	None
Reserve	No legal meaning	None

4.4.11.3   Oregon

Oregon's Columbia River Basin and the Willamette Valley produce the majority of grapes in the state. All grapes grown commercially are *V. vinifera* varieties including Pinot noir, Riesling, Pinot gris, and Chardonnay.

4.4.11.4   California

The majority of grapes are grown in the Central Valley and costal areas of California. Nearly all wine made in California are from *V. vinifera*. Red grapes represent 50% of the production, with the majority of those grapes going into white Zinfandel (juice only). In addition to wine grapes, California is a large producer of raisins and table grapes.

Generically or semigenerically labeled wine represents the majority of wine produced in the state. Only about one-third of the wine is labeled by varietal. As in other wine-producing countries, it is typical that the more precise the appellation, the higher the quality of the wine.

The Central Valley extends southward from Sacramento to Bakersfield, with the largest grape production in the San Joaquin Valley. Table grapes and raisins are primarily produced in the San Joaquin Valley because wine grapes tend to be high in sugar and low in acidity. Typically, wine grapes tend to be of lesser quality due to the hot summer temperatures and high vine yields. This area is broken down into Northern San Joaquin and Southern San Joaquin Valley.

The North Coast and Central Coast of California produced the largest supply of quality wine grapes. The North Coast extends southward to Oakland and produces 15% of California's wine grapes (45% of the total value). AVAs includes Mendocino, Lake, Napa, Sonoma, Solano, Monterey, and Marin counties. Mendocino County includes the AVA Anderson Valley. Sonoma County includes the AVAs Russian River Valley, Dry Creek Valley, Alexander Valley, Bennett Valley, and Knights Valley. Napa County includes Napa Valley and Carneros AVAs.

The Central Coast region extends southward to Santa Barbara, with AVAs that include Livermore Valley, Santa Cruz Mountains, Edna Valley, Santa Maria Valley, Paso Robles, and Santa Ynez Valley.

California wine AVAs are as extensive as the geographical characteristics. The economics of the areas and the wines produced vary tremendously. Figure 4.2 illustrates three California wine labels with different appellations.

4.5   Wine Production Overview

Winemaking is an extremely involved science in which there is constant change and setting priorities. To understand the role of the laboratory in winemaking, knowledge of the grape to wine process is necessary. The winemaking process and where you are in that process will determine the wine chemistry. In order for an analyst to know whether the test results are reasonable, they must understand wine chemistry at each step.

There are two distinct classifications of wines, still and sparkling, each having their own production procedures. Sparkling wines in the United States are defined as a wine containing greater than 0.392 g carbon dioxide (CO_2) per 100 mL, or CO_2 at 1 atm of pressure.

Still wine refers to wines with no carbonation (no CO_2). Still wine production will be discussed in more detail because still wine production far exceeds sparkling wine production.

Upcoming chapters will concentrate on specific wine chemistry and enology at each stage of wine development; for now, we will look at the general categories of production.

4.5.1   Vineyard

Wine grapes have a growth cycle that is different for every country and region. In California the growing season follows this general pattern:

April 1: New shoots emerge from the dormant bud on the vine (bud break).
May 15: Vine flowers, pollination, and grapes (berries) begin to form.
July 15: Veraison begins (ripening of berries by color change and softening), sugar concentration rise from around 4° Brix to 20–24° Brix, and acid declines from 3% to 1%.
September 15: Harvest, ° Brix levels 19–24, less for sparking wines.

The color of the wine comes from the pigments found in the skin of the berry. Wine astringency and bitterness develops from the tannins (complex phenolic substances) and seeds. Depending on the variety of grape and the winemaking style, grapes will be harvested and divided into those that will have the juice pressed out and those allowed to stay in contact with the skins and seeds. White wines are usually pressed out, whereas most red wines stay in contact with the skins and seeds.

4.5.2   Harvest

The sugar level and the level of acidity are the primary factors indicating when to pick the grapes. These parameters are checked frequently in the last weeks of ripening to pinpoint the time to harvest. Other tests might be conducted to find the level of color and tannins present in the berries.

When the decision is made to harvest, the grapes will either be hand-picked or machine-harvested. Higher-quality wine grapes are normally hand-picked. Hand-picking is performed most often in the early morning while the grapes are cool and hydrated. Machine-picking is performed most often during the night for the same reasons. Grape deliveries to the wineries can vary from small bins to large trucks that can carry several tons (see Fig. 4.4). Depending on the quality of the grapes, they might be hand-sorted for quality (see Fig. 4.5).

The winemaker might choose to ferment the grapes in whole bunches (clusters) or partially crushed. Whole clusters are transferred directly to a fer-

FIGURE 4.4. Winery grape harvest deliveries via bins and trucks.

FIGURE 4.5. Hand-sorting grape clusters.

mentation tank. Grapes that are sent through a destemmer (Fig. 4.6) have the stems removed and the berries are lightly crushed (broken berries are called must). Red wine grapes are most often sent through the destemmer. Red must is sent to a tank, where yeast and nutrients will be added, and the stems, leaves, and MOG (MOG = material other than grapes) are discarded. Red must is most often fermented in tanks. There are several different red must fermentation methods, each having a particular size of bin or tank.

White wine grapes are sent to a press, such as a bladder press, to have the juice extracted from the berries. After pressing, white juice will be sent to a tank, where yeasts and other nutrients are added. Depending on the wine-making style, the juice might remain in a tank or be pumped into barrels for primary fermentation (conversion of sugar to alcohol). Movement of the juice or must is most often done using pumps and large hoses. The filling of the barrels can be done by hand, by pump and hoses, or by automated barrel fillers often found in larger wineries (see Fig. 4.7).

4.5.3 Fermentation

White and red wines are monitored daily for temperature and sugar content (°Brix) until they reach dryness. Dryness is defined as 0.2 g/100 mL, or 2 g/L of residual reducing sugar (amount of sugar remaining after fermentation). When juice has reached the end of fermentation, it becomes a wine.

FIGURE 4.6. Grape cluster destemmer **(top)**; hopper receiving de-stemmed fruit **(bottom)**.

During this time, the red juice is extracting color and tannins from the seeds and skins. The skins tend to rise to the top of the tank (cap) and it is imperative for good color extraction to get the juice and skins back in touch with each other. This is done in several ways:

- Punch down: where the cap is pushed back down into the juice; done several times a day.
- Pump over: where the juice is taken from the bottom of the tank and sprinkled over the cap, this is done several times a day.
- Rotary fermentor: a rotating tank that has a set rotational pattern keeping the juice and cap in contact.

FIGURE 4.7. Manual and automated barrel filling.

Figure 4.8 illustrates the variety of red must fermentation tanks such as open top fermentor, rotary fermentor, tank fermentor, and bins.

Newly fermented red and white wine, depending on winemaking style, might be subjected to a second fermentation called "malic acid fermentation or malolactic fermentation" (ML). This secondary ML fermentation begins with the introduction of a bacterium (most often *Oenococcus*) into the wine. This bacterium lives on sugars and organic acids, such as malic acid. The malic acid is converted into lactic acid, which is a milder acid than malic acid, producing a softer, less acid taste to the wine. Diacetyl is a chemical by-product of this conversion that gives a distinct butter flavor to the wine. With the creation of diacetyl and other compounds during ML fermentation, secondary fermentation has its advantages by adding complexity to the wine.

Before secondary fermentation, red wines are most often placed in a press such as a basket press (Fig. 4.9) to extract the wine. The wine is sent to a tank or put into barrels and the skins and seeds are discarded. Red and white wines might be placed in barrels for secondary fermentation.

FIGURE 4.8. Red grape fermentation methods: **(A)** open top tank, **(B)** rotary fermentor, **(C, D)** tanks.

FIGURE 4.9. Typical basket press used to press the wine from the must.

Upon completion of secondary fermentation, wines are protected from spoilage organisms that can ruin the wine. Wines are protected with low doses of sulfur dioxide (SO_2). Depending on the variety and winemaking style, wines might be pumped into clean barrels and put into the cellar to mature, or simply remain in a tank to prepare for bottling.

4.5.4 Maturation

Each variety of wine and each winemaker sets the criteria for the length of time a wine is left in barrels to mature. Maturing in barrels adds different elements and flavor components to the wine. Flavor components can range from "vanilla" to "heavy smoke." Barrels interiors are charred (toast) to various degrees from a slight toast to a very heavy toast. Barrels are chosen totally by the flavor profile a winemaker desires. The longer a wine stays in a barrel, the more extraction of substances and the more complex the flavor profile becomes. Most premium wines are barrel matured 6 months to 2 years.

Barrels require continual additions of wine to keep them full (topping) as evaporation takes place. Topping a barrel reduces possible oxidation of the wine. Every few months, for the first few years, the wine is taken out of the barrels (racked), leaving the residue (lees) to be washed out of the barrels. The wine is then checked for SO_2 content to ensure that it is protected from spoilage organisms and returned to the barrel to continue the maturation process. Racking helps to gradually clarify the wine.

Long-term barrel maturation over several years is highly labor-intensive and the production costs are high. Only the very best wines are matured for long periods of time.

4.5.5   Blending, Fining, and Stabilization

Prior to bottling and after the maturation process, wines will have a final evaluation by the winemaker to determine the quality of each barrel group or tank. The winemaker will decide if wines are to be blended together or if they will stand on their own merits.

Blending involves taking a percentage of two or more wines and putting them together to create a better tasting, or better balanced product. The wine chemistries are important in choosing the wines to blend, as they will give additional information to the winemaker in making blending determinations.

Adding a substance to a wine to reduce cloudiness or haze is called fining. Only the absolute minimum of TTB-approved fining agents are added to the wine to prevent any flavor components from being removed or "stripped" from the wine.

White wines contain substances (potassium and tartaric acid) that can form potassium acid tartrate (KHT) crystals when the wine is cold. The crystalline precipitate is completely harmless, but consumers (especially in the United States) find the crystals unpleasant; therefore, most white wine producers will go to great lengths to remove these tartrates from the wine. Cold stabilization is a process where the wine is chilled in tanks for several weeks to allow the KHT crystals to drop to the bottom of the tanks. The wine is then removed, leaving the crystals, and might be filtered at this time. Another process called ion exchange can be used. Ion exchange replaces the potassium ions with sodium or hydrogen ions, which will not precipitate out of the wine when it is chilled.

4.5.6   Filtration and Bottling

Wines that contain sugar are most often sterile filtered (0.45-μm filter) to remove any microbes or yeast that could cause an unwanted fermentation in the bottle. Filtration of perceived impurities out of the wine is a stylistic decision. Typically, premium wines that are dry and have proceeded through secondary fermentation are not filtered, to allow for bottle aging. Bottle aging allows all the tannins and other compounds to continue interfacing with the wine. Wine filtration is only done when necessary to prevent stripping the wines.

Bottling wine involves putting a wine into a sanitized bottle sparged with nitrogen (or other inert gas) to remove any oxygen, putting a sterilized cork in the top, and allowing only a small amount of headspace (area between the top of the wine and the bottom of the cork).

Depending on the varietal, the bottled wine might be shipped out for immediate consumption or bottle aged for 6 months (whites) up to 2 years (reds) before being released for sale.

Once purchased, wines can be consumed immediately, as is the case for most sterile filtered or highly filtered wines, or stored (cellared) for years. Wines cellared for years usually have had minimal fining and minimal or no filtration, allowing the bottle aging to improve the wine quality.

Figure 4.10 illustrates the winemaking process for red and white wines.

4.5.7 Sparkling Wine

Sparkling wine production is quite different than for still wines. The methods used to make sparkling wine are méthode champenoise, transfer process, and Charmat process. The term "champagne" refers to the city of Champagne in France, where sparkling wine was discovered in the 1600s. Using the word Champagne in the United States requires the addition of the geographic area of origin to distinguish the difference, such as New York Champagne.

The méthode champenoise begins with a base wine, or cuvée, made from Chardonnay, Pinot noir, Meunier, French Colombard, or Chenin blanc. The cuvée is low in alcohol and high in acid. Sparkling wines made with 100% Chardonnay are referred to as "blanc de blancs" or white from whites; using red grapes are referred to as "blanc de noirs" or white from blacks.

Triage is the next step and involves mixing the cuvée with a syrup mixture containing yeast and sugar. The sugar determines the amount of CO_2 produced (approximately 1 atm per 4 g); normally 6 atm are required per bottle, equating to 6 g of sugar. The cuvée/syrup mixture is then placed in a bottle. Alcohol will increase approximately 1.3% due to the sugar fermentation. The yeast is CO_2 and cold tolerant.

Bottles are capped and placed in a cellar while they undergo this second sugar fermentation, which takes approximately 1–3 months. After fermentation, the bottles are allowed to age for 1 year before any action is taken. As the year approaches, riddling takes place. This is a procedure to rid the wine of the yeast lees. Each bottle is gradually tilted from horizontal to vertical and rotated each day. This process takes 1–3 months to accomplish.

When the wine has cleared and the sediments have accumulated in the neck of the bottle; the neck is dipped into a subfreezing solution, freezing the sed-

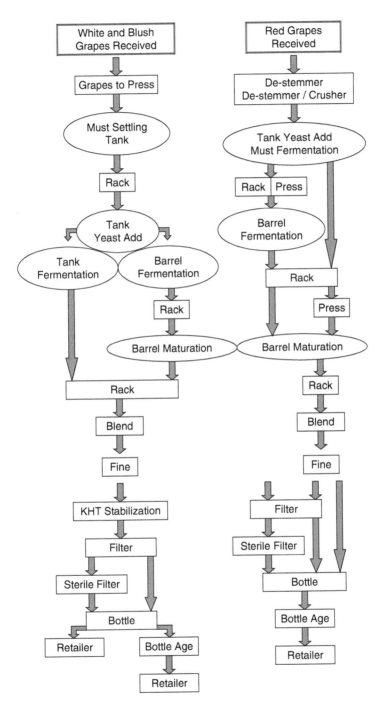

FIGURE 4.10. White and red wine production overview.

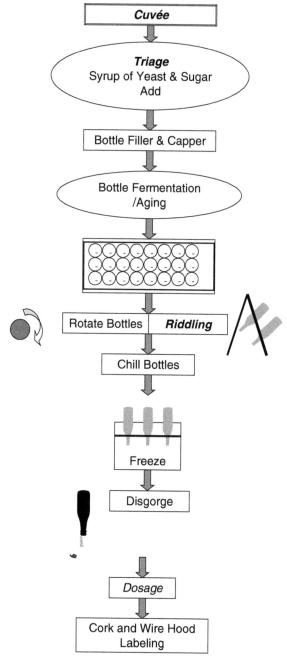

FIGURE 4.11. Méthode champenoise production.

iments in the neck of the bottle. The bottles are quickly opened and the CO_2 pressure that has accumulated pushes the frozen sediment plug out of the bottle. This process is termed *disgorgement*.

More sugar (dosage) is added to the wine and the bottle is topped and corked. The sugar additions are dependent on the acid levels and the style of sparkling wine desired:

- Sugar level of 5–15 g = brut
- Sugar level of 25–45 g = sec
- Sugar level of 50 g = demi-sec
- Sugar level of 100 g = doux

The dosage will not ferment because the yeast has been removed. If any yeast have remained in the wine, they are likely dead. Additionally, the high CO_2 level and lack of nutrients would limit any yeast growth.

Wines bottles are ready for the wire hoods to be added and labels attached. Figure 4.11 illustrates the winemaking process for sparkling wines.

4.5.8 Money Matters

There are several markets for the hundreds and thousand of wines produced every year. Marketing and selling wine is beyond the scope of this book, but it is important to understand the typical price levels for wines in the United States:

- Jug wine: < $3.00 per bottle, produced with very inexpensive grapes with a fast turnover (no secondary fermentation, maturation, or barrel time)
- Popular premium wine: $3.00 to $7.00 per bottle, slightly more expensive grapes with a fast turnover, also termed *fighting varietals* because of increased competition in the category
- Super premium: $7.00 to $14.00 per bottle, expensive grapes and/or more production costs due to barrel maturation
- Ultra premium: > $14.00 per bottle, expensive high-quality grapes and/or higher production costs due to extended barrel maturation

For most wines produced, there is little profit margin. The cost of the fruit, harvest, maturation, production, bottling, marketing, and sales greatly reduces the amount of profit a winery makes.

The laboratory is a contributing factor to keeping costs low and improving the profit margin. Accurate testing methods to determine the wine chemistry will reduce the amount of rework, protect the wine from spoilage, and ensure that the highest-quality wine is produced.

5

Vineyard to Harvest

5.1 Introduction

There are hundreds of wine grape varietals, and each varietal has its own climatic requirements, maturation timetable, nutritional needs, and chemistry. The growing cycle, climatic changes, viticulture practices, and winemaking practices will directly affect the outcome of each vintage (the wine produced in a given year). During the growing cycle, many of the elements and compounds that make good wines are developed. Winemaking techniques and procedures will stimulate the creation of additional components, resulting in a totally unique wine.

Testing of the new vintage begins before it ever leaves the vineyard. Several weeks after véraison (onset of berry ripening indicated by color change and softening), the viticulturist, grower, or winemaker will begin to monitor the sugar content of the berries in the vineyard. As harvest approaches, typically 6 to 8 weeks after vérasion. berry samples will begin coming into the wine laboratory for additional testing, including acid levels and pH. Analyses and tasting of the fruit will determine when the time is right for harvesting.

In this section we will begin with the physiology and anatomy of the mature grape berry and end at the winery, postharvest, and ready for fermentation. The processes and types of analysis required at each stage will be discussed. Fasten your seatbelts.

5.2 The Grape

Berries are formed after flowering and pollination (nouaison or berry set). The weather plays a large part in the berry set during flowering. Rain and cold weather can affect pollination, resulting in poor fertilization. Flowering can occur over several days and, in some cases, weeks in colder climates. Poorly fertilized berries will produce a lower number of seeds or no seeds at all (normal seed production is one to four per berry). Unfertilized berries will remain small and green and are commonly called *shot*.

Each vine can biologically support 100–200 berries per cluster with sufficient sugar and water. In years when there are a large number of flowers per cluster, many of the berries will not be fertilized. These unfertilized berries will all drop to the ground after berry set; this is called *shatter*.

During the last months of maturation, the berries accumulate sugar, potassium, phenol compounds, pigmentation, flavor compounds, tannins, water, and amino acids. From véraison to harvest, the sugar can increase from around 4–26% (depending on variety). Concurrently, the concentration of malic acid and ammonium decreases. Malic acid shifts, dropping from around 3% at véraison to 1% at maturation, or approximately 2–4 g/L up to 6 g/L (Boulton et al., 1999). These decreases are partially the result of increases in the water content and size of the berry, but a portion of the malic acid is converted to carbon dioxide (CO_2) (respiration). At maturation, respiratory intensity decreases and enzymatic activity increases.

5.2.1   Mature Grape Physiology

The primary constituents that comprise the makeup of the wine grape berry are organic acids, amino acids, alkaline metals, nitrogen compounds, phenolic compounds, anthocyanins, aromatic substances, water, and sugars. Each of these categories has many associated chemical compounds, but only a few will be discussed in the scope of this chapter.

The composition can be broken down into approximate percentages: water, 79%; sugars, 20%; organic acids, 0.6%; inorganic material, 0.2%; with a miscellaneous group of 0.5% (Margalit, 1997).

5.2.1.1   Organic Acids

Acid levels directly affect the fermentation, color extraction, microbial stability, maturation, and the stability of tartrates and proteins in wine. There are many different acids; some come from the grape (tartaric, malic, and citric acids) and some come from the winemaking process (lactic, acetic, succinic, and others).

Tartaric and malic acids represent 90% of all the acids in the grape. L-Tartaric acid (isomer of tartaric acid in grapes) is found in concentrations of 1–7 g/L, whereas L-malic acid (LMA, isomer of malic acid in grapes) is in concentrations of 1–4 g/L. The concentration of tartaric acid is nearly double that of malic acid. The concentration of citric acid runs around 150–300 mg/L (Margalit, 1997).

The strength of an acid is its ability to release H^+ into solution. Most wine acids are capable of releasing two protons per molecule. Each proton will disassociate and release one H^+; thus, a greater amount of H^+ is released. Tartaric acid releases nearly three times more protons than the same quantity of malic acid and contributes more to the pH.

The pH measures the H^+ and reflects the actual proton concentration in solution. Acid contribution is determined by the concentration and the ability of the acid to disassociate. Acids in grapes are considered weak organic acids and do not disassociate 100%. This ability to disassociate is given a pK_a value. The lower the pK_a, the easier an acid can disassociate releasing protons.

Free acid in the grape results in a pH of 3.1–3.5. Grapes high in acidity most often contain high amino acids and soluble proteins.

5.2.1.2   Alkaline Metals

Grapes contain a variety of minerals and are divided into two groups: cations and anions. Cations include potassium, sodium, manganese, aluminum, zinc, iron, copper, lead, and calcium. Potassium is the most abundant and has the greatest effect on pH and tartrate stability. Sodium and magnesium concentrations are about a tenth of potassium, with calcium being slightly less. The other elements have trace concentrations.

Anions consist primarily of phosphate and sulfate, with trace concentrations of boron, silicon, chlorine, bromine, and iodine. Sulfate accounts for the largest concentration, nearly three times that of phosphate. The other elements occur in trace amounts. Phosphate is an integral part of fermentation, and if concentrations in must are too low, diammonium phosphate (DAP) can be added to enhance the fermentation process by providing more available nitrogen.

5.2.1.3   Nitrogen Compounds

Nitrogen compounds are essential nutrients for the yeast's fermentation process. These compounds consist of ammonium salts, amino acids (and peptides), proteins, and nucleic acid derivatives.

Total nitrogen consists of all available nitrogen sources and varies greatly between varietals and vineyards. Generally, total nitrogen ranges from 150 to 650 mg/L. The largest concentrations of available nitrogen are derived from the ammonium salts and amino acids. Grapes contain from 0 to 150 mg/L of nitrogen in the form of ammonium salts (Boulton et al., 1999).

The largest concentrations of amino acids in grapes and must are proline and arginine. Proline is not used by yeast during fermentation, but arginine with glutamic acid, alanine, and aspartic acid are nearly depleted at the end of fermentation.

5.2.1.4   Phenolic Compounds

Phenolics are the third largest group of compounds in grapes. Phenols contribute to the red pigmentation, the brown-forming substrates, the bitter and astringent components, and, to a small extent, the taste in grapes and wine.

Phenols give the wine their character and quality and increases in concentration as the berry matures. Each grape variety has its unique makeup that varies year to year.

The primary constituents of the phenolic compounds are flavonoids and nonflavonoids. Flavonoids make up approximately 85% of the total phenols and contain anthocyanins, 3-flavanols (monomeric flavonoids or catechins), and tannin polymers. Nonflavonoids consist of phenolic acid groups, including hydroxycinnamates and hydroxybenzoates.

The nonflavonoid phenolic acids that are more prevalent in grapes come from the hydroxycinnamates group. Derivatives of cinnamic acid (including caffeic, p-coumaric, and ferulic acids) can form an esteric (reaction between alcohol and acid) bond with tartaric acid producing caftaric, coutaric, and fertaric acids. These three primary acids are responsible for the golden color in white wines. The hydroxybenzoates group, including p-hydroxybenzoic, protocatechic, vanillic, gallic, and syringic acids, is more prevalent in maturing wines and wines that contain mold.

Flavonoids are numerous, with several classifications:

- Flavone
- Isoflavone
- Flavonol
- Flavanol
- Flavane-diol
- Flavanone
- Flavanonol
- Chalcone
- Anthocyanidin

Anthocyanins, flavonols, and flavanols are the flavonoids of grapes and wines that concerns winemakers the most. Anthocyanins are responsible for color in red grapes and wine. The primary anthocyanin is malvidin, with lesser amounts of pelargonidin, cyanidin, peonidin, delphinidin, and petunidin.

Monomeric catechins (3-flavanols) produce the astringent feel and bitter taste components in wine. Catechin, gallocatechin, and afzelechin are the predominant catechins. Polymers of catechins are called tannins and will readily precipitate out of the wine with proteins (i.e., gelatin). In some literature, the catechins and their polymers are lumped and classified as tannins. The polymers are also referred to as procyanidins because of their ability to yield cyanidin when treated with acid and oxidized.

5.2.1.5 Aromatic Substances

Aromatics contribute to the nose or smell of the wine and are primarily volatile. These substances include aldehydes, ketones, esters, and terpenoids. Aldehydes are directly involved with fermentation and are reduced during the process. They are carbonyl compounds and have a distinct nose, often smelling like hay or leaves. Wines that have their fermentation stopped before they are completely dry will have a higher concentration of aldehydes than a dry wine. Oxidation of ethanol in wine results in the formation of aldehydes.

Ketones are also carbonyl compounds formed by the fermentation process. The two primary ketones are acetoin and diacetyl. Diacetyl increases with malolactic fermentation, giving the wine a buttery aroma. Acetoin levels are higher in partially dry wines to which alcohol has been added, as in Port.

Esters are the result of the reaction between alcohols and acids. In a finished wine, there are over 300 different esters. Esters are divided into volatile esters, which produce a fruity aroma, and acid esters, which are nonvolatile and have little contribution to taste or flavor. Significant volatile esters that contribute to the fruity aromas of young wines are isoamyl acetate, ethyl hexanoate, ethyl octanoate, and ethyl decanoate. A volatile ester that is undesirable in wine production is ethyl acetate. Ethyl acetate produces the smell of vinegar and it goes hand in hand with acetic acid.

Terpenoids are the result of the metabolism of mevalonic acids. Monoterpenes give us a range of volatile aromas; a few of the monoterpenes are as follows:

- Linalool: rose, plum odor
- Geraniol: rose odor
- Nerol: rose, peach odor
- Citronellol: rose odor
- Ho-trienol: fruity odor
- Pyrazines: herbaceous, vegetative odor
- β-Carotene fruity odor
- Limonene: lemon odor
- Myrcene: floral odor
- α-Terpineol: lilac, peach odor

5.2.1.6   Sugar

Grape sugars, or carbohydrates, are monosaccharides (monomers of hexose and pentose); they include glucose, fructose, rhamnose, arabinose, sucrose, and pectin.

Glucose and fructose (primarily in the forms of D-glucose and D-fructose) are the most abundant sugars in nearly equal proportions. Fructose contributes slightly more to the total, is sweeter than glucose, and is not fermented as fast as glucose. Glucose and fructose constitute the largest portion of the total soluble solids in juice and, therefore, can be measured by density and specific gravity.

Pectin and other polysaccharides known as gums tend to precipitate during fermentation because of the increase in alcohol, reducing their concentration more than 50%. The concentration of sugars at grape maturity is approximately 150–240 g/L.

5.2.2 Grape Structure

The structural components of the mature grape berry are shown in Figure 5.1. Each section of the berry carries a combination of compounds. These compounds vary not only in quantity but also in types of compound that are inherent to each grape varietal of *Vitis vinifera*.

5.2.2.1 Seeds

Grape seeds contribute up to 6% of the total weight of the berry. A berry may have zero to four seeds. The more seeds produced, the larger the berry, which slows down the sugar accumulation and tends to prolong acid retention. Seeds contain carbohydrates, nitrogen compounds, oils (oleic and linoleic), minerals, vitamin E, and phenolic compounds. The seeds contain approximately 20–50% of the total polyphenols in the berry (Ribéreau-Gayon et al., 2000), the greatest concentration of tannins. Seed tannins reach their highest concentration at véraison and diminish throughout the maturing process.

5.2.2.2 Skin

The berry skin contributes up to 20% of the total berry weight. The skin contains the essential anthocyanins required for red wine, along with flavonols and tannins. The skin is high in citric acid and contains benzoic and cinnamic acids. Aromatic substances, aroma precursors, and a small amount of sugar complete the profile.

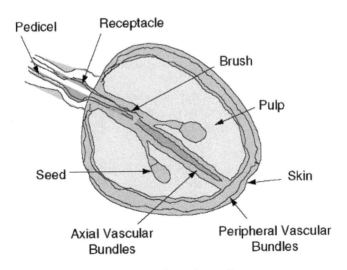

FIGURE 5.1. Grape berry diagram.

5.2.2.3 Pulp

Berry pulp is the largest component, contributing up to 85% of the total berry weight. The pulp itself has very little solid parts but has a high density and soluble solids level because of the chemical substances within. Fructose, glucose, and other sugars are the heavyweights and contribute the greatest to the soluble solids content. Soluble proteins are also a contributing factor.

The primary acids contained in the pulp are tartaric, malic, and citric, with trace amounts of other acids. The concentration of potassium predominates with lesser alkaline metal concentrations of iron, calcium, magnesium, and sodium. Pulp pH is usually in the range 2.8–3.6.

Nearly 25% of the total nitrogen is contained in the pulp as nitrogen compounds, including organic nitrogen. Pulp is high in the amino acids leucine, praline, arginine, threonine, and glutamic acid. There is a significant concentration of aromatic compounds such as aldehydes, esters, and terpenoids.

5.3 Preharvest

Harvest can be fairly well predicted when one understands the growth cycles of the vine and grapes. The type of varietal and weather changes dictate the length of each growing phase, making each year unique. The growth cycle has three phases:

- Phase 1: Approximately 45–65 days; rapid herbaceous growth; berry cellular growth begins around 2 weeks postfertilization; and berries accumulate acid coinciding with an increase in respiration. In California, bud break is around April 1 and flowering occurs around May 15.
- Phase 2: Approximately 8–15 days; slowed growth and appearance of color or translucency in the berry (véraison). In California, véraison begins around July 15.
- Phase 3: Approximately 35–55 days; increased cellular growth; decrease of respiration; increase of enzymatic activity; accumulation of sugar, potassium, amino acids, and phenolic compounds; and a decrease of malic acid and ammonium.

Knowing the vineyard history helps predict approximate dates, but keeping a close watch on the vineyard is the key to the successful timing of harvest. Approximately 4 weeks prior to the anticipated harvest date, berry samples will be gathered and the sugar content, aroma, and phenolic maturation will be evaluated in the vineyard. Vineyard sampling and testing should occur every 5–7 days.

To determine the maturity of the fruit, several factors are taken into consideration:

- Sugar concentration
- pH
- Potassium and other mineral levels
- Phenolic compound concentrations
- Nitrogen concentration
- Turbidity
- Aroma compounds
- Insoluble solids
- Titratable acidity (TA)
- Potential alcohol

Laboratory analyses can evaluate the majority of these maturation factors. Aroma evaluation, potassium content, and phenolic concentrations can be analyzed in the laboratory with sophisticated equipment and extraction methods. For most day-to-day operations, smelling and tasting the berries and seeds in the vineyard might not be the most accurate, but it is a practical evaluation method. There are several companies researching new technologies that in the future will enable better evaluation of the berries in the vineyard.

Depending on the varietal and the type of wine being made, the acceptable sugar and acid levels for harvest should be predetermined. As the sugar level nears the targeted level, grape samples will be taken to the wine laboratory for additional testing of TA, pH, juice volume, precise sugar determination, and potential alcohol (both potential alcohol and alcohol determinations will be covered in this section). The Maturation Index (sugar/acid ratio) is used by many winemakers as an aid to determine harvest. This index is used with caution because there is no direct relationship between sugar gain and acid loss.

At maturity, berries will contain approximately 230–260 g/L of sugar (higher levels could indicate dehydration in the berries). Acid levels normally range between 4 and 9 g/L tartaric acid. The phenolic compounds will accumulate to a certain level and then stop.

When the moment of harvest has been determined, the grapes will be harvested by hand or machine and transported to the winery.

5.3.1 Vineyard Sampling

Obtaining a grape or juice sample that is representative of the entire vineyard area is imperative. The analysis results are based on this sample; the decision when to harvest is based on this sample. Improper sampling of a vineyard area could lead to premature harvesting or late harvesting that might have a significant economical impact on the grower and/or winery.

Many sampling techniques have been tested, and as you would expect, the higher the sample population, the more accurate the results when compared to the actual content of the must. For practical purposes, most samples from a vineyard area consist of a sample population of 100–200 individual berries taken from various clusters throughout the vineyard.

Vineyards should be divided into manageable sections and sampled. The more sections, the more accurate the harvest picture. It is important to note the variables that affect the sugar content of the berries:

- No two clusters of grapes are the same because of uneven flowering, fertilization, and climate conditions.
- Clusters closer to the trunk contain higher sugar levels.
- Berries ripen faster further from the ground.
- Berries from shaded areas will differ from nonshaded areas.
- Irrigated vineyards will have greater variability because of water distribution.
- Raisined berries are high in sugar because of dehydration.
- Berry chemistry will vary throughout the day; it is best to sample each area at the same time of day, preferable early morning.

Collect berry samples from all over the vines throughout the vineyard area. Keep track of the locations to ensure even sampling. Evaluate and note the percentage of shot and raisins.

Samples tested in the field should be destemmed and have the juice expressed by means of a small press, or auger. The pomace (seeds, skin, and remaining solids) is discarded.

Samples obtained for laboratory analyses should be as fresh as possible. Laboratory samples can be expressed in the vineyard, placed in clean containers, stored in a cooler (not directly on ice), and transported to the laboratory as soon as possible. Whole berry samples can also be collected, placed in a cooler to prevent dehydration (not directly on ice), and transported as soon as possible to the laboratory, where the juice will be expressed from the intact berries prior to testing. The chemistry of the berry sample begins to change the moment the berry is picked. The longer a sample is delayed prior to analysis, the larger the margin of error.

The most common error when transporting samples to the laboratory is misidentification of the sample. Sample documentation must include the exact location sampled, vineyard name, time sample obtained, temperature, and any other relevant details such as weather conditions (rain, fog, dewy, etc.) or if the vines were recently watered.

5.3.2 Sugar Determinations

Berries contain several different sugars, but only two are used primarily for fermentation: glucose and fructose. Other nonfermentable sugars are pentose, arabinose, and rhamnose in concentrations of approximately 0.2–0.3%, which prevents wine from becoming completely void of sugars, or totally dry. Sugars in berries account for nearly all of the soluble solids (90–94%) in the pulp and juice, so with reasonable accuracy we can measure the soluble solids in the juice as density or specific gravity and obtain an estimated sugar value. Density and specific gravity are discussed in more detail in Chapter 1. Density

(*D*) is expressed as mass per volume (g/cc) and reported with the measurement temperature of the solution [i.e., *D*(20)]. Specific gravity (SG) is the ratio of the density of a solution relative to the density of a standard, such as water at the same temperature, and reported specifying the measurement temperature in degrees Centigrade of the solution over the temperature of the comparable standard [i.e., SG(20/20)].

Specific gravity and density can be measured by a hydrometer (Fig. 3.3) or metering devices (see Fig. 5.2). Sugar refracts light (bends light as it passes through a solution), so we can employ a device that measures the degree of refraction called a refractometer (manual or digital) to obtain an estimated sugar value, which has been adapted to the apparent °Brix scale (see Fig. 5.3). Sugar values from refractometers are slightly lower when compared to other methods because of the difference between density and light refraction.

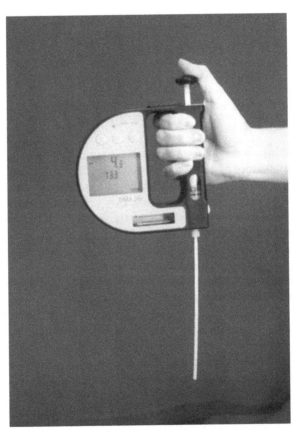

FIGURE 5.2. Anton Paar DMA 35n™ hand-held density meter with permanently stored concentration tables for °Brix, °Baumé, specific gravity, % alcohol, proof, °Plato, API gravity, and % H_2SO_4. The units store 1000 data points and optionally downloads through an RS232 port.

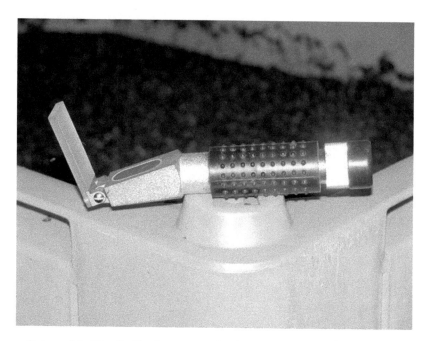

FIGURE 5.3. Handheld refractometer for the measurement of ˚Brix in juice.

Physical measurements are taken at 20°C with each instrument. Units are assigned to the values based on the particular method's calculations and a scale is then developed. The first density scale was created by Antoine Baumé in 1768. To date, there are three scales based on density in addition to refractive index used in the wine industry around the world to determine sugar levels: °Oechslé used in Germany, Austria, and Switzerland; Baumé used in France, Italy, Spain, Portugal, Australia, and New Zealand; and apparent °Brix (or °Balling) used in most other areas of the world.

It should be noted that the accuracy of any liquid measurement or determination of content is dependent on the temperature of the liquid. It does not matter if you are measuring a volume of liquid or determining the density or specific gravity. Liquids are brought to measuring temperature by heating or cooling, or the values are mathematically corrected to the proper measuring temperature. For example, specific gravity is measured at 20°C (68°F); if the temperature of the solution being measured is below 20°C, 0.0002 units are subtracted from the reading for each degree below 20°C. If the temperature of the solution being measured is above 20°C, 0.0002 units are added to the reading for each °C above 20°C.

Table 5.1 lists the conversions for these various scales and includes refractometer measurements, sugar concentration, and potential alcohol as listed in *Handbook of Enology Vol. 1,* (Ribéreau-Gayon, et al., 2000, p. 241).

TABLE 5.1. Conversion table for methods of sugar measurement.

Relative Apparent Density (20°C)	Degree Oechslé	Degree Baumé	Degree Brix	Refracto-metric Measure (in Percentage Weight of Sucrose)	Sugar Concentra-tion (g/L)	Potential Alcohol (16.83 g of Sugar per Liter for 1% Alcohol)
1.0371	37.1	5.2	9.1	10	82.3	4.9
1.0412	41.2	5.7	10.1	11	92.9	5.5
1.0454	45.4	6.3	11.1	12	103.6	6.2
1.0495	49.5	6.8	12.0	13	114.3	6.8
1.0538	53.8	7.4	13.0	14	125.1	7.4
1.0580	58.0	7.9	14.0	15	136.0	8.1
1.0623	62.3	8.5	15.0	16	147.0	8.7
1.0666	66.6	9.0	16.0	17	158.1	9.4
1.0710	71.0	9.6	17.0	18	169.3	10.1
1.0754	75.4	10.1	18.0	19	180.5	10.7
1.0798	79.8	10.7	19.0	20	191.9	11.4
1.0842	84.2	11.2	20.1	21	203.3	12.1
1.0886	88.6	11.8	21.1	22	214.8	12.8
1.0932	93.2	12.3	22.1	23	226.4	13.5
1.0978	97.8	12.9	23.2	24	238.2	14.2
1.1029	102.9	13.5	24.4	25	249.7	14.8
1.1075	107.5	14.0	25.5	26	261.1	15.5
1.1124	112.4	14.6	26.6	27	273.2	16.2
1.1170	117.0	15.1	27.7	28	284.6	16.9
1.1219	121.9	15.7	28.8	29	296.7	17.6
1.1268	126.8	16.2	29.9	30	308.8	18.4
1.1316	131.6	16.8	31.1	31	320.8	19.1
1.1365	136.5	17.3	32.2	32	332.9	19.8
1.1416	141.6	17.9	33.4	33	345.7	20.5
1.1465	146.5	18.4	34.5	34	357.7	21.3

5.3.2.1 °Oechslé

Degree Oechslé is based on the density contribution of the soluble solids in solute over that of water, or relative apparent density (D), calculated by the formula

$$D = SG(20/20), \tag{1}$$
$$°Oechslé = (D - 1) \times 1000 \tag{2}$$

Degree Oechslé was originally calculated at 15.5°C reference temperature and later changed to 20°C, which is used today.

5.3.2.2 °Baumé

The Baumé scale was originally developed based on the concentration of salt in solution at 12.5°C. The scale has been recalculated for 20°C, which is used today. The Baumé scale corresponds to the potential ethanol concentration in percent by volume and is most accurate in the 10–12% alcohol range. Based on apparent density (D), Baumé is approximately 1.8 degrees of the Brix scale and is calculated with the formula

$$D = SG(20/20),$$
$$\text{Baumé} = 145.0\left(1 - \frac{1}{D}\right),$$

where 145.0 is the modulus (constant) (Boulton et al., 1999). The general temperature correction for the Baumé reading based on 20°C:

- For temperatures over the 20° C reference, add (+) 0.05 Baumé per degree Centigrade over.
- For temperatures below the 20° C reference, subtract (−) 0.05 Baumé per degree Centigrade below.

Conversion of °Brix to Baumé = °Brix ÷ 1.8.

5.3.2.3   °Balling and °Brix

Karl Balling developed a density scale based on the gram weight of sucrose per 100 g of solution at 17.5°C. Each °Brix is equivalent to 1% w/w of pure sucrose. Since Balling's time, Antoine Brix recalculated the scale to 15.5°C in the mid-1800s, which made the Balling scale obsolete. The scale has since been updated and now is calculated to 20°C reference temperature. The remainder of the book will use the term °Brix only to avoid confusion.

The Brix scale reflects the apparent °Brix because the majority of sugar in juice is fructose and glucose, not sucrose, and exists as soluble solids in the juice. More accurate sugar determinations can be made using enzymatic analysis, which we will discuss in the next section. The formula for the calculation of apparent °Brix is

$$D = SG(20/20),$$
$$\text{Apparent °Brix} = 261.3 - \left(\frac{261.3}{D}\right)$$

where 261.3 is the modulus (constant) (Boulton et al., 1999).

From the above information, it is clear that the reference temperatures must be maintained to obtain correct determinations. The instrumentation is calibrated at these temperatures and any change will affect the results. Complete listing of temperature correction tables can be obtained from the National Bureau of Standards. For our purposes, there is a simple formula used for temperature correction for the Brix scale:

- For temperatures over the 20° C reference, add (+) 0.06 Brix per degree Centigrade over.
- For temperatures below the 20° C reference, subtract (−) 0.06 Brix per degree Centigrade below.

Conversion of Baumé to °Brix = Baumé × 1.8.

Hydrometers, refractometers, and digital density meters with programmed °Brix calculations are used to determine °Brix. Measurements of sugar in immature grapes with less than a 15 °Brix is inaccurate when using a refractometer or densitometer because of interference from organic acids, amino acids, and other materials that effect the refractive indexes and measurements (Ribéreau-Gayon et al., 2000a).

The °Brix of postharvest musts will also vary in accuracy. Red must in contact with the berries has more solid materials suspended that will affect the density of the juice, where as white grapes are usually pressed and do not have the degree of suspended solids. It is common to see a rise in the °Brix reading in red must 24 h after harvest once the skins and raisins have been soaked and release more sugar into the must. Rotten grapes and late harvest grapes will also have an increase in suspended solids and reflect an inaccurately high °Brix.

The drop in soluble solids due to the reduction of sugar and the production of alcohol will decrease density. The accuracy of °Brix measurements via hydrometer will be affected and will often show numbers in the negative as the juice reaches the end of primary fermentation, but they can still be used to monitor °Brix throughout fermentation with verification of dryness by other means. Refractometer measurements are more adversely affected because of increases of the refractive index by alcohol and are not used for monitoring fermentation.

The laboratory maintains closer quality control and most often has more sensitive equipment than is found in the field. Vineyard berry samples are analyzed by the laboratory to verify sugar content prior to harvest.

Vineyard owners who sell their grapes to wineries usually have a contractual agreement with the winery. These contracts most often specify the targeted sugar content of the grapes and could have penalties for grapes that fall above or below this set level. Many wineries in California will have government officials on site to test the grapes for sugar content upon arrival from the vineyard. Sugar content of arriving grapes should be as accurate as possible, with samples analyzed and the results documented by the laboratory regardless of government involvement.

Additional information concerning hydrometers and density meters can be found in Chapter 6. Chapter 9 will delineate procedures for the use of hydrometers, refractometers, and density meters for measuring soluble solids and sugar.

5.3.3   Glucose and Fructose Determinations

Glucose and fructose can be isolated and measured using a variety of methods, including enzymatic, high-performance liquid chromatography (HPLC), and gas chromatography (GC). The most practical method for the majority of laboratories is the enzymatic method (see Chapter 9).

Reducing sugar is the amount of fermentable sugar, or the sugar that is reduced by fermentation in a juice or wine. As mentioned earlier, glucose and fructose are the fermentable sugars. Residual sugar is all of the sugars that remain in the wine after fermentation, including the nonfermentable sugars. Today, the term "residual sugar" is used interchangeably with "reducing sugar" to mean the amount of glucose and fructose in the juice or wine. Perhaps the best term would be residual reducing sugar (Boulton et al., 1999).

5.3.3.1   Enzymatic Method

Enzymatic analyses are based on the measurement of the increase or decrease in absorbance of nicotinamide–adenine dinucleotide (NADH) or nicoti-namide–adenine dinucleotide phosphate (NADPH), which are coenzymes found in plants, animals, and microbes. NADH and NADPH absorb light in the long-wavelength region of the light spectra. Measurement of the light absorbance is often referred to as the UV (ultraviolet) method. The measurement for the presence of these coenzymes is made at 340 nm using a spectrophotometer (mercury-lamps will use 365 or 334 nm).

Spectrophotometers (spectrometers) utilize optical glass or clear plastic cuvettes with a path length of 1.0 cm (smaller cuvettes can be used depending on the instrument, with readjustment of the calculations). The reaction temperature of this method as well as the temperature of sample and reagents is optimal between 20°C and 25°C (the higher the temperature, the faster the reaction).

Enzymatic analysis can be performed to determine many chemical analyses, including lactic acid, acetic acid, ammonia, citric acid, ethanol, LMA, sulfite, and D-glucose/D-fructose, which are what we are interested in at this point of development. Further sections and chapters will cover several of the above analyses.

Enzymatic analysis identifies and quantifies glucose and fructose in a sample. The method uses several different enzymes to break down glucose and fructose into gluconic acid-6-phosphate + NADPH + H^+. The NADPH is measured because of its increased light absorbance at a wavelength of 340 nm and is stoichiometric to the amount of glucose. Fructose must first be converted to glucose and the first step is the addition of adenosine triphosphate (ATP) in the presence of hexokinase (HK) while glucose is converted to glucose-6-phosphate + ADP (Mannheim, Boehringer, 1997):

$$\text{Fructose} + \text{ATP} \xrightarrow{\text{HK}} \text{Fructose-6-phosphate} + \text{ADP,} \tag{1}$$

$$\text{Glucose} + \text{ATP} \xrightarrow{\text{HK}} \text{Glucose-6-phosphate} + \text{ADP.} \tag{2}$$

The second step to convert the fructose to glucose is the addition of phosphoglucose isomerase (PGI), which converts the fructose-6-phosphate into glucose-6-phosphate:

$$\text{Fructose-6-phosphate} \xrightarrow{\text{PGI}} \text{Glucose-6-phosphate.} \tag{3}$$

Finally, the addition of glucose-6-phosphate dehydrogenase (G6P-DH) in the presence of oxidized–dinucleotide (NADP):

$$\text{Glucose-6-phosphate} + \text{NADP}^+ \xrightleftharpoons{\text{G6P – DH}} \text{Gluconic acid-6-phosphate} + \text{NADPH} + \text{H}^+. \qquad (4)$$

This analysis requires precise dilutions and additions of reagents (see Chapter 9). Kits containing enzymes and reagents can be purchased, or reagents can be made in the laboratory and the enzymes purchased. Reagents made in the laboratory normally do not have preservatives added and will deteriorate more quickly than purchased reagents. Purchased enzymes have a limited shelf life. This method is more economical if there are several analyses to be made throughout the fermentation process. One or two samples here and there are best taken to a professional laboratory. The investment in chemicals, glassware, and time to perform a few tests is not economical.

Errors in enzymatic UV methods are numerous. Using prepared kits reduces many sources of error, but careful and precise laboratory techniques will reduce the majority of errors encountered. The most common errors include the following:

- Improper temperature for analysis and measurement of reagents or sample (optimum 20–25°C)
- Improper pipette technique
- Failure to degas or filter to eliminate CO_2
- Expired reagents or enzymes
- Failure to adjust sample pH (buffers are included with kits)
- Incorrect dilution to within the working range of the instrument
- Air bubbles in cuvette blank (zero instruments to air, recommended)
- Wrong nanometer setting on instrument
- Lamp on instrument failing
- Air bubbles in sample cuvette
- Incorrect light path length
- Contaminated reagents or enzymes

5.3.3.2 Other Residual Reducing Sugar Methods

Laboratories with hundreds of analyses will find cost savings using automated systems that perform enzymatic spectroanalysis using microquantities of reagents and enzymes, such as the Kone™ random access analyzer. The Kone is a computerized automated robotic system that performs wet chemistry, incubates, analyzes, and calculates results. The Kone can perform all enzymatic and spectrometric analyses required in a laboratory. A manual multi-sample spectrometer can handle a low to moderate influx of samples.

High-performance liquid chromatography and GC are very accurate methods for quantifying reducing sugars (RS), but the equipment is very expensive and requires highly trained personnel. Due to the analysis time, there is also a limit to the number of samples that can be processed in a 24-h period. The major advantage of GC is no reagent and enzyme costs.

A method that is being tested currently is Fourier transform infrared spectroscopy (FTIR) for determination of various juices, must, and wine parameters. FTIR, very simply, analyzes a sample by taking a spectral imprint of the light absorbed by the sample (interferogram). The interferogram is collected by the spectrometer and processed through the Fourier transform calculation. The instrument does not directly measure the analytes; data are programmed into the instrument's mathematical equation using a calibration model identifying certain parts of the spectrum to reflect an analyte. The entire accuracy of this technology depends on the accuracy of the calibration model programming. At this point, the method seems to be a promising, rapid way to scan a sample for potential problems and isolating samples for analytical chemistry evaluation.

5.3.4   Potential Alcohol (Ethanol) Determinations

It is important for the winemaker to know what alcohol (ethanol) level could result during and upon completion of fermentation. The sugar content of the grapes and must are used to calculate the estimated or potential ethanol content of the finished wine. Formulas used to calculate potential ethanol are based on the theoretical conversion of 180 g of sugar into 92 g of ethanol (51.1% by weight) and 88 g of CO_2. Pasteur's experiments showed a 48.5% ethanol yield, attributing 46.7% to CO_2 and 4.8% utilized by the yeast cell mass and other components (Boulton et al., 1999). Others have used different percentages of the theoretical yield to obtain an ethanol yield factor (yield of ethanol per 100 mL from 1 g per liter of sugar). The ethanol yield factor is different every year and varies from region to region

High ethanol levels can stop the fermentation process prematurely, causing stuck fermentations. In the United States, wine is taxed on the ethanol content, creating economical issues for the winemaker. Other parts of the world place restrictions on the percentage of ethanol allowed in wines.

Potential alcohol estimations are conducted prior to harvest and approximately midway through fermentation. Included in this section will be the alcohol determination, which is required to estimate the mid-fermentation potential alcohol.

Prior to harvest, Oechslé and apparent °Brix readings are used to calculate the potential alcohol. It is important to remember that the soluble solids of the juice are not 100% sugar, but slightly less. This nonsugar portion (extract) consists of other substances that will vary from year to year, varietal to varietal, and fermentation to fermentation. In other words, the amount of extract in juice will vary just like the sugar, acid, and pH can vary from one vintage to another. Not taking the extract into consideration when calculating potential alcohol will result in abnormally high predictions. In addition to the amount of extract contained in the juice, consideration must be made for the amount of sugar extracted from the skins during fermentation.

Because all of these values are moving targets, there is no definitive agreed upon constants that can be inserted into an equation resulting in an accurate prediction of ethanol content. Tracking of vineyard history can shed some light on the extract volume and the ethanol conversion factor to use each year.

The most current formulas for potential ethanol determination using hydrometry and refractometry and the Oechslé and apparent °Brix scales are

$$\text{Potential ethanol (\% v/v)} = 0.059\,[(2.66 \times Oe) - 30], \qquad (1)$$

where: 0.059 is the actual ethanol yield factor (in mL), 2.66 is a constant, Oe is the Oechslé reading, and 30 = 30 g/L of extract (Boulton et al., 1999), and

$$\text{Potential ethanol (\% v/v)} = 0.592\,[(\text{apparent Brix w/w \%}) - 3.0], \qquad (2)$$

where 0.592 is the ethanol yield factor (0.47 g/100 mL = 92% of the theoretical conversion of 0.511) divided by the density of ethanol (0.794), and 3.0 is 3% (30.3 g/L) of the extract (Boulton et al., 1999).

Common use of a standard ethanol yield factors such as 0.50 g/100 mL for red grapes and 0.66 g/100 mL for white grapes multiplied by the apparent °Brix or Oechslé does not take into consideration the extract and the nonfermentable sugars.

To get a more accurate potential alcohol, we can use actual glucose and fructose measurements, which eliminate the variables of extract and nonfermentable sugars. Using the European Economic Community (EEC) sugar to ethanol conversion factor of 16.83 g/L, equaling 1% ethanol, and assuming 100% of the glucose and fructose will be utilized, we should be able to use the formula

$$\text{Potential ethanol (\% v/v)} = \frac{\text{Glucose g/L} + \text{Fructose g/L}}{16.83\,\text{g/L}} \qquad (1)$$

What this formula does not consider is the yearly variation in actual ethanol conversion, but it can supply the winemaker with the highest possible alcohol. Using a historical ethanol yield factor could provide a more realistic picture:

$$\text{Potential ethanol (\%v/v)} = 0.055\,(\text{Glucose g/L} + \text{Fructose g/L}), \qquad (2)$$

where 0.055 is an example of a historical ethanol yield factor for a particular region.

If we assume that the EEC conversion factor does not change but that the amount of glucose (G) and fructose (F) metabolized is variable, we could use the formula

$$\text{Potential ethanol (v/v \%)} = \frac{(\text{Glucose g/L} + \text{Fructose g/L})\,(94.5\%)}{16.83} \qquad (3)$$

Or

$$\text{Potential ethanol}(v/v\%) = \frac{(G + F \text{ g/L})\, 94.5\%}{16.83},$$

where: 94.5% is the averaged percentage of the theoretical ethanol yield factor compared to the findings of Pasteur.

Mid-way through fermentation, potential ethanol estimates are made to give a more accurate picture of the ethanol content postfermentation, because it is taking into consideration the conversion rate of sugar to ethanol currently under way in the must. A sample is obtained and immediately filtered (to stop any fermentation) and analyzed for alcohol/ethanol content and glucose/fructose levels. The results are plugged into an equation to estimate the final ethanol content. Once again, there is no hard and fast formulation and each winemaker uses his own particular interpretation.

Theoretically, if an original glucose/fructose level was obtained on the prefermented juice, you could compare the drop in glucose/fructose and divide that number by the determined ethanol percentage to obtain the amount of glucose/fructose converted to obtain the 1% v/v ethanol conversion yield. Once the new conversion factor is determined, divide the current glucose/fructose level by the factor and add this predicted ethanol percentage to the determined ethanol percentage to yield a more precise potential ethanol. The formula assumes the sugar to ethanol conversion continues at the same pace until dry:

$$\text{Potential ethanol}(v/v\%) = \left(\frac{CGF}{(OGF - CGF) \div \%DE} \right) + \%DE, \qquad (1)$$

where OGF is original glucose fructose level (in g/L), CGF is the current glucose fructose level (in g/L), and %DE is the % v/v of determined ethanol. For example, if OGF equals 238.2 g/L, CGF equals 113 g/L, and the %DE is 7.4, the potential ethanol at mid-fermentation would be

$$\text{Potential ethanol}(\%v/v) = \left(\frac{113\,\text{g/L}}{(238.2\,\text{g/L} - 113\,\text{g/L}) \div 7.40\%} \right) + 7.40\% \quad (1)$$

$$= \left(\frac{113\,\text{g/L}}{16.92} \right) + 7.40\% \qquad (2)$$

$$= 6.68\% + 7.40\% \qquad (3)$$

$$= 14.08\%$$

Estimating the original potential alcohol based on the above OGF and the EEC conversion factor of 16.83, the potential alcohol would have been 14.2%. Using 94.5% of the OGF and the EEC factor, the predicted percentage would be 13.4%. In the equation using the ethanol conversion factor of 0.0592 and the above OGF, the potential ethanol would be 14.1%. Finally, if you use a common conversion of 0.55 g/100 mL (or 0.66 g/100 mL) and the above OGF, the result would be 13.1%. As you can see, there is a significant difference in the results depending on what conversion factor you use.

To add insult to injury, additional errors are picked up in the variations associated with the various analytical methods. Most often, high-sugar samples must be diluted to get within a detectable spectrometric range, and, as we have learned earlier, with each dilution the margin of error increases. Improper analysis of the original sample or obtaining the red must sample too early will also add to the degree of error. A small source of error is the amount of ethanol that is lost during fermentation.

5.3.5 Alcohol (Ethanol) Determinations

When we talk about alcohol, we are referring to ethanol (EtOH). Ethanol is the major alcohol in wine, with very small quantities of higher alcohols (fusel alcohols) such as methanol, propanol, butanol, and glycerol to name a few. Normal alcohol levels in wine vary from 7% upward to 24% in desert wines. In the United States, the Alcohol and Tobacco Tax and Trade Bureau (TTB) regulates alcoholic beverages, and taxes the wine according to the alcohol content. Alcohol levels are divided into five categories:

1. Not more than 14.0% ethanol
2. 14% to less than 21% ethanol
3. 21% to less than 24% ethanol (greater than 24% = distilled spirits)
4. Sparkling wines artificially carbonated
5. Sparkling wines

Alcohol content must be listed on the wine label and the TTB has applied certain tolerance levels depending on the EtOH level. There is a ± 1.5% tolerance for wines between 7% and 14.0%. For wines over 14.0%, the tolerance drops to ± 1.0%, and for wines less than 7%, the tolerance is ± 0.75%.

Alcohol determination can be accomplished using several methods, but only a few are accepted by the regulatory bodies. The TTB recognizes ebulliometry, distillation, pycnometry, densitometry, and GC for alcohol determination. The most common method used around the world is ebulliometry.

Gas chromatography measures ethanol only, but ebulliometry measures all alcohols. Near-infrared (NIR) instruments calculate the amount of light absorbed by a sample in the near-infrared spectra at certain wavelengths. The point is that there are several methods to quantify ethanol, and it is important to know how the methods vary in order to understand the differences between the individual method results. The following methods are available to measure alcohol:

- Boiling-point depression (ebulliometry)
- Distillation with specific gravity
- Distillation with hydrometry
- Distillation with pycnometry
- Densitometry
- Refractometry
- Oxidation
- GC
- HPLC
- Enzymatic spectrometry

5.3.5.1 Ebulliometry Method

Ebulliometry is the most common method for alcohol determination in the world. The method is fairly user-friendly, but ebulliometry is the least accurate of all listed, with an accuracy of ± 0.5% v/v. As technology advances, ebulliometry will most likely become obsolete, replaced with faster and more accurate methods that will be cost-effective for small wineries.

Ebulliometry is based on the Churchward technique: the depression of the boiling point in a water and alcohol mixture. Higher alcohol concentrations reduce the boiling point of the mixture. The method is accurate up to 16% v/v alcohol. Water and wine samples are each brought to their boiling points. The water boiling point sets the zero and the wine boiling point determines the alcohol content.

An ebulliometer is the instrument used for this measurement. An ebulliometer has a boiling chamber equipped with a thermometer that lies just above the liquid level in the vapor space. A lined condenser is placed above it to collect, cool, and return liquid back into the boiling chamber. The boiling chamber is heated using a small Bunsen burner, an alcohol burner, or one can use an electric ebulliometer (Fig. 5.4).

Barometric pressure affects the boiling point of a liquid, so the first step is to obtain and make note of the barometric pressure reading prior to the test, with no more than a 2-h window between readings. Take the pressure reading from a calibrated barometer or access the current weather conditions via telephone or the Internet. The condenser is filled with cold deionized (DI) water. A DI water sample is always analyzed prior to the wine sample (normally 50-mL samples). The burner is lit and placed directly underneath the arm of the ebulliometer. Once the water boils and the temperature reading remains steady, note the temperature to within 0.02 degrees. This will be the zero point. Ebulliometers (Dujardin & Salleron ebulliometer) come with a wheel calculator to assist in the calculations. Rotate the wheel and place the temperature reading at the "0" percent alcohol mark located on the outer rim of the wheel.

The process is repeated with the wine sample. Note the temperature at which the reading remains steady. Go back to your calculator and locate the

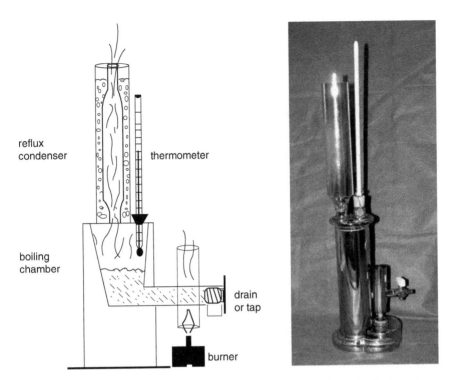

reflux
condenser

thermometer

boiling
chamber

drain
or tap

burner

FIGURE 5.4. Typical ebulliometer exaggerated for illustration and a Dujardin & Salleron ebulliometer.

wine boiling-point temperature on the wheel. The alcohol determination will be read on the inside of the wheel across from the wine boiling-point temperature. If the wine has been diluted, the alcohol determination should be multiplied by the dilution factor (see Chapter 9).

The Churchward technique uses the difference between the water and the wine (ebulliometer degree) and applies the results to a table with corresponding alcohol percentages.

The key to success with ebulliometry is as follows:

- Use a functional instrument that has been cleaned with a 2-g/100 mL solution of sodium hydroxide.
- Because sugar can alter the boiling point of the wine; dilute wines with sugar content of >2 % or high alcohol content >16% v/v.
- Centrifuge excessively turbid samples such as juices to reduce the solid content.
- Use the smallest possible dilution to reduce the amount of error.
- Keep the ebulliometer away from drafts, which could cause a fluctuation in temperature. Use precise sample measurement for both the water and the wine.

- Never leave the ebulliometer unattended at anytime during the test, and always use fire safety procedures. Electric ebulliometers eliminates the fire danger, but they are more costly.

5.3.5.2   Other Alcohol Determination Methods

The distillation method incorporates boiling the wine sample, condensing the steam in a cooling chamber thereby creating a distillate, and analyzing the distillate by any of several different methods—pycnometry being the international standard. The distillation process is carried out using precision hand-blown glassware with heating elements and cooling condensers. Distillation is an accurate method that reduces the amount of interference from sugars and solid materials. The downside to this method is the amount of time involved to distill the sample and analyze the distillate. Distillation is an approved method by TTB and Association of Official Analytical Chemist International (AOAC).

Another AOAC-approved method for determining ethanol is GC. GC has an accuracy of around ± 0.2% v/v. The GC principle is based on the movement of molecules through a detector. The larger, or heavier, a molecule, the longer it takes the molecule to reach the detector (retention time). Elements and compounds are identified by their molecular weights and retention time.

Gas chromatography requires precise dilution of the samples prior to injection. A 0.5-μL sample is injected into a chamber heated to 120 – 125°C and immediately changed into its gaseous phase. A carrier gas such as helium moves these particles through a heated (80°C) column, giving the different elements and molecules a chance to separate according to their molecular weight. The molecules will finally pass through a heated (125°C) detector, which will mark the time and the quantity. GC ethanol determination is a reliable automated system geared for a moderate to large winery.

Densitometry is an approved TTB and AOAC method that is used with or without predistillation of the wine sample. This method has an accuracy range of approximately ± 0.10 – 0.15% v/v. The densitometer uses harmonic motion to determine the density of a liquid in a U-shaped tube. The liquid is vibrated electromagnetically to its natural frequency. The change in the U-tube vibrating frequency is proportional to the density of the liquid at 20°C. Earlier in the chapter, we learned that alcohol is less dense than water; the lower the density, the higher the alcohol percentage. Densitometry is a fast, accurate, and user-friendly method for alcohol determination.

Another fast, accurate, and user-friendly method is Near Infrared (NIR). To date, NIR has not been approved by the TTB for ethanol determination. NIR utilizes spectrometry based on the Lambert–Beer law to determine alcohol content via light absorption at different wavelengths. The spectral range is designed to be very narrow and the detector has the ability to measure low absorbencies of 10^{-1} to 10^{-3}. In addition, the instrument maintains precise temperature control for more accurate readings. The accuracy of NIR determinations is ± 0.01% v/v. An example of a NIR spectrometer is the Anton Paar Alcolyzer® shown in Figure 5.5.

Enzymatic analysis of wine breaks down the alcohol's NAD^+ in the presence of alcohol dehydrogenase to form NADH. NADH is easily detected using a spectrometer at 340 nm. The more NADH present, the higher the alcohol concentration. It is bit more time-consuming and labor-intensive than some of the other methods.

Fourier transform infrared spectometry has also shown good repeatability in alcohol determination and is being used for monitoring alcohol levels.

5.3.6   pH and Acidity Determinations

In Section 5.2.1.1, we discussed how acids are determined to be strong or weak by their ability to disassociate and release all their protons and hydrogen ions (free H^+). Most organic acids in grapes and wine are considered weak because they do not dissociate 100%.

The acidity of a juice, must, or wine is articulated in several different ways:

- Total acidity
- TA
- Fixed acidity
- Volatile acidity (VA)
- pH

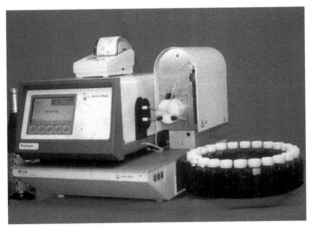

FIGURE 5.5. Anton Paar Alcolyzer® NIR spectrometer for the determination of alcohol % v/v (based on ethanol tables). The Alcolyzer may be teamed up with the Anton Paar DMA 4500® or DMA 5000® density meter for determination of alcohol % w/w. The automatic sampler with bar-code reader holds up to 60 samples for attendant-free use.

The pH reflects the hydrogen ion concentration of the juice, must, or wine. Total acidity is the sum of all organic acids, including their salts. TA (titratable acidity) is the proton concentration of the total acidity. These two terms often get confused. Total acidity considers the proton content and the buffering effect of potassium and sodium (Margalit, 1997):

$$Total\ acidity = TA + K + Na\ (molar). \qquad (1)$$

Because of the buffering (neutralizing) effects produced primarily by potassium and to a lesser extent sodium, TA is always less than the total acidity.

For practical production purposes, TA is often referred to as the total acidity because it comprises the majority of the total acidity.

Titratable, fixed, and volatile acidity are all part of the total acidity:

- TA expressed as tartaric acid = Total acidity – Buffered acids
- VA expressed as acetic acid = Total acidity – Fixed acids
- Fixed acidity expressed as tartaric = Total acidity – (VA × 1.25)

Many countries regulate the amount of acids in wine. TTB regulations state the fixed acid concentration should not be less than 5 g/L. Volatile and fixed acids will be discussed in more detail in the next chapter.

5.3.6.1   pH

The pH is the measurement of free-hydrogen-ion activity in a solution. The *lower* the pH of a solution the more *acidic* the solution because of the increased free-hydrogen-ion activity; the *higher* the pH of a solution, the more *alkaline* the solution and the less hydrogen ion activity. The activity (concentration) is measured with the use of a pH meter and electrodes that measures the free-hydrogen ions and the voltage potential created by those ions. There exists a linear relationship between the pH of a solution and the voltage potential created; the greater the voltage potential, the higher the hydrogen ion concentration.

The pH meter includes a device that measures electrical potential, called a potentiometer, and it is calibrated in pH units from pH 0.0 to 14.0. Each pH unit is equal to a 10 times increase in H^+ concentration from pH 14.0 (i.e., a measurement of 0.0001 moles equals a pH of 4.0, a 10 times increase equals 0.001 or pH 3.0, and so on). As the number of moles decrease, the pH rises, indicating a less acidic solution. The potentiometer is connected to two electrodes; one electrode is equipped with a membrane sensitive to hydrogen ions and the other is the reference electrode. As hydrogen-ion levels increase around the ion-sensitive membrane, a voltage change occurs. This voltage change compared to the reference electrode is read by the potentiometer. The meter then calculates the pH by using the Nernst equation (Iland et al., 2000):

$$pH = pH_T^0 - (\frac{E}{R} \times S \times T), \tag{1}$$

where: pH_T^0 is the 0 pH value where there exists zero potential at a certain temperature (neutral), E is the measured voltage (in mV), R is the gas content, S is the electrode sensitivity, and T is the temperature.

Meter electrodes are calibrated using two standards (buffers): one with a pH 7.0 and the other with a pH of 4.0. Electrodes are sensitive and have slight inherent variations between electrodes, and in time, they decrease in efficiency. These differences must be taken into consideration for accurate measurement. The pH 7.0 buffer response sets the zero point, whereas the pH 4 buffer response sets the slope of the linear line (mV reading) that passes through the zero point. The theoretical response for the pH 4.0 buffer is 59.2 mV at 25°C.

The pH calculations are temperature dependent and the equation calculations are set at 25°C. Any change in the solution temperature above or below this theoretical will shift the slope from the ideal 59.2 mV, producing inaccurate calculations. In most laboratories, it is impractical to bring buffers and samples to 25°C for pH measurement. The object of calibration is to make proper measurements in order to set the slope as close as possible to the ideal 59.2 mV. Most meters are equipped with temperature-detecting devices (temperature compensated) that measure the temperature of the solution, insert that temperature into the equation, and allow the meter to recalculate the equation during calibration, producing the ideal 59.2-mV slope.

Calibration buffers should be at room temperature, with proper temperature compensation of the meter. Samples tested using this calibration should also be brought to room temperature. The sample temperature is automatically noted and the pH will be adjusted accordingly. The larger the variance of sample temperature from the calibration temperature, the greater the shift of the slope occurs and the larger the error. Meters that do not have temperature compensation will need to have the results calculated by hand or the solutions and buffers brought to 25°C. Calibrations should be performed frequently or daily. Older electrodes and samples containing juice might require additional cleaning and calibration.

Electrodes are purchased separately or purchased as a combined unit of reference and sensing. The sensitivity and accuracy of a meter and electrode should also be considered. There are many different types of meters and electrodes on the market and it is advisable to thoroughly read the manuals that accompany the meter and electrodes for proper care and use.

5.3.6.2 Titratable Acidity

Titratable acidity (TA) is quantified by titration of a strong base into the sample to an end point of pH 8.2 (in the United States), noting the amount of base used (titer), and calculating the amount of acid, expressed as tartaric acid. The more base used, the more acidic the sample; the higher the proton level and the more H⁺ present, the higher the TA and the lower the pH. The French use an end point of pH 7.0, with the TA being expressed as sulfuric acid. Ough et al. (1988) pointed out that the true end point would be greater than pH 7.0 up to a possible pH 8.3.

The pH of the juice without the transfer of protons and cations (buffering) would be around pH 2.2, whereas completely buffered juice would have a pH of 7.5 (Boulton et al., 1999). The buffering capacity increases as the grape matures reducing the acidity by maturation.

A common base used to determine TA is NaOH; the amount of NaOH used to reach the pH of 8.2 is the titer. The formula for determining TA is

$$TA\,(g/L\text{ as }H_2T) = 75 \times N \times \left(\frac{T}{S}\right), \tag{1}$$

where N is the normality of NaOH, T is the titer volume (in mL), S is the sample volume (in mL), and 75 is a constant.

Acidity levels are expressed differently in some countries. Acidity is also expressed as the volume of a $0.1N$ base needed to titrate 100 mL. The conversion factors for expressions of acidity are (Ough et al., 1988) as follows:

- Tartaric: 1.00
- Sulfuric: 0.653
- Malic: 0.893
- Acetic: 0.80
- Citric: 0.853
- Lactic: 1.20

When working with must and wine, it should be noted that must will most often have a lower pH and higher TA due to the increase in malic acid. As fermentation progresses and the malic acid converts to lactic acid (weaker acid), the TA will decrease and the pH will increase.

It is important to note that unbound gases such as carbon dioxide and sulfur dioxide can add to the TA in the form of carbonic acid and sulfuric acid; for this purpose, samples must have the gases removed prior to testing.

Normal TA ranges for grapes and wine is between 6 and 9 g/L, but can climb to 12 g/L in botrytized wines or grapes grown in cool regions (Boulton et al., 1999; Ough et al., 1988).

Titratable acidity determinations utilize a pH meter and a manual titration device such as a glass burette. Computerized automated systems are also readily available. The automated systems come with sample changers that can accommodate many samples at one time and are computer-driven to provide the user with calculated results. The procedures are listed in Chapter 9.

5.3.7   Potassium, Calcium, and Other Alkaline Metals Determinations

The cations of potassium, calcium, sodium, iron, and copper are the primary alkaline metals quantified for winemaking because of their effect on pH, involvement in reactions and oxidation, and their contribution to wine appearance. Excess quantities of potassium, calcium, iron, and copper contribute to the cloudiness of wines and/or precipitate as tartrates in wine, whereas excess sodium has health implications for hypertensive people. Each country sets their allowable limits of these elements in finished wine.

There are five primary methods for quantifying different metals in wine: atomic absorption spectrometry (AAS), flame emission photometry, inductively coupled plasma–mass spectrometry (ICP-MS), ion-selective electrode (ISE), and colorimetric. Sophisticated laboratories producing high-volume metals testing will utilize the AAS, flame photometry, and the ICP-MS, and small to medium wine laboratories might use the ISE and colorimetric methods. AAS and ICP-MS are very accurate but are expensive pieces of equipment that require some expertise to operate and interpret the results, whereas the ISE method is not as accurate, but much simpler. Commercial laboratories are a great resource for the occasional metal testing, a bit pricey but the most cost-effective way to get the results needed.

The very basic theory of AAS, flame photometry, and ICP-MS centers on the color spectrum wavelengths, intensity of light, or absorption of light from metal-ion solutions when exposed to a flame or heat source.

When burned, metals emit color at certain wavelengths. The flame photometer introduces a sample containing metal ions to a flame and measures the intensity of light emitted at a wavelength known to be associated with the metal of concern. AAS measures the amount of light absorbed from a sample containing metal ions, introduced to a flame at a wavelength known to be associated with the metal of concern. The method uses standard additions of a known metal-ion concentration to a sample, in a graduated series of aliquots. The standard samples are analyzed along with the original sample and the results are plotted according to their known concentration, which forms a linear line. The metal-ion concentration of the sample is extrapolated from that graph. Ionization suppressants are used to reduce the effects of other ions in the samples and standards. Cesium is used as an ionization suppressant in the determination of potassium and sodium. Lanthanum or strontium is used as an ionization suppressant for calcium detection.

Inductively coupled plasma–mass spectrometry is a more precise and sophisticated method. A sample is aspirated into a nebulizer and sprayed as an aerosol into a chamber. A carrier gas moves the aerosol into the plasma created by argon gas and high radio-frequency (RF) energy at a temperature of 6000–10,000 K. The aerosol is immediately decomposed into analyte atoms and ionized. A vacuum takes the ions into the mass spectrometer, which then separates them by mass/charge ratio and systematically draws them into the detector to be quantified.

Ion-selective electrode methods (ISE) are available for calcium, potassium, sodium, iron, and copper. ISE utilize electrodes with ion-selective membranes sensitive to particular ions and a reference electrode utilizing the same theoretical principles as discussed in the section on pH. To allow for accurate measurements and reduce interference, samples and standards are diluted 1:1 with a total ionic strength adjuster and buffer (TISAB). The TISAB consists of $1M$ NaCl to adjust the ionic strength, acetic acid and acetate buffer to control pH, and a metal complexing agent. ISE methods are similar in theory; refer to other ISE methods listed in Chapter 9 (ammonia, fluoride).

5.3.8   Total Phenols and Anthocyanin Determinations

Wine color depends on the amount of anthocyanins in the grape, the pH, fermentation reduction, and sulfur effects. Depending on the varietal and the growing season, grapes will contain a certain amount of anthocyanins in their skins at maturity. White grapes normally have their juice expressed and have little to no contact with the skins, but flavonoids, hydroxycinnamates, and brown pigments will affect the wine as it ages.

Red grapes are in contact with their skins after harvest for the primary reason of extracting as many anthocyanins and other phenolic compounds as possible. After several days of skin contact, the juice will have extracted the majority of the phenolic compounds. At this early stage, sulfur addition will result in a bleaching effect, causing a decrease in color as a result of the binding of a small portion of the sulfur to the anthocyanins.

The pH greatly affects the color in juice and wine. A decrease in pH (more acidic) will intensify the color and shift the color forms of free anthocyanins to the red scale.

Fermentation reduces color because of the gradual increase in alcohol content. This rise in alcohol decreases the hydrogen-ion bonds. The removal of alcohol will restore the color (Margalit, 1997). Red pigments are formed throughout fermentation and maturation and are less sensitive to pH changes and sulfur additions. Not all of the red color in wine comes from the pigment; there are other red-colored compounds that contribute to the total color profile. Compounds of yellow/brown are formed during fermentation and maturation, contributing to the red/brown color of aged red wines and the golden brown color of aged white wines.

Quantification of these colors, tannins, and other phenolic components is difficult, laborious, and does not fit well into most winery laboratories. Spectrometry can be utilized to quantify pigments, flavonoids, hydroxycinnamates, and total phenolics. These are fairly simple, but obtaining quantification of tannins from skin and seeds, free and bleached anthocyanins, and phenol content involves serial dilutions and extractions. The Folin–Ciocalteu method, LA method, and the grape maturity assays for the above can be found in various texts such as *The Handbook of Enology Volume 2, The Chemistry of Wine Stabilization and Treatments* (Ribéreau-Gayon et al., 2000b).

For the purpose of this book, we will look at spectrometer determinations. Samples are placed in the spectrometer (red wines and some white wines are diluted, with unknown concentrations up to 10 times), a light is passed through the liquid at a certain wavelength (using a 1-mm or 10-mm light path), and the light is collected and compared to the original light source. Different substances absorb light at different wavelengths, and the wavelengths that are absorbed by substances in wine and juice are being determined or have been determined. Absorbance is also called optical density (OD).

Because pH affects the color, it is wise to set a precedence to analyze all samples at a constant pH, such as pH 3.50. Samples will need the pH adjusted prior to analysis. Winemakers might want to have the wines tested at their current pH to evaluate the phenolics at that pH, but either way, the pH should be indicated on the report.

Sugars tend to interfere with the readings; therefore, to obtain accurate readings, dry wines should be tested. Juice test results should include the sugar measurement for reference. Always read the equipment directions closely to ensure proper usage and do not forget to zero your machines before and after the readings.

Various wavelengths (in nanometers) are used in the estimated determinations:

- 280 nm = red-colored anthocyanins and compounds, flavonoids, total phenolics
- 320 nm = hydroxycinnamates
- 420 nm = yellow/brown pigments
- 520 nm = red pigments

The addition of 10% w/v acetaldehyde (CH_3CHO) will pull the sulfur away from the anthocyanins and create a stronger bond, thus releasing the anthocyanins into the free state and negating the bleaching effect of the sulfur. A sulfur addition of around 2000 mg/L (25% w/v) added to the sample will bond with the free anthocyanins, leaving the red-colored pigments resistant to sulfur bleaching. Total red color can be measured by adding $1M$ HCL to the sample to reduce the pH to the point where all the anthocyanins and pigments go to red.

With the above information, we can make determinations of estimated values for total phenolics, wine color intensity, hue, total red pigments, and sulfur-resistant pigments, according to these formulas. Refer to Chapter 9 for method details.

5.4   Harvest

The grape maturity has been determined and the harvest date set. Picking the grapes is done manually or by machine. The manual method allows more selective picking because all grapes do not mature at the same rate. Grapes with *Botrytis* fungus, used for late harvest wines, can be harvested berry by berry if desired. Second harvesting is often done in the same vineyard when there is a considerable amount of fruit that is unripe at the time that the majority of grapes are harvested. There is less damage to the vineyard trellis and irrigation systems than machine-harvested fruit. Picking by hand allows the fruit to remain intact and brought to the winery in whole clusters that can be sorted for quality.

Machine-harvesting grapes has many advantages and some disadvantages. Machine harvesting is much more economical and faster than the manual method. Reducing skipped rows increases overall tonnage and harvesting at night keeps the grapes cool and reduces dehydration before they arrive at the winery. Some disadvantages include the damage to the vineyard, an increase in material other than grapes (MOG = stems, leaves, parts of the vine, small animals, etc.), inability to sort, loss of juice due to breaking of the berries (approximate 6–8% loss), and the inability to selectively harvest *Botrytis* grapes. All in all, machine harvesting is improving at leaps and bounds, and despite the negative aspects, it is still a more cost-effective harvesting method. Not everyone agrees with machine-harvesting of fruit. Some winemakers say it ruins the quality of the fruit and negatively impacts the agricultural labor force. Machine-harvesting is prohibited in some areas, such as Champagne, France.

As the grapes arrive at the winery and the receiving process begins, it is a common winemaking practice to add small quantities of sulfur dioxide to the fresh red and white must as an antimicrobial, to reduce browning, and to partially delay fermentation.

White grapes are received at the winery as whole clusters (if hand-picked) or as juice that has been pressed from grapes harvested by machine. The grapes are transferred to a press of some sort (bladder press, basket press, etc.) to express the juice (Fig. 5.6). The pomace, which includes mostly skin and seeds, is discarded. The juice (or must) is moved into tanks (stainless steel is preferred), most often cooled, allowed to settle, and prepared for fermentation. Some white varietals will not be pressed but be sent to a destemmer to have the MOG removed and then transferred to a tank to await yeast inoculation and nutrients.

Red grapes are received at the winery in whole clusters but might arrive partially crushed due to machine-harvesting. For premium wines, the clusters might be sorted by hand according to quality and then sent to the destemmer, whereas other wineries send the clusters directly to the destemmer for the removal of MOG. The crushed fruit (whole must) is transferred to a tank to await yeast inoculation and nutrients for fermentation. There

FIGURE 5.6. Movement of grapes from delivery bin to press.

are those winemakers who prefer to ferment their grapes in whole clusters or in small vats.

Winemaking, although a science, is not an exact science when it comes to grape processing. Many winemakers are convinced that their methods and style are the best for the varietal of grapes they process. Please note that the processing of the fruit is strictly a winemaking decision and there are many ways of handling the fruit depending on varietal, style, and winemaker.

This is the beginning of intense analyses, record keeping, and monitoring of each tank filled. Once the juice or must is in the tank, several types of analysis, in addition to the sugar levels, TA, and pH, are needed to make winemaking decisions. Analyses performed could include the following:

- TA
- pH
- Fluoride
- Ammonia
- Apparent °Brix
- Potential alcohol
- Insoluble solids
- Total sulfur dioxide
- Free sulfur dioxide

Analyses begin with confirmation of the sugar levels, TA, and pH. These parameters verify the anticipated preharvest predictions and reflect the actual must content for the entire harvested area. If the vineyard sampling was done well, the vineyard analyses should be close to the actual harvest analyses. Keep in mind that the sugar content of red must will rise after the berries have been soaking in the juice.

White juice requires insoluble solids measurement, and red and white must requires assessment of the available nitrogen via ammonia measurements. Areas that are suspected of using chemical sprays too close to harvest might require fluoride determinations or more involved analyses that indicate the use of certain prohibited chemicals. Determination of total sulfur dioxide and, in some cases, free sulfur dioxide content is necessary because increased sulfur dioxide levels can hinder fermentation.

5.4.1   Nitrogen and Amino Acid Determinations

Nitrogen is a necessary nutrient for *Saccharomyces cerevisiae* yeast during sugar fermentation. Ammonia (NH_3), ammonium (NH_4), and amino acids are the primary sources of nitrogen for the yeast. The three primary sources of nitrogen are glutamate, glutamine, and ammonia. The yeasts use these three primary nitrogen sources first; when the nitrogen is depleted, the yeast will begin to attack other amino acids. Amino acids that are utilized by the yeast are aspartic acid, alanine, and, when ammonia has been exhausted, arginine.

Typical must ammonia is around 100 mg/L (ppm) (Margalit, 1997). Winemaking techniques, varietal differences, and weather will dictate the nitrogen concentration at harvest. The amount of sugar in the must will dictate the amount of nitrogen required to ensure adequate yeast growth for complete fermentation. There are several different opinions as to what the required level of ammonia should be to have a successful fermentation. Some reports indicate 25–50 mg/L or 70–140 mg/L free amino nitrogen (Ribéreau-Gayon et al., 2000a), whereas other sources report 100–120 mg/L NH_3 (Boulton et al., 1999), and 150–200 mg/L NH_3 (Margalit, 1996).

Winery laboratories use two different methods to quantify the nitrogen (ammonia) levels in must: ISE with a known addition (see Chapter 9) and determination of free amino nitrogen (FAN). There is a third method that has been used in the past that involves a distillation process—a bit time-consuming and more dangerous. The ISE method is relatively inexpensive and much more user-friendly. Depending on the results and the winemaker, additional ammonia could be added to the must in the form of ammonia salts in preparation for fermentation. There are several theories on adding nitrogen and when, but most additions are made prior to the onset of fermentation. TTB allows up to 96 g/100 L (hL) additions of ammonium phosphate.

Ammonia salts such as diammonium phosphate, $(NH_4)_2HPO_4$ (DAP), and diammonium sulfate, $(NH_4)_2SO_4$ (DAS) contains approximately 27% NH_3 or 27 mg in 100 mg of DAP or DAS (Ribéreau-Gayon et al., 2000a). The rate of addition works out to be 10–20 g/hL, or 100–200 mg/L for a 27–54 mg/L increase in NH_3. Ammonium salt additions will increase the acidity of the must or wine by approximately 0.005 g/L tartaric acid per 100 mg/L added.

Nitrogen deficiency can result in the catabolism of other amino acids that contain sulfur, such as cysteine. This process releases the sulfur that bonds with hydrogen ions, creating hydrogen sulfide (H_2S), an unpleasant smelling compound comparable to the smell of sulfur or skunk (many other descriptors have been used).

Too much ammonia will increase the amount of amino acids in the finished wine (because they are not being used for nitrogen), which can modify the aroma character in the wine and also promote the development of ethyl carbamate, which has carcinogenic properties.

5.4.2 *Insoluble Solids Determinations*

Insoluble solids (also referred to as solids) consist of small particles (<2 mm) of cellulose, hemicellulose, pectin, mineral salts, lipids, and insoluble proteins. The presence of solids in white juice might promote elimination of carbon dioxide during fermentation, promote yeast multiplication, absorb toxic fatty acids that impede fermentation, supply yeast nutrients, and absorb certain metabolic inhibitors (Ribéreau-Gayon et al., 2000a).

The proper concentration of solids in juice is necessary for the fermentation process to proceed to completion, with the added benefit of improving the fruity flavor and quality of white wines. Too high a concentration of solids will produce a heavy, herbaceous aroma and bitter taste due to higher concentrations of phenolic compounds. The wine will have increased color, but the color will be unstable. Another danger is creation of postfermentation off-odors. Too few solids can affect the yeast's nutrients, resulting in an increase in acetic acid, which will decrease the fruity aromas desired in most white wines.

Juice in tanks are settled or clarified by several different methods, including cold settling, centrifuging, addition of enzymes, or microfiltration. The cold-settling method cools the juice, allowing the grape pectinase to act on the colloidal structures of the solids and naturally dropping to the bottom of the tank. Spinning the juice in large centrifuges is another way of removing solids. In recent years, some winemakers have begun adding enzymes to the juice, which promotes rapid destruction of colloidal structures speeding up the settling time. The need for these methods to remove solids in white juice is decreasing as presses evolve, thus reducing the amount of solids in the pressed juice.

Currently, there are three common procedures for measuring the quantity of insoluble solids in juice: nephelometry, turbidity, and percent solids. A nephelometer and turbidimeter work on the principle that small particles will scatter UV or visible light. A strong light beam at 90° is passed through a glass tube; the nephelometer detects and measures the amount of scattered light. With a higher percentage of solids to sample volume, light is scattered to a greater extent, resulting in higher nephelometer readings (NTU).

A turbidimetric system passes a straight beam of light through the sample tube, employing a detector that measures the remaining light after scattering (turbidity). The higher the percentage of solids to sample volume, the more light is scattered and lost and the higher the turbidity reading.

Turbidity readings are recommended for samples containing a higher level of solids, whereas nephelometry is recommended for samples containing a lower percentage of solids. Nephelometric readings in the range 100–250 NTU indicate good insoluble solid levels (Ribéreau-Gayon et al., 2000).

Centrifuging involves measuring a precise volume of juice in a graduated tube, placing the tube in the centrifuge, and spinning for several minutes. The percentage of solids is determined by measuring the amount of solid material in the bottom of the centrifuge tube, divided by the initial volume of juice, and reported as % solids. The range of % solids measurements indicating good levels of insoluble solids is between 1% and 2%. This method is neither as fast nor as precise as the nephelometric or turbidimetric reading.

The solids that form at the bottom of the tank are referred to as lees. When the juice has settled for a period of time, the clarified juice will be pumped or moved to another tank (racking) and the lees will be sent to a press to express as much juice as possible. The clarified juice is analyzed for % solids, and if the results indicate the solids level is still too high, the juice will be allowed to settle and racked again. If the results indicated too few solids, lees can be added back into the juice.

5.4.3   Total, Free, and Bound Sulfur Dioxide Determinations

The use of sulfur dioxide (SO_2) as a food preservative began hundreds of years ago, most likely with the Romans and Egyptians. By the Middle Ages, the benefits of SO_2 additions to wine were known, but the use of SO_2 was frowned upon and outlawed in most areas. Common use of SO_2 in wine did not come about until the 1700s. Today, SO_2 is in general use throughout the food and beverage industry and is clearly one of the best and safest preservatives available when used responsibly within specified ranges. The TTB has set the limitation of SO_2 in wine at no greater than 350 mg/L (ppm).

At the end of natural yeast fermentation, SO_2 levels have been found up to 30 mg/L depending on the varietal, temperature, pH, and growing region (Ribéreau-Gayon et al., 2000). The SO_2 is produced by the yeast during the fermentation process and will be found in some degree in all wines.

In finished wine, the majority of the SO_2 content has been added during the winemaking process. The known benefits of SO_2 additions to wine are numerous and research continues to explore the effects and interaction of SO_2 in wine.

With good winemaking techniques, the benefits of SO_2 can be utilized with a minimum risk to health:

- First and foremost are the antimicrobial effects of SO_2. Low levels of SO_2 will eliminate most strains of bacteria, including lactobacillus, pediococcus, and oenococcus, but it will only inhibit the growth of yeast for a few days. Higher levels of SO_2 will eliminate all bacteria, molds, and yeasts.
- Sulfur dioxide is an antioxidant agent and can react by direct action, binding with dissolved oxygen and forming sulfite, or indirectly by inhibiting nonenzymatic oxidation by reacting with carbonyl compounds such as acetaldehyde. The acetaldehyde reaction forms ethane sulfonic acid, which accounts for the majority of the bound sulfur. The balance of bound sulfur is made up from the SO_2 reactions with phenols, pigments, and other components in the juice or wine. The general effect is a reduction in browning and oxidation of the wine.
- Inhibition of enzymatic oxidation with SO_2 basically hinders the function of oxidation enzymes such as tyrosinase and laccase (Ribéreau-Gayon et al., 2000a).
- By binding with ethanal, SO_2 helps protect wine aromas.

5.4.3.1   Forms of Sulfur Dioxide

Sulfur dioxide in solution exists in three forms: hydrated sulfur dioxide (the molecular form) and two ionized forms. In the presence of water, SO_2 produces sulfonic acid (H_2SO_3). This acid molecule does not exist in solution but produces two ionic forms [bisulfite (HSO_3^-) and sulfite (SO_3^{2-})] plus hydrogen ions (Ribéreau-Gayon et al., 2000a). These three forms of SO_2 found in wine and juice exist at different temperatures and pH ranges, with different dissociation (pK_a) values (Margalit, 1997):

- Molecular SO_2 exists primarily at pH range of 0.0 to 1.81 with a pK_a value of 1.81. The higher the pH, the lower the concentration of SO_2 exists in solution. SO_2 represents approximately 1–7% of the SO_2 in the wine pH range of 3.0–4.0.
- HSO_3^- exists primarily at a pH range 1.81–7.2 with a pK_a 1.81–7.2. HSO_3^- represents the largest percentage, 90–99%, of SO_2 in the wine pH range of 3.0–4.0.
- SO_3^{2-} exists primarily in the pH range 7.2–10.0 with a pK_a of 7.2. SO_3^{2-} represents 0.01–0.1% of the SO_2 in the wine pH range of 3.0–4.0.

Molecular SO_2 is of the greatest interest to winemaking. SO_2 produces nearly all of the beneficial effects in juices and wines, especially the antimicrobial effects. In order to tap into the benefits of SO_2, it must be unbound and available in solution, or free, to interact with the various mechanisms to produce the desired effects. The temperature, pH, and pK_a value will dictate which form of SO_2 is providing what percentage of the free SO_2 and how much of that free SO_2 will exist as SO_2. Free SO_2 plus the amount of SO_2 that is bound comprises the total SO_2 content, but only a small percentage will actually exist in the desired form of SO_2.

The SO_2 form exists 100% at pH 1.0. At pH 1.81, the percentage is reduced to 50%; at pH 3.0, the amount drops to around 6%; and at pH 4.0, the SO_2 concentration drops to less than 0.7% (Margalit, 1997).

The HSO_3^- form is nonexistent at a pH of 0.0 but gradually begins to form as the pH increases. At pH 1.81, HSO_3^- exists at approximately 50%; at pH 3.0, the concentration rises to nearly 94%; and at pH 4.0, HSO_3^- exists at its highest level (99%). As the pH continues to rise, the HSO_3^- form decreases, dropping to 50% at pH 7.2 and is nonexistent at pH 10.0 (Margalit, 1997).

The last form, SO_3^{2-}, exists in the alkaline pH range. It is basically nonexistent at pH 3.0 (very small percentage). At pH 4.0, SO_3^{2-} increases to 0.06%, rising to 50% at pH 7.2 and reaching 100% at pH 10.0 (Margalit, 1997).

Note the percentage of each form at the pK_a levels of pH 1.81 and pH 7.2. At pH 1.81, SO_2 represents 50% and HSO_3^- represents 50% of the forms of SO_2 present. This is called the first point of equilibrium, where the forms SO_2 and HSO_3^- equally contribute to the SO_2 content. The pH 7.2 is the second point of equilibrium, where the forms HSO_3^- and SO_3^{2-} each exist at 50%. In the first equilibrium, as the pH raises, the form SO_2 declines and the greater the amount of form HSO_3^- will be available as a SO_2 source.

As stated earlier, at pH 3.0, approximately 6% of the form SO_2 is present and 94% is HSO_3^-, with a trace amount of form SO_3^{2-}. For practical purposes, we can define free SO_2 as the sum of the forms SO_2 and HSO_3^-, which are in equilibrium and unbound (Margalit, 1997). Total SO_2 would then equal the bound and the free portions of SO_2 and HSO_3^-. At pH 4.0, form HSO_3^- exists at 99.2%, SO_2 exists at 0.6%, and SO_3^{2-} makes up the remainder of the forms present; the free SO_2 and total SO_2 will be comprised of HSO_3^- and SO_3^{2-}. Figure 5.7 illustrates the availability of the various forms of SO_2 from pH 0.0 to pH 10.0 according to Margalit (1997) and Boulton et al. (1999).

According to Margalit (1997), to calculate the amount of SO_2 and free SO_2 in a solution the following formulas can be used:

$$SO_2 = \frac{SO2\,free}{[1 + 10^{(pH - 1.81)}]}, \tag{1}$$

$$SO_{2\,free} = [1 \times 10^{(pH-1.81)}] \times SO_2, \tag{2}$$

where 1 is the value of water, $10^{(pH - 1.81)}$ = the log 10 of the pH minus the pK_a value.

These formulas are derived from the first equilibrium formula (Margalit, 1997):

$$\log\left(\frac{SO_2}{HSO_3^-}\right) = pH - pK_a. \tag{2}$$

For example, let us say we have a free SO_2 of 40 mg/L (ppm), at pH 3.5. As the pH increases, the ability of the solution to support the existence of SO_2

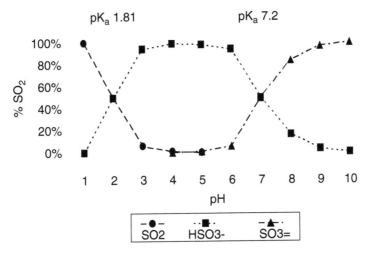

FIGURE 5.7. Relationship of pK_a disassociation values and pH to approximate percentage of SO_2 forms. ●: SO_2; ■, HSO_3^-; ▲, SO_3^{2-}.

decreases dramatically, and at a pH of 3.8, the free SO_2 of 40 mg/L has been reduced to nearly zero. To maintain a 0.8 mg/L moleculor SO_2 level at the new pH of 3.8, 80 mg/L of free So_2 must be present. For every increase in pH, increased amounts of free SO_2 are required to maintain a particular level of molecular SO_2 in solution.

When calculating additions of free SO_2 to a wine, temperature and alcohol content must be taken into consideration. In the presence of ethanol (EtOH), the first equilibrium pK_a value increases by approximately 0.1 for every 5% EtOH present and the temperature increases will raise the pK_a value (Ribéreau-Gayon et al., 2000a). The pK_a value of 1.81 is based on a non-EtOH solution at 20°C. In a solution of 14% v/v EtOH at 19°C, the pK_a value will be approximately 2.06. If the EtOH level does not change but the temperature drops, the pK_a will also decrease, increasing the amount of free sulfur required to reach the desired molecular level.

The majority of the bound SO_2 is the union of bisulfite and carbonyl compounds mainly acetaldehyde, keto acids, and glucose. To some extent, SO_2 will bind with certain monomeric anthocyanins, creating a bleaching effect as well as binding with ethanal. As free SO_2 is utilized, oxidized, or vaporized, the bound SO_2 will supplement the free SO_2 by dissociating (to some extent) to maintain equilibrium.

5.4.3.2   Maintaining SO_2 Levels

The concentration of molecular SO_2 required as an antimicrobial varies depending on the stage of wine development and the type of organisms to be controlled or killed.

Sulfur dioxide could be added to must and juice in preparation for fermentation depending on the winemaking style. Free SO_2 additions at low levels, around 5–8 mg/L, provides several benefits:

- Delays fermentation until the juice can be cooled and settled
- Blocks development of wild yeasts
- Prevents oxidation
- Eliminates most bacteria

Juices that will undergo malolactic fermentation might be affected by the presence of free SO_2 in concentrations as low as 5 mg/L and bound sulfur in concentrations of 20–60 mg/L. Low concentrations can significantly delay malolactic fermentation or prevent the fermentation. Care is taken to make SO_2 additions to must low enough to derive the benefit, but allow for reduction of the free SO_2 by the end of alcohol fermentation.

When wines finish fermentation, SO_2 additions are made to kill the remaining yeast, bacteria, and basically stop all growth activity. Most yeast are eliminated in the presence of 0.825 mg/L molecular SO_2 over several hours. The amount of free SO_2 to be added is calculated according to the pH of the wine. During the first SO_2 addition, a large percentage of the free SO_2 will bind with compounds and yeasts in the wine, some of which will precipitate out of the wine. This could reduce the amount of free SO_2 by 30–50%; therefore, a greater addition is made to compensate. Free SO_2 levels should be reanalyzed a few days after the addition, allowing time for the free SO_2 to interact with the wine components and yeast.

Sweet wines, for which cessation of fermentation is desired, might require free additions up to 100 mg/L to compensate for binding of SO_2 with the high glucose content. Late harvest wines made with grapes containing *Botrytis cinerea* require very high SO_2 additions to arrest the fungus and allow for binding.

Maturing wines will bind less and less of the free SO_2 as the wine becomes more stable, and the components that bind sulfur dioxide are oxidized or precipitated from the wine.

Molecular SO_2 levels are usually maintained at 0.825 mg/L throughout the winemaking process to protect the wine from possible contamination by spoilage organisms. Figure 5.8 reflects the estimated amount of free SO_2 required for a 0.825-mg/L molecular level based on the formulas found in the previous subsection. Always obtain the wine temperature and EtOH content to calculate precise free SO_2 additions. For general calculation, using the standard 1.81 pK_a seems to work well for most wines with alcohol contents around 14% v/v at cellar temperatures of 13–15°C. Tanks can be maintained at higher temperatures around 19°C, where the pK_a value for a 14% v/v wine would be approximately 2.06.

Sulfur dioxide is added to the wine or must in the form of potassium metabisulfite ($K_2S_2O_5$) containing approximately 58% SO_2, sodium metabisulfite ($Na_2S_2O_5$) containing approximately 67% SO_2, or SO_2 gas or liquid

Free SO2 mg/liter

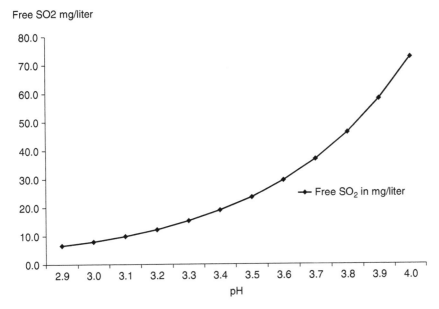

FIGURE 5.8. Free SO$_2$ levels needed for approximately 0.825 mg/L molecular SO$_2$ at various pH values in a 14.0% v/v EtOH solution, at 19°C.

SO$_2$ solution containing 100% SO$_2$. Typical additions are as follows (Margalit, 1997):

- K$_2$S$_2$O$_5$ = 0.044 g/L or 44 g/1000L for a 25-ppm addition; 0.174 g/L or 174 g/1000L for a 100-ppm addition.
- Na$_2$S$_2$O$_5$ = 0.037 g/L or 37 g/1000L for a 25-ppm addition; 0.148 g/1000L or 148 g/1000L for a 100-ppm addition.
- SO$_2$ = 0.025 g/L or 25 g/1000L for a 25-ppm addition; 0.1 g/1000L or 100 g/1000L for a 100-ppm addition.

5.4.3.3   Aeration–Oxidation Method

Determination of free and bound SO$_2$ continues throughout the winemaking process until the wine is bottled. Aeration–oxidation is an AOAC- and TTB-approved method for determining free and total SO$_2$ and the most widely used. To determine free SO$_2$, a small amount of H$_3$PO$_4$ (dilute phosphoric acid) is added to a sample to aid in the release of free SO$_2$ from the sample. The sample vessel is kept in an ice bath to reduce evaporation of other volatile compounds and prevent the dissociation of bound SO$_2$. A continuous volume of air is drawn through the sample, carrying the volatile free SO$_2$ with it. This airflow continues through a volume of H$_2$O$_2$ (hydrogen peroxide), where the SO$_2$ is oxidized into sulfuric acid, SO$_4^{2-}$ + 2H$^+$.

The $SO_4^{2-} + 2H^+$ solution is titrated with NaOH back to the H_2O_2 pH (see Fig. 5.9). The calculation for free SO_2 is (Ough et al., 1988)

$$\text{Free } SO_2 \text{ (mg/L)} = \frac{v}{N \times V \times 32 \times 1000}, \tag{1}$$

where N is the normality of NaOH, V is the volume of the NaOH titer (mL), and v is the volume of the wine sample (mL), or

$$\text{Free } SO_2 \text{ (mg/L)} = V \times 16, \tag{2}$$

where V is the volume of NaOH titrated and 16 is the conversion factor obtained using $0.01N$ NaOH and a sample volume of 20 mL.

Total SO_2 can also be determined using the aeration–oxidation method. Rather than placing the acidified sample in an ice bath, place the sample vessel on a heating element and bring the sample to a boil. Boiling the acidified sample releases the bound sulfites, which will evaporate and be carried to the peroxide. A good cold water condenser should be in place to eliminate volatile acids from being drawn into the peroxide (see Fig. 5.9). The calculation will be the same. The total minus the free will equal the bound SO_2.

Errors in this method are approximately 2.5–3% (Margalit, 1997). Possible errors include the following:

- Measurements of sample and reagents
- Inadequate cooling of sample or poor condenser function, which allows volatiles to be carried to the peroxide
- Too high or too low airflow rate, which affects the contact time of the SO_2 with the peroxide and recovery rate (flow rate of 700–1200 cc/pmin depending on the setup)
- Loose connections
- H_2O_2 concentration too weak, old, or contaminated
- Shift to the left of the equilibrium due to the addition of acid, which will release some of the bound as free (Margalit, 1997).
- Dirty glassware

Refer to Chapter 9 for more details.

5.4.3.4   Ripper Method

The Ripper method is based on oxidation–reduction reactions. It is fast and easy, but not the most accurate method of determining free and total SO_2 with an estimated error of approximately 10% (Margalit, 1997). To determine free SO_2, a sample is acidified with 25% H_2SO_4 to reduce the oxidation of polyphenols by the iodine (Ough et al., 1988). Starch solution is used as an indicator. Iodine ($0.02N$) is used as an oxidizing agent and titrated into the acidified sample. The iodine will oxidize the free SO_2, creating iodide ions, which will turn the starch blue:

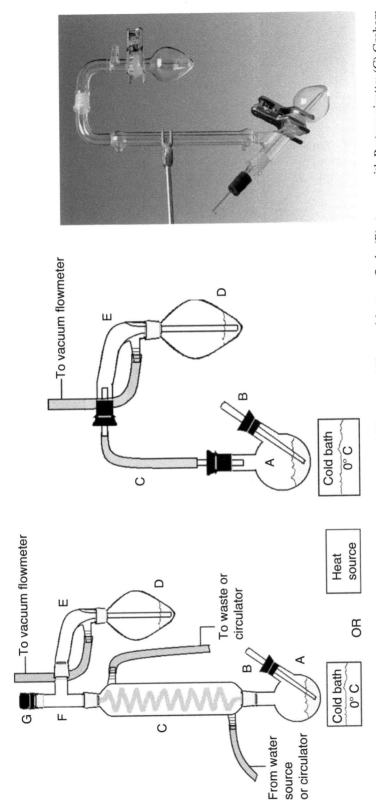

FIGURE 5.9. **(A)** Aeration–oxidation system for free, total, and bound SO_2 assays: (A) round-bottom flask; **(B)** stopper with Pasteur pipette; **(C)** Graham condenser; **(D)** pear flask; **(E)** vacuum connecting tube with stem; **(F)** three-joint 90° (or 75°) tube; **(G)** stopper (not to scale). **(B)** Aeration–oxidation system for free SO_2 assay only: (A) round-bottom flask; **(B)** stopper with Pasteur pipette; **(C)** stoppers with glass connecting tubes; **(D)** pear flask; **(E)** vacuum connecting tube with stem (105° or 90° angle) (not to scale). **(C)** Research and Development Glass Products & Equipment's apparatus for SO_2 determination by aeration–oxidation.

$$SO_2 + I_3^- + H_2O \rightarrow SO_3 + 3I^- + 2H^+.$$

The calculation is (Ough et al., 1988)

$$\text{Free SO}_2 \text{ (mg/L)} = \frac{v}{V \times N \times 32 \times 1000}, \tag{1}$$

where V is the volume of iodine solution titrated (mL), N is the normality of the iodine solution, and v is the volume of wine sample (mL). When $0.02N$ iodine is used with a 25-mL sample,

$$\text{Free SO}_2 \text{ (ppm)} = V \times 25.6. \tag{2}$$

Total SO_2 determination uses the same method, but just before titration, NaOH ($1N$) is added to hydrolyze the bound sulfites and raise the pH value, which releases the bound as free SO_2. The calculation is the same as above.

End-point determination is the greatest source of error in this method. End-point determinations can vary person to person depending on their ability to detect the color changes, especially in red juice and wine. It is helpful to have a light box with either a yellow or a white lens and examples of end-point color for analysts to use for comparison. Theatrical light gels work well as color comparisons and come in a wide array of colors.

Other sources of error include using the wrong concentration of reagents, incorrect sample volume, failure to standardize reagents, not enough reaction time, letting the sample sit too long before titration, presence of ascorbic acid in the sample (ascorbic acid also reacts with iodine), and titrating samples that are over 20°C.

Refer to Chapter 9 for method details.

5.4.3.5   Other Methods

Free and total SO_2 determination is one, if not the most, frequently performed analyses in the wine laboratory. The aeration – oxidation method as well as the ripper method are used most often because they are fairly fast. As explained earlier, these methods have limited accuracy, but more accurate methods usually call for detailed training and special equipment or take additional time to perform the analysis.

Continuous flow is a technology being used in some production laboratories. The method used for analyses of free and total SO_2 is similar to the wet chemistry methods described in the previous subsections. This automated device continually aspirates microquantities of sample or water at precise intervals, each separated by an air bubble. Pumps continue to move the liquid through small bore tubing at a methodical pace, directing each sample to a separate area (cartridge) where the desired test will commence. In the designated cartridge chemicals are added, mixed, heated (totals), and more chemicals added, as the sample continues its movement through the system. The last movement is into the spectrometer, where the absorbance is read and the sample is directed to the waste container.

The advantage of continuous flow is the number of tests (up to 60) that can be performed in 1 h. Each test takes 15–20 min to makes its way through the instrument, but being a continuous system, samples will be continually analyzed one after another (Astoria-Pacific, 2003). Additional advantages are the microliter quantities of reagents used, the ability to perform several different tests at the same time from the original sample, and the autocalculation of results.

Free SO_2 can be determined by enzymatic reactions, which measure the NADH disappearance. This method is a bit tricky and time-consuming. Perfection of this method would allow utilization of the same equipment and could easily be automated creating a more cost-effective method for a laboratory.

Fourier transform infrared spectrometry is utilized to scan for free SO_2, but, again, the accuracy depends on the data programmed into the instrument.

High-performance liquid chromatography, GC, capillary electrophoresis, and amperometric procedures have been developed, but these methods are not yet in common practice.

5.5   Summary

Great grape development, preharvest monitoring, proper harvested fruit evaluation, and preparation of the must are a recipe for great wine. The next major step to producing a great wine is fermentation. Poor fermentations are troublesome, costly, and most often produce inferior wines. Chapter 6 will discuss alcohol fermentation and fermentation to reduce malic acid in the wine. Analyses determination of those procedures not yet covered will continue to be explained with each stage of wine production.

6

Fermentation × 2

6.1 Introduction

There are two types of fermentation covered in this chapter: primary, or alcohol fermentation, and secondary, or malic acid fermentation (MLF). Fermentations can occur consecutively or concurrently depending on the varietal and wine-making style. Both fermentation processes are involved for clarity; I will cover them separately in the order they most often occur.

Chapter 5 ended with the juice and must in tanks, cooling, settling, and/or awaiting additions to prepare them for primary fermentation. Several analyses have been completed and the results evaluated to determine what, if any, additions or adjustments are to be made to the must and juice prior to inoculation with cultured yeast. Depending on those results, additions of nitrogen, solids, water, or sugar can be made. Adjustments to the acid levels could be made or additional racking might be indicated.

Additional analyses might be needed at this point in time to obtain fermentation baseline levels for acetic acid or volatile acidity (VA) and L-malic acid (LMA). If enough lactic acid bacteria (LAB) are present (without inoculation with cultured LAB), a reduction in the LMA level can occur during primary fermentation, increasing the acetic acid and VA levels.

Prior to the arrival of the fruit, winemakers will decide if wild yeasts or cultured yeast will be used for the fermentation. Wild yeasts are of a non-*Saccharomyces* genus, including *Kloeckera, Brettanomyces, Dekkera, Hanseniaspora, and Hansenula*. They exist naturally in the vineyard and arrive at the winery on the grapes. Wild natural fermentation is a very common and traditional method in Europe, especially in France. The downside to this natural fermentation is its unpredictable behavior. Some years might have an abundance of wild yeasts, others might not, and different strains can dominate the fermentation year to year. Most winemakers prefer to rid the grapes of wild yeast by rinsing them or by the addition of SO_2 to kill them and inoculate the must or juice with commercial cultured yeast of a specific strain of the genus *Saccharomyces*.

There are many types of commercial yeast on the market, all geared for different varietals. Yeast can impact the fermentation speed, temperature, and, ultimately, the taste profile of the wine. Each yeast type performs differently depending on temperature, sugar level, pH, and alcohol level, making the correct choice in yeast imperative.

Once the must has been inoculated, a daily regimen of monitoring the fermentation takes place until it reaches dryness. Verification of dryness will require accurate reducing sugar (RS) analysis.

The varietal and winemaking style might require the now young wine to proceed through MLF. MLF describes its function: changing harsh malic acid into a milder lactic acid. If the young wine is not to proceed with MLF, SO_2 will be added to stop any microbial growth or oxidation. The young wine will then proceed to a further stage of development: maturation.

This chapter will begin with the final preparation for primary fermentation and progress through MLF.

6.2   Preparation for Fermentation

The key factors effecting a good fermentation and producing a wine with varietal characteristics include the following:

- Insoluble solids
- Available nitrogen
- Acid level
- Sugar level
- Appropriate yeast selection
- Temperature control

Adjustments to the solids, nitrogen, sugar, and acid, plus the addition of nutrients are made at this point. Chapter 5 discussed insoluble solids and their importance to fermentation. The results of the solids analysis will determine whether the juice requires additional racking to reduce the solids or the addition of solids to a level of 1–2% solids. Yeast hulls can be added to supplement the solids, with their ability to activate fermentation. Recheck the solids level after any addition or racking.

Nitrogen levels between 100 and 200 mg/L, discussed in the previous chapter, are necessary for yeast metabolism. The amount of diammonium phosphate (DAP) addition to the juice or must will depend on the results from the ammonia determination and/or free amino nitrogen determination. DAP additions may require a follow-up ammonia or free amino nitrogen (FAN) analysis to confirm the addition.

Sugar determinations and the acidity of the juice and must have been determined. The sugar will indicate the potential alcohol, which, if high, could lead to a difficult fermentation or, if too low, might not produce the varietal characteristics desired.

We will now move on through the process to make the final additions prior to the addition of yeast.

6.2.1   Sugar Adjustments

Sugar levels can be too high or too low. High sugar levels produce wines that are high in alcohol. High alcohol levels can prevent the completion of fermentation by killing the yeast. Low sugar levels will produce low-alcohol wines, which might not be desirable for particular varietals.

Most sugar adjustments are made before the start of fermentation with the exceptions of sparkling wine and some northern European varietals. Each country regulates the amount of sugar added (chaptalization) and/or the increase in potential alcohol by either dehydration or sugar additions to the juice or must. These methods are available for areas where there are poor ripening conditions and are not meant to complete the maturation process because of early harvesting techniques or overcropping.

Dehydration of the must or juice will help concentrate the sugars and improve the quality (increase of phenolic compounds in red must), but, unfortunately, there is a drop in yield and the process takes time. The European Community (EC) allows a 20% maximum volume decrease or a 2% maximum increase in potential alcohol (Ribéreau-Gayon et al., 2000a).

Chaptalization in the form of 99% pure saccharose (or sucrose) and concentrated must (or juice) are the most common methods of sugar adjustment. The EC allows additions of saccharose to red must of 18 g/L and 17 g/L for white must (Ribéreau-Gayon et al., 2000a). Additions of 17–19 g/L sucrose will raise the alcohol approximately 1%.

In the United States, sugar additions are not to exceed an apparent °Brix of 25 according to the Alcohol and Tobacco Tax and Trade Bureau (TTB). California does not allow chaptalization in still wine production but does allow the addition of grape concentrate.

For those cases in which there is an excess of sugar, water additions (amelioration) are made prior to fermentation. Many countries do not allow amelioration, which can reduce the quality of the wine. The TTB allows limited amelioration to drop the apparent °Brix to no less than 22 °Brix. California does not allow amelioration but states that a minimum of water can be added prior to fermentation, which will facilitate a normal fermentation.

After any addition, always reconfirm the sugar levels. If dehydration or amelioration methods are employed, the titratable acidity (TA), pH, alcohol (when applicable), plus the sugar levels should be analyzed. These methods will change the wine chemistry due to dilution or concentration of compounds.

Neither amelioration nor chaptalization are desired in quality winemaking. The arrival of the fruit at the winery within the desired level of sugar is always optimal, but sometimes Mother Nature has different ideas.

6.2.2   Acid Adjustments

The perfect wine pH is thought to be around pH 3.6 (depending on varietal) and the perfect pH for yeast and LAB is around pH 4.5. At pH 4.5, not only do yeast and LAB grow well, but the spoilage bacteria are enjoying the conditions and thrive also. The good news is that spoilage bacteria do not do well below pH 3.6. Wine yeasts and some LAB can still do their job in a pH range of 3.3–3.6. The lower the pH, the more prolonged the fermentation due to the slower growth rate of yeast and LAB.

Another pH consideration is MLF. MLF will usually increase the pH 0.1–0.2 units depending on the beginning pH value (the higher the pH, the greater the effect). The TA will decrease (1–3 g/L depending on initial pH) as the malic acid is converted into the milder lactic acid, resulting in the pH increase.

After reviewing the above, it appears that the best pH range for primary fermentation is slightly below pH 3.6. For musts that are to complete MLF, the initial fermentation pH range of 3.3–3.4 should result in a final pH around 3.6.

6.2.2.1   Acid Additions

Low acid levels can result from prolonged hot weather. Malic acid degrades through respiration as the fruit matures, increasing the pH. Traditional white wines have lower pH values, which gives the wine the crisp, tart mouth feel especially in German and Burgundy varietals. Acid adjustments must always consider the varietal characteristics as well as the effects of the acid addition on total acid, including succinic, acetic, and lactic acids, which can increase the total acidity 10–20% during fermentation (Margalit, 1997).

The most common acid used for increasing the acidity of must or juice is tartaric acid. Tartaric has the greatest effect on lowering pH with a lesser increase in total acidity (potassium becomes insoluble in the presence of tartaric). A certain amount of the tartaric will precipitate during fermentation and allowances should be made for this loss of tartaric. Additions of 1 g/L tartaric acid will result in a decrease of pH by 0.1 units (Margalit, 1997). Acid additions will require a follow-up TA and pH analyses.

Cation exchange is used in some countries to reduce the pH. This method does not reduce pH more than 0.2 units. Citric acid has been used, but during MLF, the bacteria can react with the citric to form increased levels of diacetyl or acetic acid. Malic acid can be used to acidify but is not used in wines that will go through MLF. Gypsum (calcium sulfate) will reduce the pH without affecting the TA, but much more gypsum must be used than with a tartaric acid addition. This method is referred to as plastering and the maximum addition in the United States is not to exceed 2.0 g/L.

6.2.2.2   Deacidification

Areas with short growing seasons or poor ripening seasons produce immature fruit containing a higher acid content (higher concentrations of malic acid). High-acid, low-pH juice or must is not conducive to the yeast or bacteria involved in fermentations, and corrective measures are usually taken to reduce the acid level.

Additions of calcium carbonate ($CaCO_3$), potassium bicarbonate ($KHCO_3$), amelioration, neutral potassium tartrate, and double salt precipitation can be used to reduce the acid content. Adjustment level calculations should consider the effects of MLF on the pH and total acid levels of the juice or must if MLF is slated (see Section 6.2.2).

Calcium carbonate is used for wines that will not go through MLF and is approved in many countries. The downside to $CaCO_3$ use is the reaction of Ca^+ with tartaric acid, which causes the tartaric to precipitate out of solution over a long period of time. According to Margalit (1997), 0.66 g/L $CaCO_3$ reduces the TA by 1 g/L which, in effect, increases the pH by approximately 0.1 pH units and is more effective in wine than juice or must.

Potassium bicarbonate or K_2CO_3 (potassium carbonate) is effective for slight acid adjustments and will precipitate tartaric to a certain point but does not continue to do so. Approximately 0.66 g/L will reduce the TA by 1 g/L and increase pH by 0.1 units (Margalit, 1997).

Neutral potassium tartrate precipitates potassium hydrogentartrate, which effects acidity to a lesser extent. This method is costly and has a lower deacidifying ability.

Double salt precipitation incorporates $CaCO_3$ and calcium tartrate malate, which removes quantities of tartaric and malic acid. This method is used most often in cold winemaking regions like Germany.

All of these carbonates and bicarbonates will be decomposed in wine to carbonic acid, releasing CO_2. The carbon or potassium form salts with the tartaric, which precipitates out of solution. There are no limits on the amount of additions of these chemicals, but there are some limits established as to how low the acidity can be reduced. The TTB states that the fixed acid level cannot drop below 5.0 g/L. Analyses for TA and pH are performed after any deacidulation.

Amelioration will reduce the acid concentration but does not affect the pH to any great extent. With a water addition prior to fermentation, sugar levels should be retested, in addition to TA and pH levels.

6.3   Baseline Analyses

When the juice and must are prepared and ready to begin fermentation, baseline analyses for reducing sugar, TA, pH, acetic acid, volatile acidity (VA), and LMA is helpful to track the progress of the fermentations, as we will learn further in the chapter. These parameters provide useful information concerning any spoilage problems that might be occurring.

6.3.1   Volatile Acidity and Acetic Acid Determinations

In the previous chapter, we discussed total acidity (TA), fixed acidity, and volatile acidity (VA). Volatile acidity consists of volatile fatty acids, primarily acetic acid and, to a much lesser degree, formic, propionic, and butyric acids. These acids can be steam-distilled and detected by smell and taste (organoleptically).

Because acetic acid comprises the majority of the VA, it is expressed as acetic acid. Acetic acid is formed during yeast fermentation from the side reaction of acetaldehyde oxidation. Normal acetic acid levels formed during the fermentation process are approximately 0.2–0.4 g/L. Botrytized wine, Sauternes, and late harvest wines can contain much more.

Small amounts of acetic acid can be formed during MLF by the malolactic bacterial decomposition of citric acid (Ough and Amerine, 1988). Concurrent MLF and primary fermentation can also create small amounts of acetic acid due to competition of fermentation yeasts and LAB (Boulton et al., 1999).

Increases in acetic acid after fermentation is directly attributed to spoilage organisms such as *Acetobacter*. It is believed that *Acetobacter,* in an aerobic environment and depending on exposure time, oxidizes EtOH, producing acetic acid (Margalit, 1997). Acetic acid and VA are direct indicators of wine spoilage. Acetic acid content can exist in higher levels in late harvest and Sauterne wines. Concentrations greater than 2–3 g/L are considered spoiled, or vinegar. *Acetobacter* will be discussed later in this chapter.

Another concern indicated by high acetic acid levels is high ethyl acetate levels. Ethyl acetate is produced proportionately to acetic acid. It is not only the acetic acid but the ethyl acetate that you are able to organoleptically detect. Levels starting around 0.6–0.9 g/L for acetic acid can be detected by the vinegar smell and taste. Ethyl acetate levels of 0.15–0.2 g/L give off the odor often described as similar to fingernail polish remover (Margalit, 1997).

Legal limits for VA calculated as acetic acid, exclusive of sulfur dioxide, in the United States is 1.2 g/L in white wine and 1.4 g/L in red wine. These limits are for wines made from grapes below 28 °Brix. For wines made from juice with greater than 28 °Brix and unameliorated, the maximum VA for white wines is 1.5 g/mL and 1.7 g/mL for red wines [Section 27 CFR 4.21 (a)(1)(iv) Federal Regulations]. German wines range from 1.2 g/L in white wines to 2.5 g/L in some other varietals as a comparison.

Fixed acidity calculations (Section 5.3.6) utilizing the VA results are used in Europe to detect amelioration in wine (expressed as tartaric). This detection method assumes that the normal fixed acidity of the wine is fairly close to that of the must. Abnormal changes in the fixed acidity from the predicted fermentation changes might indicate amelioration. Analyses for acetic acid and VA are performed throughout the development of the wine to act as indicators of potential spoilage problems.

Volatile acidity is measured by steam-distillation, whereas direct acetic acid is measured via enzymatic analysis, high-performance liquid chromatography (HPLC), or gas chromatography (GC). Measurements of acetic acid and VA will never be equal, but acetic acid levels correlate very closely to VA levels in juice, must, and young wines. As the wine ages and the acetic acid levels reach the 0.5–0.7-g/L range, a divergence occurs where the VA and acetic acid levels differ (possibly due to extraction of volatile acids from the wood). The greatest divergence occurs in Pinot noir wine. At this point, VA analysis alone is relied upon as the spoilage indicator.

In the early stages of production, acetic acid analysis via an enzymatic method is more cost-effective because a larger volume of samples can be analyzed in a given time frame. Very small wineries will often forego acetic acid analysis and use the VA analysis throughout wine production.

6.3.1.1   Acetic Acid Enzymatic Method

Acetic acid determination is performed via enzymatic analysis utilizing a spectrophotometer, random access analyzer, or continuous-flow analyzer in addition to other methods, including GC and HPLC (these technique principles are discussed in Chapter 5). In the majority of wine laboratories, enzymatic analysis, utilizing a spectrophotometer, is the most common method. As the sample load increases, more automated instrumentation is required. The Alcohol Tobacco Tax and Trade Bureau (TTB) utilizes the AOAC methods of GC and HPLC.

The chemical principle is based on the light absorbance at 340 nm of the coenzyme NADH produced from the following reaction according to the procedure stated by Boehringer Mannheim (1997):

$$\text{Acetic acid} + \text{ATP} + \text{CoA} \xrightarrow{\text{ACS}} \text{Acetyl–CoA} + \text{AMP} + \text{Pyrophosphate}, \quad (1)$$

$$\text{Acetyl-CoA} + \text{Oxaloacetate} + \text{H}_2\text{O} \xrightarrow{\text{CS}} \text{Citrate} + \text{CoA}. \quad (2)$$

The oxaloacetate for the above reaction is formed from L-malate and NAD^+, where NAD^+ is reduced to NADH:

$$\text{L-Malate} + \text{NAD}^+ \xleftarrow{\text{L-MDH}} \text{Oxaloacetate} + \text{NADH} + \text{H}^+, \quad (3)$$

where CoA is coenzyme A, AMP is adenosine-5-monophosphate, CS is citrate synthase, and L-MDH is L-malate dehydrogenase.

The production of measurable NADH in this reaction is not proportional to the acetic acid concentration in the sample and must have additional calculations. Calculation of results is based on the Beer–Lambert law. The chemicals and kits that can be purchased have detailed instructions on the proper procedures. The errors that can be occur during enzymatic ultraviolet (UV) analysis are numerous. Using prepared kits reduces many errors, but careful and precise laboratory techniques will reduce the majority of errors encountered. The most common errors include the following:

- Improper temperature for analysis and measurement of reagents or sample (optimum 20–25°C)
- Improper pipette technique
- Failure to degas or filter sample to eliminate CO_2
- Expired reagents or enzymes
- Failure to adjust sample pH (buffers are included with kits)
- Incorrect dilution to within the working range of the instrument
- Air bubbles in cuvette blank (zero instruments to air, recommended)
- Wrong nanometer setting on instrument
- Instrument lamp failing
- High alcohol content (slows reaction time)
- Air bubbles in sample cuvette
- Incorrect light path length
- Contaminated reagents or enzymes
- High levels of ethyl acetate
- Decolorization of red wines with anticipated acetic acid levels less than 0.1 g/L

Consult Chapter 9 for more precise information on acetic acid analysis via the enzymatic UV method.

6.3.1.2   Volatile Acidity Distillation Method

Volatile acidity determination is performed via steam-distillation using stills including the Cash still, Markham still, and Sellier tube. The TTB and AOAC official methods are Cash still distillation and segmented flow (similar to continuous flow described in Chapter 5, Section 5.4.3.5). Segmented-flow and continuous-flow instruments actually incorporate a tiny glass steam-distillation component designed for very small samples.

In the United States, the Cash still and variations of the Cash still are more commonly used. The principle of steam distillation involves passing a flow of steam through a sample and heating it, allowing the volatile components to be swept up with the steam into a cooling condenser where the steam cools to a liquid distillate. The distillate is collected in a container that is titrated with a base such as NaOH. The amount of NaOH used to neutralize the acidic sample is proportional to the amount of acid in the sample. A typical Cash still setup is made of hand-blown glass and consists of a boiling chamber for water; an inner chamber (Sellier tube) for the sample, and a water-cooled condenser. Figure 6.1 illustrates a typical Cash still setup.

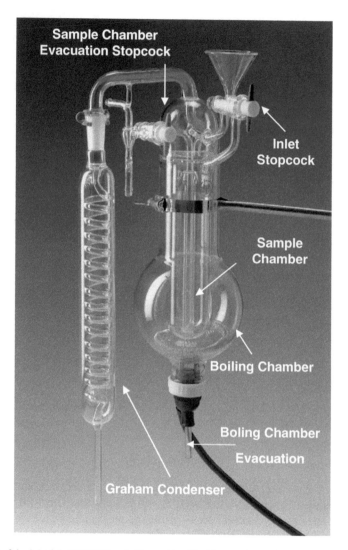

FIGURE 6.1. Model RD80™ volatile acid still (Research and Development Glass Products & Equipment, Berkeley, CA).

The calculation of VA is as follows:

$$VA\,(g/L) = 60 \times \frac{(V_T \times N)}{(V_w \times F)} \tag{1}$$

or

$$VA\,(g/100\,mL) = \frac{(60 \times 100)(V_T \times N)}{(V_w \times F)} \times 1000,$$

where V_T is milliliters of NaOH titrated, N is the normality of the NaOH (0.1N), V_w is the milliliters of sample, and F is the normality factor of 0.1N or the normality of the NaOH used. If the sample size is 10 mL and the NaOH normality is 0.1N, the equation can be simplified to

$$VA\ (g/L) = 0.6 \times V_T. \tag{2}$$

The standard deviation range of the distillation method can vary from ± 0.05 g/L upward. The success of VA analysis is dependent on the elimination of as many variables as possible. Variables are numerous and associated with technique, equipment, procedure, and carryover of other volatile substance.

Carbon dioxide can distill over as carbonic acid, which will add acid to the distillate, producing an abnormally high VA. High SO_2 levels in wine will distill over as sulfurous acid, again increasing the acid content of the distillate, resulting in an abnormally high VA.

Juice, must, and wine samples must be degassed to remove the CO_2 content prior to testing. The water used in the still to produce the steam must also be degassed for CO_2 or boiled to release the CO_2. Bottled sparkling wines have a high content of CO_2, and more rigorous methods such as double distillation might be required to remove the CO_2 (Ough et al., 1988).

Sulfured juice, must, or wine is introduced into the Cash still, followed by the addition of 0.3% hydrogen peroxide (H_2O_2) to bind the majority of SO_2. Very high SO_2 levels might require the addition of a mercuric oxide solution that can bind up to 250 ppm SO_2, or the addition of sodium arsenite in conjunction with H_2O_2 has also been used in high SO_2 wines (Ough et al., 1988). Unfortunately, both of these solutions are a health hazard. The common method in use today is back-titration of the distillate with iodine.

Legal limits for VA in the United States are listed as exclusive of SO_2, which means the SO_2 component is removed. The majority of wines have a VA below the legal limit, so most wine laboratories do not remove the excess SO_2 contribution unless the VA is close to the legal limits. In addition to the above, other volatile fatty acids can distill over. Remember, that acetic acid is the most abundant volatile acid, not the only volatile acid.

Vigorous boiling and steam production in the still can cause lactic acid to volatize, thus erroneously increasing the VA. Lactic acid is not considered a spoilage acid; therefore, it should be eliminated from the VA. This is especially true for wines that have gone through malolactic fermentation, which have higher levels of lactic acid. Preventing vigorous boiling will decrease the amount of lactic acid distilled over. The greater the volume of heat and steam, the greater the amount of lactic acid distilled over.

Each distillation unit has small differences. Testing the distillation unit using various standards containing CO_2, SO_2, lactic acid, and water blanks can give the operator information on the correct boiling temperature and steam production. Testing the units can identify problem areas that will need to be corrected.

The most common equipment problems include loose connections, worn O-rings, condenser water temperature that is too cold or too warm, water level in the boiling chamber that is too high or too low, dirty condenser and/or still, and leaking drainage hoses.

Recirculation pumps containing water or a chilling agent are often used to maintain a constant condenser temperature. Tap water or deionized (DI) water can vary in temperature during the day, weeks, months, and years, so good temperature and flow control are necessary for consistency in testing. A condenser temperature of 15–17°C is a good place to start.

A meticulous preventative maintenance program and scheduled cleaning of the stills will eliminate many of the encountered variables. A dilute NaOH solution is boiled in the still to clean the sample chamber, followed by thorough rinsing. Condensers might require careful dismantling to clean properly.

We have discussed titrations associated with previous methods and know that there are many technician and reagent errors that can occur. There are three common errors that most affect the consistency and validity of VA results: discerning the end point of the distillate titration, reagent problems, and poor degassing of the samples. The amount of sample and titrated volume should be exact measurements. Excellent pipette and burette techniques are mandatory.

After pipetting the sample into the sample chamber followed by the addition of H_2O_2, a very small amount of CO_2-free distilled water is added to flush all the chemical and sample into the sample chamber. After closing off the sample chamber, add a small amount of water above the stopcock to ensure a good seal.

For consistency in VA analysis, an exact preset volume of distillate must be collected for each sample. In my experience, 100 mL of distillate seems to work the best and is recommended by Ough and Amerine (1988), but the volume of distillate collected will depend on the functionality of the still. Analyze several wine samples of a known VA that have been spiked with a known concentration of acetic acid. Including a control, analyze each sample repeatedly at various distillate levels and ascertain the percent of recovery in order to set the appropriate analysis time and distillate volume. It is recommended that each unit should be tested minimally once a week with an acetic acid standard. The analyzed standard results are recorded to ensure the recovery and validity of the method.

Titration of sample distillate is carried out immediately upon completion of the distillation process. If a titration must wait a few minutes, stopper the distillate flask. Phenolphthalein indicator solution is added to the distillate and gently stirred or swirled while performing the titration. This allows mixing of the distillate with the NaOH but prevents introduction of CO_2 by vigorous agitation. The distillate is quickly titrated with NaOH to a pale pink end point that should persist for a minimum of 15 s up to 30 s before it completely fades. Slow titration will produce abnormally high results. All laboratories need to standardize the rate of titration and length of persistence, and apply it to all samples. A light box under the distillate flask is helpful in detecting the end point and the length of persistence.

Chemical and reagent problems contribute as a source of error and are attributed to the following:

• Using the wrong chemical or normality of the chemical
• Using expired chemicals
• Using contaminated chemicals
• Errors in standardization
• Uncalibrated burettes
• Using improperly stored chemicals

It is prudent to always double check the normality or percentage of the chemicals you use, empty the chemical container before refilling with a stock solution, read and know storage requirements of the chemicals used, and maintain the integrity of the chemicals by correct and frequent restandardization.

Many times, a sample is degassed and left to sit on the workbench until it can be distilled, or a sample has been degassed, sealed, but will sit for a long period of time. Fermenting juice and must samples can generate a great deal of CO_2 in a short period of time, enough to affect the VA results. These samples must be distilled as soon as possible after degassing. If a fermenting sample sits for 15 min, repeat the degassing step. Wine samples can sit for approximately 30 min before redegassing. Degassed samples left exposed to the air will reabsorb CO_2 and require degassing prior to distillation. It is good laboratory practice to seal the sample after degassing and note the time on the container to alert the analyst to a potential problem.

Other errors associated with the distillation method are as follows:

• Sample contamination
• Contaminated glassware
• Poor lighting
• Calculation errors
• Allowing too much steam to escape prior to closing system

6.4   Primary Alcohol Fermentation

Yeast fermentation of sugar into ethanol is a very complex process involving many biochemical reactions. The scope of this section is to give an overview of fermentation, the reader is well advised to seek further information for a more complete understanding of this life process.

Yeasts (fungi), bacterium, molds, and so forth are micro-organisms; in other words, unable to be seen by the naked eye. Yeasts are classified as a fungus, and because of their life cycle and the by-products from that life cycle, they are used in the food industry to create certain positive effects when introduced into various food media.

Many varieties of yeasts occur in nature and on most any naturally grown food. These yeasts are referred to as wild yeasts, or yeasts that have not been cultured (for the intent of introduction into a food media). Wild yeasts and cultured yeast can be found on food processing equipment that has not been sanitized. A winery cellar most often contains a variety of yeasts at harvest. Some wineries choose to sanitize and destroy the wild yeasts, opting for a single species of cultured yeast; yet other areas of the world welcome wild yeasts and the geographical flavor profiles they add to the wine.

The cultured yeast most often used for alcohol conversion in grape juice is *Saccharomyces cerevisiae* (Fig. 6.2), one of the seven wine-related species from the genus *Saccharomyces. Saccharomyces cerevisiae* have been isolated and cultured further by their ability to tolerate alcohol, temperature, pH, and SO_2, in addition to their speed of fermentation and foam production. These subcategories include Pasteur-Champagne, Pasteur-Red, Eperny, Montrachet, Prise-de-mousse, Tokay, Steinberger, and California-Champagne.

6.4.1   Yeast Metabolism

Living organisms require a source of energy to maintain their biological functions, growth, mobility, and reproduction. Plants containing chlorophyll collect solar energy; they are called phototrophs. Certain bacteria, chemolithotrophs, oxidize minerals for energy. Chemoorganotrophs are the majority of yeasts (fungi), bacteria, and animals that break down organic nutrients as their source of energy. The process of breaking down organic nutrients is known as catabolism; the energy created is transferred through a chain of synthesis reactions and is known as anabolism.

Only a portion of the energy created by catabolism is used; the rest is dissipated as heat. The energy that is to be used is called free energy and supplies the cell with the energy for transport, movement, or synthesis. This free energy is transported throughout the cell by a chemical called adenosine triphosphate (ATP). ATP contains two phosphoanhydride bonds that when hydrolyzed (phosphorylation) into adenosine diphosphate (ADP) release a large quantity of the free energy for biosynthesis and the active transport system of metabolites. Yeast growth is directly related to the quantity of ATP produced. In wine yeast, as in all living cells, there are two processes that produce ATP: substrate phosphorylation (aerobic or anaerobic) and oxidative phosphorylation (aerobic).

According to Ribéreau-Gayon et al. (2000a), substrate phosphorylation (fermentation) involves electron loss via oxidation and the creation of an ester–phosphoric bond between the oxidized carbon of the substrate and a molecule of inorganic phosphate. This bond is transferred to ADP (transphosphorylation), forming ATP during glycolysis (transformation of glucose into pyruvate plus ATP). The oxidative phosphorylation process (respiration) produces ATP by the transport of electrons to an oxygen molecule via the cytochromic respiratory chain in the mitochondria.

Ethanol is a by-product of glycolysis that is directly followed by alcoholic fermentation. Glycolysis provides the yeast with two molecules of ATP per molecule of degraded glucose and pyruvate via a complex system of enzymes

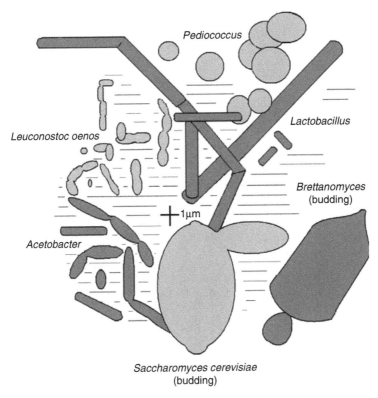

Pediococcus

Lactobacillus

Leuconostoc oenos

Brettanomyces
(budding)

1μm

Acetobacter

Saccharomyces cerevisiae
(budding)

FIGURE 6.2. Comparative drawing of typical wine bacteria and yeast.

and cofactors of those enzymes. Alcoholic fermentation begins as glycolysis ends, breaking down the pyruvate into acetaldehyde and CO_2, and further into ethanol via enzymatic activity. Figure 6.3 illustrates this complex process followed by the reactions that take place during alcoholic fermentation. To simplify Figure 6.3, the following reaction takes place (Margalit, 1997):

$$C_6H_{12}O_6 + 2ADP + 2(H_2PO_3) \rightarrow 2CH_3 - CH_2 - OH + 2CO_2 + 2ATP.$$

Other sugar fermentation processes can take place in addition to, or in lieu of, alcoholic fermentation. All fermentations are derived from the glycolysis process and differ due to the aerobic conditions that produce different by products. Glyceropyruvic fermentation produces glycerol, CO_2, and a bisulfitic form of acetaldehyde; the acetaldehyde combines with sulfite, preventing the reduction to ethanol. The fermentation takes place in the very first stages of fermentation, when the cultured yeast is not yet acclimated to the anaerobic conditions. Respiration (oxidative phosphorylation) results in the production of CO_2 and water via the Krebs cycle. The Krebs cycle is a sequence

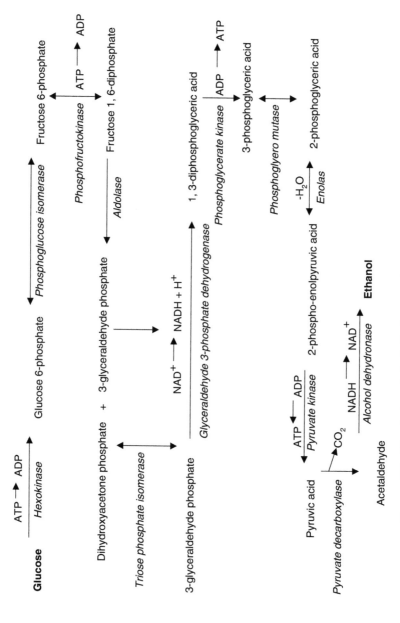

FIGURE 6.3. Glycolysis and alcoholic fermentation diagram. Enzymes are italicized.

of reactions oxidizing acetic acid or acetyl equivalents, providing energy that is stored in phosphate bonds.

Theoretically, under ideal conditions, 1 mol of sugar (180 g) fermented produces the following (Margalit, 1997):

- 180 g sugar
  - = 1 mol
  - = 56.4 kcal/mol of free energy—
  - 32.4 kcal/mole of energy utilized by yeast and 24 kcal/mole of energy released as heat
  - = 2 mole of CO_2 produced or 48 liters at 20°C
  - = 84.6 g EtOH
  - = 107.2 mL EtOH
- Addition of 1 liter H_2O
  - = 108 mL/L EtOH
  - = 10.8% v/v
- 180 g/L sugar
  - = 20 °Brix
- °Brix
  - = 9 g/L sugar
  - = 1.14 kcal/mol of energy released as heat
  - = 0.1 mol CO_2 produced or 2.4 liters
  - = 4.23 g EtOH
  - = 5.36 mL EtOH
- Addition of 1 liter H_2O
  - = 5.4 mL/L EtOH
  - = 0.54% v/v EtOH

The heat generated from excess energy not used for cellular metabolism is of great concern during fermentation. Vigorous fermentations produce a large amount of heat, raising the temperature of the must or juice to the point that it can kill the yeast, halting fermentation. Temperature control of fermentation is necessary to protect the yeast and control the rate of the fermentation for winemaking purposes.

6.4.2   Cultured Yeast and Inoculation

Yeasts used for inoculation are of a single-species type chosen for the character it will impart to the finished wine and its adaptability to certain fermentation conditions. Commercially prepared yeast is available as dry active yeast and as liquid culture.

Dry yeasts are vacuum-packed freeze-dried yeast. This type of yeast requires hydration with very warm water to activate the yeast. The temperature of the hydrated yeast (inoculum) should equal the temperature of the receiving juice or must. This is accomplished by small periodic additions of juice or must to the inoculum until temperature equilibrium is reached. Dry yeasts come packaged in large quantities as well as small 500-g packages.

Liquid cultures are live, viable yeasts in a juice medium prepared from either a live "mother" culture or dry yeast. Liquid cultures can be purchased from a reputable laboratory by the liter or in 400-mL vials (slants). Liquid culture is usually more expensive than dry yeast, but it is ideal for small inoculations.

The key to a successful fermentation is to provide the yeast with the best growing conditions, including temperature, pH, nutrients, adequate sugar levels, timely oxygenation, and CO_2 elimination. Inoculation with a sufficient or more than sufficient number of yeast colonies is critical.

Fermentation should optimally begin at 20°C to allow the yeast to acclimate. White wines are normally fermented at cooler temperatures between 8°C and 15°C (46–59°F). The cooler temperatures and slower fermentation tend to preserve the fruity taste often desired in a white wine. Red wines are fermented at much higher temperatures between 25°C and 30°C (77–86°F) (Margalit, 1997). The higher temperatures allow for a faster fermentation, but more important, the higher temperatures are required to extract optimum color and phenols from the grape skins. Temperatures that are too low will limit the growth of the yeast and reduce the population; conversely, too high temperatures can kill the yeast, as discussed earlier.

The optimum pH range for fermentation is between 3.0 and 4.0 pH units. The closer to pH 4.0, the less time is required for the yeast to acclimate. As discussed earlier in this chapter, most micro-organisms prefer a higher pH, not only the desired microbes but the undesirable microbes as well.

High sugar content can adversely affect fermentation by the production of high levels of EtOH in excess of 16% v/v. EtOH levels in this range will kill the yeast by the destruction of the cell membrane.

Timely exposure to oxygen can stimulate fermentation in the first few days. Oxygen improves the permeability of the yeast cell membrane allowing glucose penetration (Boulton et al., 1999). Oxygenation of white juice is not normally done because of the browning effect of oxidation, which is more of a danger than a slower fermentation. Red must is aerated via several different methods; the two most common methods are "pump over" and "punch down." The pump over method simply takes must or juice from the bottom of a tank and pumps it up and over the top of the tank, sprinkling it over the must cap. This method helps aerate the must; keeps the cap cool by dissipating heat, and reduces bacterial growth in the cap. A punch down is exactly that, physically or mechanically pushing the cap down into the must. Small-tank punch downs use a device that has a flat perforated plate at the end of a long pole. The cellar worker pushes sections of the cap down into the juice. Automated systems are available for larger tanks. These automated systems mechanically move a large perforated disk over the top of a large tank and then proceeds to slowly lower the disk pushing the cap down into the must (Fig. 4.8A).

We have discussed the need of nutrients such as amino acids including ammonia, phosphate, minerals, and vitamins. Figure 6.4 illustrates the primary nutrient requirements of the yeast and a few of the primary by-products of the fermentation.

The number of yeast cells added to a substrate is called the inoculation rate. The fermentation of juice and must has a recommended inoculation rate of 10^6–10^7 yeast cells/mL (1 million –10 million yeast cells/mL). The optimum yeast population at the height of fermentation is 10^8. A low inoculation rate will not reach the desired optimum population and fermentation will end without all of the sugar being fermented. A higher initial inoculation rate will speed up the fermentation but will also increase the cost (yeast are not inexpensive) per gallon produced. Inoculation of juice and must is most successful when the yeasts are in the middle of intensive budding. Dry yeast inoculation is typically 120–240 mg/L (Boulton et al., 1999). Yeast cell counts (titer) are determined using a microscope and a counting chamber (refer to Chapter 9, Section 9.2.20). Yeast counts in juice can also be monitored with spectrophotometry. A 10-mL inoculated juice sample and 1.0 mL of water are placed in a cuvette and measured at optical density (OD) 600 nm. Readings of 0.1 equate to 10^6 cells/mL.

Yeast growth has a specific pattern that is broken down into four phases:

1. *Lag phase*: Period of time when the yeast cells are acclimating to the pH, temperature, and sugar content of the juice or must. Yeast cell membranes are being built and there is no significant multiplication. The lag phase can be as short as a few hours to several days. The shortest lag phase occurs in warm must or juice with a higher pH.
2. *Growth phase*: Period of time when there is exponential growth in the yeast population. The growth phase can last several days with a significant drop in sugar levels, as much as 50% of the initial. During this phase, metabolism becomes anaerobic due to the increase in CO_2.
3. *Stationary phase*: Period of time when the yeast having attained their maximum population growth and continue to ferment the remaining sugar with little reproduction. This phase lasts only a day or two.
4. *Death phase*: Period of time when the population begins to die due to lack of nutrients and an increase in toxic by-products. Very little reproduction occurs during this phase as the population steadily falls to below a viable level of 10^5 cells/ml (Ribéreau-Gayon et al., 2000a). The length of this phase can be weeks.

Venting the fermentation tanks or barrels is a necessity to avoid a buildup of CO_2 gas pressure within the fermenting vessel. Excessive CO_2 pressure buildup can lead to tank and barrel ruptures or overflowing juice and must due to excessive foam created by high levels of CO_2 gas.

Open-top fermentors have no restrictions, but closed tanks require a pressure-relief system to allow the gas to escape. There are several types of pressure-relief system available for tanks.

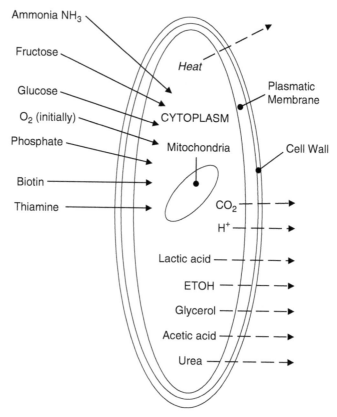

FIGURE 6.4. Selection of yeast nutritional requirements and by-products of fermentation (simplified).

During barrel fermentation, the barrel hole is closed with a fermentation bung. A fermentation bung is a large stopper with a simple pop-up valve to release gas pressure. There are several devices available that allow gas to escape from the barrel. Solid bungs are installed in the barrels postfermentation. Figure 6.5 shows a simple fermentation bung compared to a solid bung.

If the conditions are good, most cool white fermentations can be completed in 2–3 weeks. Red fermentations can be completed in a few days.

6.4.3 Monitoring Fermentation

Fermentation progress in all juice and must should be closely monitored. Winery laboratory staff will monitor fermentations once or twice a day, calculating the sugar content and noting the temperature of the fermentation. The results produce a picture of the fermentation that allows the winemaking staff to analyze the trends, note potential problems, and anticipate and plan their next production step.

Monitoring fermentation can be accomplished several ways, including density (soluble solids), volume of CO_2 produced, amount of heat released, and weight changes in the fermenting juice or must. The most common method used in wineries is the measurement of the soluble solids via hydrometer or electronic density meters. The other methods require extensive instrumentation and specially designed fermentation containers.

6.4.3.1 Soluble Solids (Apparent °Brix) via Hydrometer

Chapter 5 discussed sugar analysis and the theory behind the methods. In most parts of the world, apparent °Brix (which will be referred to as Brix) is used to express sugar content during fermentation. Although Brix is not an exact measurement of the sugar content, it does allow one to follow the trend of the fermentation. Confirmation of dryness should always be performed via enzymatic analysis for residual reducing sugar (RS).

Brix (also °Baumé and Oechslé) readings are obtained by floating a hydrometer in a column of juice. The hydrometer has been designed to float at the same levels that correlate with sugar content at various densities. The hydrometer method can be successfully accomplished when the variables in the method have been addressed, thus reducing errors.

FIGURE 6.5. Barrel fermentation bung on the left and solid wine barrel bung on the right.

It is important for anyone working in the cellar during fermentation to remember the dangers and safety rules associated with cellar work (see Chapter 2), especially those concerning CO_2 gas, mouth suctioning, barrels, tanks, climbing, forklift traffic, and splinters.

Prudent laboratory practices and techniques and following the listed guidelines are keys for successful Brix evaluation:

- *Clean equipment.* Clean and sanitize all equipment between every measurement with a 70% EtOH solution, followed by a minimum of three rinses. If time permits, allow the equipment to soak in the EtOH solution and rinse just before use; this will improve the sanitizing effects. Residual solids will carry over to the next measurement when dirty equipment is used, producing inaccurate readings. Clean tank valves after obtaining the sample and rinse the area with water. Cleanliness reduces the production of fruit flies that can carry bacteria and protects juice or must samples that are returned to their tank or barrel.
- *Sampling sites.* Your results are only as good as your sample. This test does not discern between soluble solids and insoluble solids. Brix measures soluble solids and you want to obtain the best representative sample for the measurement. Samples should be obtained from mid-barrel or as close to mid-tank (includes bins and open fermentors) as possible. Avoid the bottom of the barrel or tank and the cap area, where insoluble solids are at high levels. It is best to avoid taking a sample during or directly after a tank pump over or after filling a barrel. The agitation mixes the insoluble solids back into the juice or must and will result in abnormally high Brix readings. Take samples prior to any movements. Sampling tanks containing whole berries, seeds, or other materials requires the juice or must to be poured through a sieve to remove debris prior to testing. The temperature of the walls of the tank will affect the fermentation rate of the juice or wine in that area (usually cooler with slower fermentation) and the hottest and fastest fermentation is happening in the center of the tank. When obtaining a sample from any valve for any reason, it is important to flush the valve area with 1–2 liters of juice, must, or wine, bringing more of the inner contents to the sampling valve. The sample will better represent the Brix and temperature (or other analyses) of the majority of the volume in the tank. Working with tank valves requires the utmost safety precautions. The pressure of the juice, must, or wine in a tank is significant, and if released accidentally, it can cause severe injury. Valves should be opened very slowly to avoid an uncontrollable rush of juice, must, or wine. The most dangerous valves to use are the must and racking valves located at the bottom of the tank where the pressure is the greatest. A sampling valve (also referred to as the tasting valve) located closer to the tanks mid-line is the best location to obtain the sample on most tanks, as shown in Figure 6.6. If a tank sample is unable to be obtained from the sampling valve, a sample can be taken from the top of the tank. If you choose to obtain the sample from a must or racking

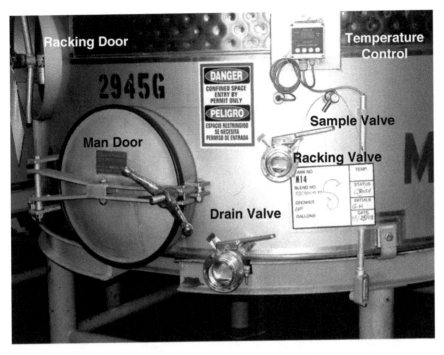

FIGURE 6.6. Typical stainless-steel tank front.

valve, a notation should be made alongside the results to alert the review-er. Tank-top samples are usually obtained using a heavy stainless-steel metal cylinder attached to a long stainless-steel chain (Fig. 6.7). With the chain secured, the cylinder is inverted and dropped through the middle of the tank. As the cylinder reaches the end of the chain, a quick jerk will upright the cylinder, allowing it to fill; it is then pulled back to the top. Barrel samples can be obtained using a barrel thief (large-volume, non-graduated pipette with a handle) preferable made of plastic (for wine that will not be kept for tasting) or glass. Small hoses equipped with a suction bulb can also be used, but they are more difficult to clean properly. Automatic pipetting devices are available that will draw up a preset volume of sample using a large-volume pipette (sterile or sanitized). The advantage to the automatic pipette is precise sample volume, it is safer, faster, and cleaner, and there is less waste of juice or must. Juice or must used for flushing valves and for testing can be returned to their individual fermen-tors *if* good sanitizing practices are in place *and* nothing is added, such a defoam. It is always good laboratory practice to discard all used juice and must. Contamination of the fermentation by introducing bacteria is not worth the few dollars saved in product.
- *Sample preparation*. Fermentation CO_2 displaces volume, decreasing the density of the sample and giving an erroneously low Brix reading. The

FIGURE 6.7. Sampling from the top of a tank.

CO_2 needs to be eliminated from the sample directly before testing because the fermentation continues to produce CO_2. In the cellar, the fastest and cleanest way to degas the sample is by driving off the gas by aggressive agitation of the sample. Pouring a sample from one container to another several times, shaking the sample, and vigorous stirring are the most common methods. Non-battery-operated electrical devices should not be used in the cellar because of the wet conditions. If the sample contains too much foam to obtain a good hydrometer reading, add a few drops of antifoam to reduce the bubbles. Only use what you need. If obtaining a sample that is to be analyzed at a later time, ensure that the container is well vented. Samples left in closed glass bottles can explode.

• *Brix determination.* The basic tools required for Brix determination are a hydrometer, a 250-mL plastic cylinder, a thermometer, and a graduated 2-liter plastic beaker with a handle. Brix hydrometers come in a variety of scale ranges, the three most commonly used in the wine industry are 0–12 °Brix, 9–21 °Brix, and 19–31 °Brix. Past determinations of the Brix readings, sugar levels, or anticipated sugar levels will dictate the correct Brix hydrometer to use. The best readings are obtained using a Brix hydrometer with the shortest scale range calibrated at 20°C (certified). A laboratory-grade non-mercury-filled thermometer (certified) is recommended to obtain the temperature of the juice or must (Fig. 6.8). Temperature variation from the ideal 20°C will require correction of the Brix reading (refer to Chapter 5 and Table 5.1). Protective metal sleeves can be purchased to protect the thermometer from breakage. Combination units of Brix hydrometers with an integral thermometer (thermohydrometers) are most convenient, but are twice as expensive as the hydrometer alone and require twice the sample size. The analyst should be skilled at reading a meniscus when using a hydrometer as illustrated in Figure 6.9. The degassed sample is poured into the cylinder set on a level surface. Take the temperature reading first and then immediately after, slowly inserted the hydrometer into the sample. An easy spin of the hydrometer stem will dislodge any gas bubbles that might be trapped on the bottom of the hydrometer. Before the reading is obtained, carefully lift the hydrometer slightly and wipe the stem clean, leaving no juice or grape material to weigh down the hydrometer that could result in

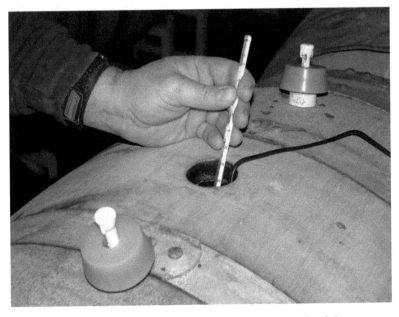

FIGURE 6.8. Obtaining the temperature of fermenting juice.

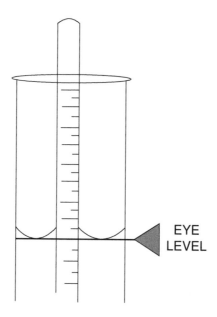

FIGURE 6.9. Reading a meniscus.

a false reading. Take your readings to the nearest 0.1 °Brix, make temperature corrections, and compare the results to a previous result. If the Brix and temperature are in line with the phase of yeast development, all is well; however, sharp changes in temperature, Brix readings that drop sharply or increase, or Brix readings that have not moved could indicate a poor sample or a problem with the fermentation. Always repeat the Brix and temperature with a fresh sample if there are any obvious discrepancies. If the anomaly persists, notify the winemaking staff. See Chapter 9, Section 9.2.4 for more details.

- *Cellar work*. Obtaining samples or monitoring Brix in the cellar is a dirty job. Wear clothing that can be thrown away at the end of harvest or use a washable smock (not white) to protect your clothing. Wear comfortable shoes that can either be cleaned very well or thrown away after harvest. If there is forklift traffic, always wear a safety vest to draw attention to your position, especially when working between barrel stacks.

6.4.3.2 Soluble Solids (Apparent °Brix) via Density Meter

A density meter programmed with °Brix calculations such as the Anton Paar DMA 35n (Fig. 5.2), is a more accurate tool for measuring Brix. These units are so quick and easy to operate that they significantly cut down the analysis time. The sampling and sample preparations are identical to the hydrometer method, with the exception of much smaller sample size. These units require only a few milliliters of degassed sample, making cellar degassing easier to manage. Cleaning the units between tests is fast, but it must be stressed that these units require very thorough cleaning to operate properly.

A degassed sample is aspirated and expelled several times to rinse the inner chamber. The final sample volume is aspirated; the meter will take its measurements and digitally read out the temperature and the Brix reading. If the sample is actively fermenting, the meter is turned on its side to allow gas bubbles to escape the measuring chamber. Always check with the manufacturer's recommendations for gas elimination.

Density meters might, or might not, be temperature compensated. Temperature-compensated devices can automatically recalculate the Brix (or any measurement that is based on temperature as a variable) from one initial reading. Temperature-compensated equipment is faster with less computation errors. The DMA 35n will flash the temperature reading if it falls out of the measurement range, alerting the technician. The instruments will continually monitor the temperature of the sample and cease flashing when the proper measuring temperature is reached and the Brix reading can be noted. Samples can also be placed in a water bath until the measuring temperature is achieved and resampled. The DMA 35n offers several different options for setting measuring temperatures, thus making correct calculations automatic.

The portable battery-operated density meter is checked daily with water or a standard sample for accuracy and cleaned routinely. Little maintenance is required. Both hydrometers and meters are subject to breaking. The hydrometer stems frequently snap off with any rough handling. The meters are not as fragile and the inner glass chambers can withstand a few bumps, but if the unit is dropped and the chamber breaks, the cost to repair the unit is expensive. Securing the meter to a cart or attaching it to a wrist band or belt will help prevent it from hitting the ground, improving the longevity of the units. It is prudent lab practice to always have a backup supply of hydrometers with or without the use of meters, especially at harvest.

Most wineries using meters for monitoring Brix are very pleased with them and have experienced the cost savings in labor with the added benefit of improved accuracy. Density meters can be utilized throughout the year, not only for Brix, but for specific gravity and density readings (see Fig. 5.2). The procedure for use of the Anton Paar 35n can be found in Chapter 9, Section 9.2.5.

6.4.3.3   Verification of Dryness

As the fermentation slows and stops, the Brix readings will be zero or below zero due to the alcohol content of the wine (alcohol is less dense than water). It is common practice to observe three consecutive zero or below zero readings before performing analysis to determine dryness. The laboratory will perform enzymatic analysis to confirm the level of remaining RS is less than

2 g/L. An alcohol analysis is performed to set a baseline and to verify that the potential alcohol was achieved. Refer to Chapters 5 and 9.

The juice and must become a young wine upon completion of primary fermentation. The wine will either be prepared for secondary MLF, maturation, or bottling depending on the varietal and winemaking style.

6.5   Secondary Malolactic Fermentation

Malolactic fermentation is a process whereby a LAB *Oenococcus oeni* (previously called *Leuconostoc oenos*) converts LMA, containing two acid groups, into L-lactic acid, containing one acid group. The catabolic pathway utilizes the enzyme malate carboxy lyase for the decarboxylation of the malic acid into lactic acid and carbon dioxide:

$$COOH-CH_2-CHOH-COOH \xrightarrow{\text{Malate carboxylyase}} CO_2$$
$$+ CH_3-CHOH-COOH. \tag{1}$$

Although this reaction is called fermentation, it is not a true fermentation because it does not produce energy (Margalit, 1997).

The major results of this conversion are an increase in pH of approximately 0.3 pH units, a decrease of TA in the range of 1 to 3 g/L, the production of diacetyl, acetoin, acetic acid, and esters, and the creation of other preferred stylistic characteristics in the finished wine.

Malic acid fermentation is advantageous for low pH wines and in stabilizing (microbiologically) wines with higher pH and low SO_2, which will not be sterile-filtered to remove any remaining bacteria or yeast prior to bottling. The presence of yeast or bacteria in bottled wines with a high pH or low SO_2 can lead to an undesired fermentation, causing cloudiness and carbonation or fizzy wine. Wines that contain sugar are highly susceptible under the above conditions, but dry table wines also contain an adequate supply of carbohydrates to sustain MLF.

Stylistically, MLF is allowed in wine to improve the fruity aroma of chardonnays and the bouquet of white wines. The flavor profile can be altered with flavor components such as diacetyl, which imparts a buttery flavor to the wine. Wines that remain on their lees (French term *Sur lies*) during MLF are noted to obtain a yeasty, nutty flavor and wines that have undergone MLF in barrels can obtain a smoky or spicy flavor. Malic acid fermentation can add complexity and improve the mouth feel (softening, less acetic and harsh) of the wine. Red wine color can be improved simply by the increase in the pH that occurs during MLF. Because most red wines have a higher pH before MLF, they might require an acid adjustment after MLF if the pH becomes too high. Acid adjustments with malic acid are not recommended because commercial malic acid contains a racemic mixture of L (levorotatory) and D (dextrorotatory) malic acid.

In worldwide wine production, around 20% of the white wine undergoes MLF and nearly all the red wines proceed through MLF. Natural MLF (no additional bacteria culture added) can occur in must and juice from LAB found on the grapes and winery equipment and in wooden tanks and barrels. Grapes coming into the winery, on average, contain 10^2 CFU (colony-forming units)/mL of LAB. By the end of harvest, the bacteria population can infect the entire winery unless adequate sanitation practices are met.

Many wineries around the world rely on natural MLF to occur as routine practice and feel that the indigenous bacteria from their vineyards produce wine typical for the area and the varietal. Other wineries want to control if and when MLF will take place and institute strict sanitation procedures to alleviate possible inoculation of juice, must, or wine. Sulfur dioxide is frequently added to juice and must prior to alcohol fermentation to prevent MLF or delay the onset of MLF, especially in red grapes, which typically have a higher pH.

If the wine is to undergo controlled MLF, it will be inoculated with a pure culture of the LAB *Oenococcus oeni*. This species is the most widely used LAB and will produce the greatest amount of lactic acid.

There are several fermentation methods used by winemakers to produce the desired MLF organoleptic and acidic changes in the wine. Wooden tanks and barrels are often used because wood is somewhat permeable to oxygen and allows micro-oxidation, which appears to benefit the *Oenococcus* (classified as anaerobes but microaerophilic). Smaller containers such as barrels provide the growing population with consistency in temperature. Keeping cellars warm will promote a faster fermentation

Concurrent primary and secondary fermentation can be advantageous for LAB due to the decreased alcohol and the increase in nutrients such as B complex vitamins provided by yeasts autolysate. Some yeast strains can compete with the bacteria for nutrients and amino acids, in which case the yeast will always prevail reducing the LAB population. There are strains of yeast that actually produce low levels of SO_2 inhibiting or stopping LAB growth. The production of alcohol levels of 16% v/v or greater will inhibit LAB growth as the end of primary fermentation nears. LAB produce a great deal of acetic acid in the presence of sugars; a delay or increase in yeast lag time can produce high amounts of acetic and lactic acids, which, in turn, inhibit the yeast population.

Secondary fermentation conducted after primary fermentation can be advantageous in the reduction time for the fermentation due to unhindered *Oenococcus* growth and the reduced risk of high acetic acid levels. When the yeast population is low, there is an increased production of diacetyl due to the greater population of *Oenococcus* (Zoecklein et al., 1999). As stated earlier, *Oenococcus* do not thrive in high-alcohol wines; if MLF is desired the alcohol levels must be dropped below 16% v/v before inoculation. Upon completion of primary fermentation, white wines are inoculated with *Oenococcus*, whereas red wines can be racked off their lees and pressed to release all of the wine from the must before they are inoculated. Post-alcohol fermentation is conducted in tanks or preferably barrels.

Sur Lies wines are not racked before inoculation and complete MLF while still on the lees. Lees contact can provide the increased nutrients from yeast autolysate. This method is typical of Burgundy, France.

Carbonic maceration is a concurrent alcohol and ML fermentation using whole unbroken berries. The alcohol fermentation takes place inside the berry while the MLF proceeds outside the berry in an anaerobic environment. This method of fermentation is allowed to proceed naturally with no additions. Once MLF is complete, the berries are pressed to express the wine and/or juice, which will quickly finish alcohol fermentation. Carbonic maceration is a typical method used in the Beaujolais region of France.

6.5.1   Taxonomy and Morphology of Wine Microorganisms

Before we begin to discuss the propagation and development of inoculum for the wine, it is important that the reader understand and be skilled at microscopic identification of the most common bacterium and yeast. Taxonomy is the orderly classification and description of organisms, and morphology studies the form, structure, and the metabolism of organisms. A comparative drawing indicating the approximate size and shape of the most common LAB and yeast is provided in Figure 6.2. This illustration is based on a 1.0-µm scale to assist in the identification of these common microbes. Actual photographs of LAB and yeasts can be found in several enology and microbiology texts. A commercial microbiology laboratory might be able to provide literature, photographs, and additional information concerning wine-related microbes.

6.5.1.1   Lactic Acid Bacteria

Lactic acid bacteria originate from two families, Lactobacillaceae and Streptococcaceae. Species *Lactobacillus* is from the family Lactobacillaceae; they are Gram-positive, nonmobile, nonsporulating, aerotolerant anaerobes, with a regular elongated and rod-shaped cell. The cell diameter ranges between 0.5 and 1.2 µm and the length varies between 1.0 and 10.0 µm. *Lactobacilli* can be microscopically identified by their rod-shaped appearance. They will appear as a single cell, in pairs, or in chains (see Fig. 6.2).

Species *Oenococcus* is derived from the Streptococcaceae family. *Oenococcus* are Gram-positive, nonmobile, nonsporulating, aerotolerant anaerobes, with a spherical or slightly elongated cell. The cell is the smallest of the common LAB with a diameter range of 0.5–0.7 µm and a length between 0.7 and 1.2 µm. *Oenococcus* can be microscopically identified by their round to lenticular shape (similar in shape to a lentil bean). They will appear as a single cell, in pairs, or in chains (see Fig. 6.2).

Species *Pediococcus* is also derived from the Streptococcaceae family. *Pediococcus* are Gram-positive, nonmobile, nonsporulating, aerotolerant anaerobes, with a spherical shape and is never elongated. The cell is spherical with a diameter range of 1.0–2.0 μm. *Pediococcus* can be microscopically identified by their spherical shape and their tendency to form tetrads (arrangement of four cells). They will also appear as a single cell or in a pair, but they do not form chains (see Fig. 6.2).

Lactic acid bacteria are chemo-orgnotrophic, gaining their energy from the oxidation of chemical compounds. In the presence of glucose, LAB can be divided into two groups: hetero and homo fermentors and each ferment glucose differently. Homo fermentors produce 2 mol of lactic acid and 2 mol of ATP per mole of glucose fermented. Hetero fermentors lack the aldolase enzyme (see Fig. 6.3) and use two different fermentation pathways: one that results in lactic acid and CO_2 production and the one that reduces acetyl phosphate to ethanol or oxidizes acetyl phosphate to form oxygen and acetic acid (Zoecklein et al., 1999). Both groups can produce acetic acid in sugar.

During MLF, LAB utilizes three pathways to metabolize malic acid. The first is very direct, producing lactic acid and CO_2; the second involves the breakdown of malic acid into oxaloacetic acid and CO_2, to pyruvic acid, to lactic acid; and the third is the breakdown of malic acid to CO_2 and pyruvic acid, to lactic acid (Margalit, 1997).

6.5.1.2   Yeasts

Yeasts that are of primary concern to winemaking are those used for fermentation and those that create flaws in the wine. The most common yeast used in wine fermentation is from the family Saccharomyces etaceae; genus *Saccharomyces* sensu *stricto*; and species *Saccharomyces cerevisiae*. *S. cerevisiae* is Gram-positive, sporulating, and aerobic, and its reproduction is asexual with multilateral budding. This yeast is spherical to ellipsoidal in appearance, with an average width of 7 μm and length of 8.0 μm, according to Ribéreau-Gayon et al. (2000) (see Fig. 6.2). *S. cerevisiae* can be microscopically identified by its size and elliptical shape and appear as a single cell. Yeast viability can be tested using methylene blue stain because live yeast with intact cell walls will not absorb the dye. Only those yeast cells that have died and lysed will absorb the stain and turn blue. This simple viability test has led to the development of sophisticated equipment and tests that can monitor yeast during fermentation. One of these devices is a microhemacytometer using flow cytometry; it can monitor the viability of yeast.

Section 6.4.1 describes the morphology of yeast fermentation. Please refer to Chapter 9, Section 9.2.20 for information on simple yeast viability testing and counting.

The wild yeast *Brettanomyces* (commonly referred to as Brett), could be a desired or dreaded guest depending on the geographical wine-producing region and stylistic approach. Wines of the Bordeaux region of France typically contain Brett, which can metabolize sugar and produce high levels of acetic acid as well as impart a horsy, wet wool aroma to the wine, which people might, find offensive. Brett is highly contagious and spreads quickly throughout a winery. The off-odor is 4-ethyl phenol, formed as a by-product from the activity of the enzymes cinnamate decarboxylase and vinylphenol reductase with *p*-coumaric acid. The genus *Brettanomyces* is from the family Cryptococcaceae (Ribéreau-Gayon et al., 2000a). *Brettanomyces* are nonsporulating, asexually reproduce by polar budding, and are slightly smaller than *S. cerevisiae*. *Brettanomyces* can be microscopically identified by its ogival shape on one end, resembling a gothic arch, and they appear as a single cell. Occasionally, there is incomplete separation of the cells that will take on a chainlike appearance (Zoecklein et al., 1999) (refer to Fig. 6.2).

6.5.1.3 Spoilage Bacterium

Acetobacter and *Gluconobacter* bacterium use the pentose phosphate pathway to metabolize ethanol and glucose via respiration, producing acetic and lactic acids. These bacteria can greatly increase the level of volatile acid and acetic acid in the presence of oxygen and are generally referred to as spoilage organisms. *Acetobacter* and *Gluconobacter* grow on the surface of the wine, juice, or must. Some *Acetobacter* thrive in warm wine conditions between 30°C and 35°C (86–95°F), whereas *Gluconobacter* prefer cooler temperatures near 20°C (Zoecklein et al., 1999).

The genus *Acetobacter* and *Gluconobacter* belong to the family Acetobacteriaceae. *Acetobacter* contains several species: *A. aceti, A. pasteurianus, A. hansenii,* and *A. liquefaciens. Gluconobacter* contains only one species: *G. oxydans. Acetobacter* and *Gluconobacter* are Gram-negative, aerobic, and nonsporulating and may have cilia for mobility. *Acetobacter* and *Gluconobacter* are ellipsoidal to rodlike in appearance, with a diameter range of 0.6–0.8 μm and a length between 1.0 and 4.0 μm. *Acetobacter* and *Gluconobacter* can be microscopically identified by the structural difference in their cell wall compared to LAB, size, and shape. They appear as a single cell, in pairs, or chains.

Lactic acid bacteria can create spoilage problems. Acrolein production from the dehydration of glycerol can introduce a bitter taste when it reacts with the wine tannins. Mannitol production from glucose metabolism in high-pH, sweet wines can result in unpleasant concentrations of diacetyl, acetic acid, and mannitol. *Pediococcus* and, to an extent, *Oenococcus* might form polysaccharides, producing a viscous and oily character to the wine (often called "ropiness") in low-acid wines. As mentioned earlier, LAB in the presence of sugar can result in excessive acetic acid production.

6.5.1.4 Other Methods of Identification

In this age of technology, the ability to refine the taxonomy of yeasts and bacteria is improving daily. Sophisticated instrumentation uses cellular DNA and RNA to look at the chromosomal difference of organisms in order to enhance identification. Older and simpler identification methods such as the Gram tests, selective culture media, and physical identification still play a significant role, especially in the wine laboratory. More information can be found in Chapter 8.

6.5.2   Prefermentation Analyses

Prior to the start of MLF, it is prudent to analyze the juice, must, or wine for sulfur dioxide levels, acetic acid, TA, pH, ethanol, baseline LMA levels, and RS.

Perform a microscopic evaluation of the juice, must, or wine for spoilage organisms. This is especially important when using juice to create an inoculation culture or to increase the volume of a culture. Lees from actively fermenting wine are often added to a sluggish fermentation to stimulate the LAB population; these lees should always be microscopically evaluated for spoilage organisms. Use of contaminated lees will spread unwanted bacteria into an otherwise good wine.

As stated earlier, the best fermentations with the least amount of problems occur when the conditions are right. Analyses allows the winemaker to make adjustments to the juice, must, or wine to provide the proper fermentation environment.

6.5.3   Oenococcus oeni Culture Preparation

Oenococcus oeni are nutritionally fastidious organisms requiring a complex organic medium, including vitamins, amino acids, and a small amount of carbohydrates. The primary factors that affect the population growth of *O. oeni* are pH, temperature, alcohol content, elevated sugar content, lack of nutrients, and sulfur dioxide levels.

Determine the amount of culture required for the anticipated harvest. Most inoculations are performed near the end or after primary fermentation at the rate of 5–10% and contain a bacteria population (titer) of 10^6–10^8 CFU/mL. The greater the amount of inoculum added, the shorter the fermentation time. Small wine producers might need only a few liters of inoculum, whereas larger wineries might need tens of thousands of liters. Small wineries might choose to purchase their inoculum from a reputable laboratory and forgo enlarging the volume of culture. Larger wineries find that it is more economical either to isolate and grow their own cultures or to purchase initial cultures and expand them into larger quantities (buildup) to meet their needs.

Whether you use pure liquid culture or commercially freeze-dried bacteria is a matter of economics and time. Small liquid cultures are expanded more quickly, usually every 2–3 days but are more labor-intensive compared to starting a culture in a large volume of juice, which could take several days to mature.

6.5.3.1 Culture Buildup

The production of initial cultures requires aseptic techniques to prevent any contamination, including the use of sterile water to hydrate freeze-dried bacteria. All steps require the cultures to be allowed some aeration and CO_2 gas release by using sterile porous plugs (such as foam or fermentation bung) to close the containers. It is important to note that with every expansion in culture volume, the greater is the risk of contamination. It is wise to prepare several different cultures in the event that one culture is contaminated; it is always better to have too much culture than not enough! The buildup of a high-quality inoculum is paramount to a successful fermentation and development of premium wine.

Initial cultures are prepared using sterile nonsulfated grape or apple juice. They are inoculated with *O. oenos* from an isolated bacteria population grown on plates or from commercially freeze-dried bacteria at the rate of approximately 10 g/100 mL. These cultures are placed into small quantities of sterile juice having a sugar content of around 10%, a pH of 5.5 (adjusted with calcium or potassium carbonate, or tartaric acid), rich in the proper growth nutrients (commercially prepared nutrients such as Oenovit™ or yeast extracts), and placed in an incubator at 25°C (77°F). Reduce the sugar content of the juice by diluting with sterile water (bottled water works well) to the desired level. Deacidification of the juice using 0.03 g/L potassium carbonate (KCO_3) will increase the pH by 0.1 units.

Sterile juice and techniques are used to build up the smaller culture into 500-mL cultures (20% buildup). The sugar level of the juice is around 10%, the pH is adjusted to around pH 4.5, growth nutrients are added, and it is incubated at 25°C.

The next step can be accomplished using the enlarged or built-up culture prepared above or freeze-dried bacteria. Prior to using the built-up culture, there are three major steps to assess the readiness, purity, and maturity of the culture. A microscopic examination of the culture is necessary to confirm that the bacteria are growing well and can be seen in pairs or chains (old cultures will have very long chains). The second reason for the examination is to verify the purity of the culture and the absence of spoilage bacteria such as *Acetobacter* and *Lactobacillus*. Any contaminated cultures should be discarded and never used. Third, analysis of the LMA levels will indicate the timing for build up (feeding) of the culture with additional juice or the cultures readiness to be used as an inoculum.

The third buildup step is to increase the volume to 10 liters (5% buildup). The microscopically approved expanded culture above is used, or freeze-dried bacteria can be used. In a large sanitized container (sterile if possible), sterile grape juice is mixed in a ratio of 1:1 with sterile water to drop the sugar level to 50% of the original juice °Brix. The pH is adjusted to 4.0, growth nutrients added, and incubated at 20°C. At this point forward, the addition of a low-titer (10^3–10^4 CFU/mL), low-foaming yeast can be added to the culture to encourage alcohol fermentation of the grape juice (Zoecklein et al., 1999). This method allows culture to slowly acclimate to an alcohol environment that can decrease the lag time when inoculated into dry wine. Temperatures can be maintained in these larger containers using approved submersible heaters (often used in fish tanks).

Step four builds the microscopically approved culture to 200 liters (approximately 52 gal, a 5% buildup). Sanitized and well-rinsed stainless-steel 55-gal drums are filled (leaving room for the 10L of inoculum) with nonsterile grape juice that has been microscopically evaluated for spoilage organisms, adjusted to 18–20 °Brix, growth nutrients and yeast added, and adjusted to a wine pH of 3.6–3.8. The temperature is maintained at around 16°C (61°F).

For larger wineries, further expansion might be necessary, making the next buildup into 1000-gal tanks at a pH of 3.4 units. The same steps as above continue with this and further expansions.

As these volumes of culture grow and metabolize the malic acid, more juice must be added to provide the bacteria with the necessary malic acid and nutrients or the culture must be used as inoculum. Daily analysis for L-malic levels, microscopic examination, and, if possible, acetic and or volatile acid levels of each culture should be made and documented. When the culture has reached an LMA level less than 1 g/L (0.1 g/100 mL), it should be expanded or used as inoculum. Any juice added to a culture for the purpose of buildup or feeding must be approved microscopically, void of sulfur dioxide, include nutrients and yeast, and adjusted to a Brix of 18–20 unless otherwise instructed.

Oenococcus oeni in the stationary phase (chains of five cells or greater) are used for inoculations, most often because of their resistance to lytic bacteriophages (Zoecklein et al., 1999). A culture that has matured to the point where the bacteria have entered the death phase or the VA is in excess of 0.8 g/L (0.08 g/100 mL) should not be used as an inoculum. These cultures can be barreled down and allowed to ferment to dryness or discarded if spoilage bacteria have taken over the culture.

Wine can be inoculated from the culture, or a portion of the juice that is to undergo primary fermentation, can be inoculated for a concurrent fermentation. This wine becomes the inoculum for the remainder of the wine once it is dry. Concurrent inoculation methods require a longer lag time but do well in juice that is to be fermented at lower temperatures. This method of inoculation is more cost-effective for larger facilities despite the risks discussed earlier.

Several manufacturers are now producing a freeze-dried LAB that can be used for direct inoculation of wine without any prior buildup procedures, just a simple hydration step. The freeze-dried direct inoculation bacteria is expensive, but the cost is somewhat offset by the significant reduction of labor costs involved with a build up of culture.

The volume of culture needed and the type of culture used will dictate how many steps it will take to build up pure culture into a mature culture that is near the same pH and temperature of the juice that is to be inoculated. Direct bacteria inoculations into juice without a gradual acclimation to grape juice conditions lead to a very high bacteria death rate. If direct grape juice inoculation is required, a high bacteria titer must be used. Higher bacteria titer inoculations increase the cost of the process and might not always be successful. Normally, it takes approximately 38.4 liters (10 gal) of culture to inoculate 1000 gal of wine at a rate of 10% (Zoecklein et al., 1999).

6.5.4   Monitoring MLF

Once the wine has been inoculated with *O. oeni*, the fermentation is followed on a weekly basis. MLF is a more lengthy process than primary fermentation and can take weeks to reach completion. Routine monitoring of the fermentation includes obtaining a good cross-section representative sample for analyses of LMA and acetic acid or VA levels. Wines that are off-dry require analysis of the RS levels to monitor the completion of primary fermentation. As discussed earlier, off-dry wines inoculated with *O. oeni* can create large amounts of acetic acid and should be monitored more frequently. Weekly monitoring of dry inoculated wine and daily monitoring of off-dry inoculated wine is sufficient to warn the winemaking staff of pending problems created by uncontrolled growth of unwanted organisms.

Determination of free sulfur dioxide, alcohol, RS, acetic acid, VA, TA, pH, and microscopic examination has been discussed previously. Discussion of LMA determination will round out the array of analyses required for fermentation. Please refer to Chapters 5 and 9 for more in-depth information.

6.5.5   L-Malic Acid Determinations

Malic acid is a racemic mixture of L (levorotatory) and D (dextrorotatory) malic acid. During MLF, it is the LMA that is utilized and metabolized into L-lactic acid. Monitoring the LMA levels will assist the winemaker in determining the quality and speed of the fermentation. The point of completion of MLF, or secondary dryness, has not been agreed upon by the winemaking community. Opinions range from levels of 200 mg/L (2 g/100 mL) (Boulton et al., 1999) to 15 mg/L (0.015 g/100 mL) or less. According to Fugelsang and Zoecklein (1993), the majority of winemakers agreed that LMA levels of 15 mg/L in wines to be bottled is considered safe from further MLF.

Methods to quantify LMA include paper chromatography separation for post-alcohol-fermented wines, thin-layer chromatography (TLC), HPLC, electrophoresis, GC, and enzymatic analysis.

For many years, paper chromatography was used by most wineries to indicate the presence of racemic malic acid. This method is somewhat accurate and is sensitive down to 100 mg/L malic acid. Confirmation of dryness requires more sensitive analysis.

6.5.5.1   Paper Chromatography

Paper chromatography is partition chromatography. A sheet of pure cellulose paper is spotted with sample and a control; it is then placed into a special solvent that moves up the paper via capillary action. The various acids in the sample are moved up the paper with the solvent. The analytes separate based on their hydrogen-bonding or water solubility. Acids having a high solubility or hydrogen-bonding moves slower up the paper than those that are less polar, which move faster along with the solvent. Once the solvent reaches the top of the paper, it is removed from the excess solvent and allowed to dry before reading. The ratio of the distance traveled by the analyte spot to the distance traveled by the solvent front is called the relative front (Rf). The Rf ranges have been determined for several acids, including malic, tartaric, citric, lactic, and succinic acids. Malic acid has an Rf of 0.54 (based on malic acid traveling 3.25 in. up a 6-in.-high paper), which means malic acid will travel up the paper slightly over half-way on a 6-in. paper. This method is qualitative, not quantitative. Drawbacks to this method are dealing with the solvents, the 6 hs it takes to run the analysis, analysis of racemic malic acid, limitations of the sensitivity, and inaccuracy of the test. On the positive side, this method has a low cost per test, can follow the progression of MLF, and does not require expensive equipment.

6.5.5.2   Thin-Layer Chromatography

Thin-layer chromatography is based on adsorption chromatography. The method is similar to paper chromatography with the exception of the paper. A thin layer (250 mm) of adsorbent is placed on a plate made from glass, aluminum, or polyethylene. The adsorbent powder and solvent slurry is spread over the plate and allowed to dry. As in paper chromatography, the sample and control is spotted near the bottom of the plate, placed into the special solvent until the solvent reaches the top, dried, and then qualification of the presence of the acids at the corresponding Rf points. TLC has similar positive and negative points to paper chromatography, with an additional advantage of a greater throughput of samples.

6.5.5.3   Capillary Electrophoresis

The separation of organic acids via capillary electrophoresis (CE) is based on differences in the velocity of an electrically charged solute through a capillary tube in an electric field. Smaller highly charged particles will move faster than larger particles. Tartaric, malic, succinic, citric, acetic, and lactic acids (listed in order of the fastest migration times) will move through the electric field at different velocities with the aid of a UV ion solution of 5 mM sodium phthalate buffer at 5.6 pH. An absorbance detector quantifies the amount of solute that passes at different intervals.

Capillary electrophoresis quantifies racemic malic acid and is a precise tool for monitoring MLF. Confirmation of LMA dryness requires enzymatic analysis. The advantages to CE are the relative speed of analysis and the economical cost per test because it does not use many chemicals or enzymes. These systems can be expensive and difficult to operate. The throughput of samples is limited due to the long analysis time, but it works well in smaller wineries.

6.5.5.4   Enzymatic Method

The enzymatic method for determining LMA is similar to most other organic acid enzymatic methods such as acetic acid described earlier in this chapter. As with other enzymatic methods, the samples are read by a spectrometer at 340 nm. Adaptation from a single-sample spectrometer method to an automated instrument, such as the Kone mentioned in Chapter 5, makes analysis of hundreds of samples a day possible. In the enzymatic process LMA is oxidized by NAD in the presence of L-MDH to oxaloacetate:

$$\text{L-Malic acid} + \text{NAD}^+ \xrightarrow{\text{L}-\text{MDH}} \text{Oxaloacetate} + \text{NADH} + \text{H}^+. \quad (1)$$

Oxaloacetate is further broken down to allow for an equilibrium shift away from LMA. The oxaloacetate in the presence of L-glutamate is catalyzed by the enzyme glutamate–oxaloacetate transaminase (GOT) to L-aspartate and oxoglutarate:

$$\text{Oxaloacetate} + \text{L-glutamate} \xrightarrow{\text{GOT}} \text{L-aspartate} + 2\text{-oxoglutarate} \quad (2)$$

The amount of NADH produced directly correlates to the amount of LMA in the sample. This method requires precise measurement techniques and takes some skill to perform correctly. The cost per test can be reduced with batch runs that maximize the analyst's time and supplies as well as utilization of microsampling instrumentation such as the Kone automated analyzer. The accuracy of this test is approximately ±0.05 g/L.

6.5.6   Cessation of Fermentation

Confirmation of the completion of fermentation requires exact analyses for RS and LMA by the most accurate method you have available, or a sample should be sent to an outside laboratory. Once primary and/or secondary fermentations has been completed, it is important to remove or inhibit the yeast and bacteria populations. Sterile-filtering will remove all spoilage microbes, LAB, and yeasts. The addition of SO_2 at sufficient molecular levels will inhibit microbiological activity as long as the molecular level is maintained (refer to Chapter 5).

The wine is usually racked off the lees at this point and returned to clean barrels or tanks. SO_2 is added to protect the wine during maturation. If filtration is scheduled, the wine will be filtered and returned to clean barrels or tanks prior to the addition of SO_2.

Additional analyses is needed at this point in the wine's development to set the baseline for wine maturation:

- Final EtOH/alcohol content
- RS confirmation
- LMA confirmation
- VA
- Acetic acid (AC)
- TA
- pH
- Free sulfur dioxide (FSO_2) after the addition of SO_2 (do a preaddition analysis if there has been sulfur added to the wine, juice, or must)
- Total sulfur dioxide (TSO_2) after the addition of SO_2 (do a preaddition analysis if there has been sulfur added to the wine, juice, or must)

Alcohol content, RS, LMA, VA, AC, TA, and pH will change slightly during maturation but should not change dramatically unless there is a problem. TSO_2 will fluctuate in response to the concentration of FSO_2 in the wine (refer to Chapter 5).

7

Maturation Matters

7.1 Introduction

Upon verification and completion of fermentation, the winemaking and varietal styles will dictate the future of the young wines. The winemaker will create
a plan that will address specific questions: How the wine will be protected? Will
the wine be bottled immediately, as with Beaujolais Nouveau, or matured?
Should the wine be stored in tanks to maintain the fruit flavor or matured in
barrels like many Bordeaux varietals; if so, for how long? Will the wine require
blending, fining, and/or stabilization?

It is important to note that minimal intervention produces a higher-quality
wine and any manipulation of the wine is carried out only to the point where
the actions enhance the maturation process. Wines slated for immediate bottling will be processed quickly (protected with possible stabilization and filtration) with little to no storage time. For wines that are not to be immediately
bottled, the winemaker will initiate a maturation plan. Most often the plan
includes several of the following basic steps:

1. Removing unwanted substances that could adversely affect the wine
2. Placing wine in the proper storage containers
3. Protecting the wine from contamination, further fermentation, and/or
   unwanted oxidation
4. Performing baseline analyses
5. Maintaining wine during storage
6. Clarifying the wine
7. Stabilizing the wine
8. Blending wines
9. Preparing the wine for bottling

Each varietal is treated and processed differently and each winemaker
might treat that individual varietal differently. Chapter 4 includes information about wine varietals and typical regulations concerning production and
bottling. Winemaking styles are numerous and can fill their own book; this

chapter will discuss general winemaking and analyses guidelines for the maturation and bottle preparation of wines.

7.2   Maturation and Aging

According to Boulton et al. (1999), maturation occurs during the bulk storage of wine, whereas aging occurs in the bottle. Maturation is the chemical changes that develop in wine as a result of the environment, activities, and procedures affecting wine quality prior to bottling. These include additions, movements, storage conditions, fining, stabilization, and blending. Bottle aging allows much slower chemical reactions to occur in the bottle, and in some varietals, it results in desirable changes in the wine over time to a point where the quality is optimally improved. Bottle aging will be discussed in the next chapter.

Fruity white wines tend to have very short maturation time and go to bottle much sooner than other white wines, such as Chardonnay. Red fruity wines will go to bottle earlier than other red wine varietals. Red wines can be matured for a few years, such as with a Cabernet Sauvignon, or upward of 40 years in the case of some tawny ports. Typical barrel maturation periods are listed in Table 7.1.

Wine maturation requires an investment of time and money on the part of the winery. Depending on the varietal, higher-quality wines tend to be matured longer than fruity wines, or wines of lesser quality. Wines that have extended maturation time will almost always have a higher price point in the retail market.

7.2.1   Advantages of Maturation

The advantages of maturation in most young wine varietals can be broken down into four categories: subtraction, addition, multiplication, and carry-over (Boulton et al., 1999).

Subtraction is the removal or reduction of undesirable elements in the young wines that might affect the quality. Yeast, lactic acid bacteria (LAB), CO_2, and suspended solids are removed to improve clarity, promote stabilization, reduce gassiness, and reduce microbial contamination of the wine.

TABLE 7.1. Time frames commonly used for barrel maturation of wines.

Dry white wine	0–8 months
Sweet white wine	0–6 months
Dry red wine	0–3 years
Port	0–20 years
Madeira	0–3 years
Sherry	1–12 years

Reductions of acidic or alkaline elements and astringent components can significantly add to the taste and overall quality of the wine.

Addition is the introduction of substances that improve wine aroma, bouquet, flavor, taste, and color. The two most common additions come from oxygen (oxidation) and the extraction of oak compounds from barrels during storage.

Multiplication refers to the development of a more complex flavor profile, which adds depth and breadth to the wine. Duration and temperature of wine storage, maintenance of the stored wine, and blending affect the development of wine complexity.

Carryover is retaining and preserving a young wine's fruity flavors and varietal aromas for bottling. Carryover is accomplished by cold storage conditions of the wine.

The following sections will explore each step in the maturation process.

7.3   Removal of Substances

At the completion of fermentation, there is usually an accumulation of yeast, bacteria, and insoluble solids that has precipitated out of the wine. This precipitant is referred to as gross lees and lies at the bottom of the fermentation container. The wine is most often removed from the lees to stop any further fermentation, improve wine clarity, and prevent any undesirable characteristics from developing in the wine. For some varietals, the winemaker might choose to continue to mature the wine Sur lies; other varietals, if left on the lees, could become bitter.

The wine is carefully removed to another container (racking), leaving the lees in the fermentation container. The lees can be discarded, pressed, or used to stimulate fermentation in a slowly fermenting juice or wine. Depending on the varietal, the wine might be clarified further by filtration to remove yeast and bacteria, or centrifuged to remove excess solids. Lees are not just a result of fermentation, but will continue to form in the storage container as the wine matures and changes chemically. Monomeric phenols (anthocyanins and tannins) polymerize to form larger molecules that precipitate out of the wine over time, particularly during barrel storage. Fermentation produces the largest volume of gross lees, and the amount of lees produced during maturation decreases as the wine becomes more clear and stable.

Carbon dioxide created during fermentation is most often removed during the racking procedure by the agitation of the wine and subsequent release of gas. The removal of CO_2 leaves the wine unprotected from oxidation, requiring immediate addition of an inert gas to protect the wine by displacing oxygen from the wine's surface.

More information on filtration is presented in the latter part of this chapter.

7.4   Storage Containers

Typically, wine is stored in tanks or barrels. The choice of storage container is dependent on the varietal style and the winemaking technique. Wines that will not benefit from micro-oxidation or oak flavor components, further chemical changes, or require closely controlled temperatures will most often be stored in tanks.

Tank storage is much less expensive than barrels. Today, stainless-steel tanks are the most popular tank used in winemaking because of their effective, easy, and reduced labor-intensive cleaning and sanitizing process. Stainless-steel tanks can be constructed to hold very large volumes and can be temperature controlled, and the stainless steel does not react with the wine, making it practical for any size of winery. Tanks come in a variety of sizes ranging from a few hundred gallons to huge tanks with capacities of hundreds of thousands of gallons. Barrels also come in a variety of sizes but hold only a fraction of the volume. The most common barrels (often the French "barrique") used in the United States have a capacity of 55–60 gal (225 liters), with a cost in the range of $400–$700 each.

It is clear that tank storage is very economical and is reflected in the retail price paid for those varietals, such as some Sauvignon blancs, Rieslings, and other fruity wines that benefit from tank storage. The retail price of barrel matured wines increases proportionately to the time spent in barrel.

7.4.1   Cooperage

Cooperage is the product of coopers; coopers make and repair barrels. A winemaker's choice of cooperage can make or break a vintage; understanding all aspects of wood barrel construction is imperative to the winemaker. Volumes of information exist on wood types, forest locations, flavor profiles, manufacturers, styles, and economics. I will only touch on the highlights and recommend that the reader seek other, more in-depth sources.

Most barrels are made from seasoned oak heartwood and are used from 5 or 6 years (French oak) to 10 years (American Oak) depending on the winery. New barrels impart the greatest amount of wood flavor components to the wine within the first year of use and are typically used for maturing Chardonnay and many Bordeaux varietals, which benefit from the added flavor components. Each subsequent year of use results in less extraction of flavor components from the wood. The use of oak wood chips, slats, or blocks placed in tanks, allowing extraction of wood flavor components without the high cost of barrel maturation, has become popular in the past few years.

Winemakers will often use a variety of barrels made from different woods and of different ages to mature their wine. This creates varied taste profiles, which work well when blending, to give the wine more complexity.

Barrels provide the wine with a controlled slow micro-oxidation environment that chemically results in softening tannins, increasing color intensity, and improving stability. Wood consumes oxygen as it passes from the outside layers inward, protecting the wine from excessive oxygen contact. The introduction of oxygen into the wine is thought to be the result of wine evaporation (Boulton et al., 1999).

Barrel maturation promotes the extraction of phenolic compounds from the wood (e.g., nonflavonoid gallic, vanillic, and ellagic acids, which add desirable flavors and bouquet to many wine varietals). During cooperage production, the interior of the barrels can be toasted by an open fire. Toasting decomposes the ellagitannins and polysaccharides on the surface of the wood, which softens the phenol extraction in the new wood (Margalit, 1997). Toasting can be light, medium, or heavy. Light toasts affect the surface of the wood, whereas a heavy toast can penetrate the wood surface 3–4 mm. White wines are most often matured in new toasted barrels to soften the extraction and to add smoky or toasty odors and flavors.

7.4.1.1   Wood

Many different types of wood have been used for cooperage, but the most common wood used is oak, and not just any oak. Of the 300 or more species of oak (*Quercus*), only a very few are used for cooperage. According to Zoecklein et al. (1999), European oak (*Quercus rober*) contributes approximately twice the solids and phenolic compounds (mostly nonflavonoid ellagitannins) than American oak (*Quercus alba*), making it more desirable. Oak forests in Europe can be found in France, Italy, Portugal, Yugoslavia, Scandinavia, and Russia. American oak comes from forests in Kentucky, Missouri, Arkansas, Oklahoma, and Texas. Wood, being a semiporous material, will absorb compounds from the wine and other sources.

Bacteria and yeast can infect the wood and be inadvertently introduced to a clean wine. Sanitation of cooperage should not include chemicals such as sodium hydroxide, citric acid, chlorine, or iodine, which can be transferred to the wine inadvertently. Chemicals are used only when there is a bacterial or yeast infection in the barrel. Chemical cleaning is followed by neutralization of the chemical (usually with citric acid) and thorough rinsing. Some wineries are using ozone to treat infected barrels with great success.

7.4.1.2   Barrel Construction

Barrel size is an important factor the winemaker must consider. The greater the interface of wine to wood, the more extraction that takes place and the shorter the maturation time. Customary cooperage is based on traditional European style. Although there are many styles and sizes of barrel, the barrel styles from France (as with many other winemaking traditions) are most frequently used.

The Bourgogne region of France produces oak barrels called Bourgogne Tradition, which are wider and shorter than the Bordeaux Chateaux barrels produced in the Bordeaux region of France. The volume of the barrels is approximately 225 liters (approximately 59.5 gal). Both regions produce larger barrels for export use.

When working in the cellar, it is important to know barrel construction terminology and identification (see Fig. 7.1).

7.5   Protecting the Wine

There are two preventable situations that can potentially ruin a good wine: oxidation and microbial infection. Excess oxygen exposure drastically reduces the quality of the wine and leads to spoilage and browning. A microbial infection of *Acetobacter* can turn wine into vinegar in a very short time, which is a plus if you are a vinegar producer but a disaster for the winemaker. An untreated growth of the wild yeast *Brettanomyces* can impart an off-odor described as horsey or wet wool to the wine, which many people find offensive at moderate to high concentrations.

There are certain winemaking styles that embrace oxidation and microbial activity to develop desired flavor profiles in their wines, such as with Sherries. For the majority of wines, oxidation and microbial infections are

FIGURE 7.1. Chateaux barrel construction.

avoided at all costs. Daily winery practices will always include methods to prevent oxidation and microbial infection of the wine. The use of sanitized airtight containers, inert gases (to displace oxygen), and SO_2 are the weapons of defense.

Sulfur dioxide should not exceed concentrations that can easily maintain the molecular SO_2 level. High SO_2 concentrations in red wines can slow the phenolic polymerization process, reducing the benefits of maturation to soften the wine (Olsen, 1994).

7.5.1 Oxidation

Oxygen dissolves in the wine and reacts with wine components. Oxygen is an oxidizing agent, causing other compounds to lose some of their electrons, or become oxidized, creating a reduced form of the compound. The compounds that lose electrons and bind with oxygen are termed *reducing agents*. Examples of reducing agents are SO_2, EtOH, ascorbic acid, and phenols.

This ability to gain or lose electrons is defined as the oxidation–reduction potential (redox). Reducing agents lower the overall redox potential of the wine by binding with the oxygen and, conversely, oxygen increases the wine's redox potential. At a certain wine pH, equilibrium is reached where the net reduction equals the net oxidation. A higher wine pH diminishes the redox potential, improving the action of the reducing agents (Zoecklein et al., 1999).

The degree of wine oxidation is determined by exposure, pH, and temperature, as well as the buffering capacity supplied by reducing agents (antioxidants). A wine's oxygen consumption rate is proportional to its total phenol content (Boulton et al., 1999). As the wine becomes oxidized, the phenols decrease and the wine is subject to the detrimental effects of oxygen. White wines contain less phenolic compounds and quickly suffer detrimental oxygen damage.

Wine's exposure to oxygen has positive and negative results. Micro-oxidation is a necessity in red wine production to develop color, soften tannins, reduce astringency, reduce bitterness, and promote stabilization. A few white wines having higher tannin and phenolic levels obtain positive attributes from limited micro-oxidation.

Excess oxygen exposure promotes browning and the development of acetaldehyde; it supports microbial growth and results in the loss of fruit character. For most wine varietals, formation of any one of the above is deleterious. On the positive side, excessive oxidation with the development of acetaldehyde and browning is a stylistic method in Sherry production, which gives a distinct flavor character and appearance to the wines (flor, oloroso, and amontillado). Madeira wines are actually heated to accelerate this oxidation process.

The unintentional introduction of oxygen into wine results from movements and storage of the wine. Racking and moving wine from one container to another introduces oxygen during pumping out and filling of tanks or barrels. The development of headspace in tanks or barrels due to improper sealing, frequent opening, or evaporation will introduce oxygen into the wine.

The use of an inert gas such as CO_2, N_2, or argon is used extensively in wineries to flush out hose lines, tanks, barrels, tankers, and bottling lines to protect the wine from oxygen contact. Nitrogen is used most frequently because it is insoluble in wine and less toxic than CO_2. Fine bubbles of N_2 can be passed through a wine (sparging) to remove excess oxygen or CO_2. The advantages of using CO_2 are the lower cost (dry ice or compressed gas) and, because it is heavier than air, less is needed to form a protective gas layer above the wine surface (blanket).

The addition of SO_2 to the wine at the appropriate time will buffer the wine by combining with oxygen to form sulfite (SO_3) and sulfate (SO_4), lowering the redox potential of the wine. The first addition of SO_2 to a young wine will require a significantly larger addition to achieve the appropriate molecular level. The majority of SO_2 will bind with the oxygen and other components in the wine, leaving less free SO_2 available to maintain proper molecular levels. First SO_2 additions can be increased by 40–50% to compensate for the binding of the SO_2 in some varietals. Verification of the SO_2 addition will be reflected primarily in the total SO_2 analysis, which indicates the bound SO_2 (see Chapter 5). Free SO_2 analysis should be repeated until the targeted free SO_2 level is reached. After the SO_2 addition and the wine is in storage, the free SO_2 will continue to bind to oxygen and other components in the wine. Free SO_2 levels should be rechecked 1–2 weeks after the first SO_2 addition and adjustments made.

7.5.1.1   Oxygen and Wine Color

Wine color is affected by the enzymatic oxidation of phenolic compounds via molecular oxygen. Polyphenoloxidases and laccase are two enzyme groups involved in this oxidation process. Laccase (found on moldy grapes) has a greater capacity to oxidize a wider range of phenolics than polyphenoloxidases.

The white wine phenols responsible for browning are the flavonoid catechins (epicatechin, gallocatechin, and catechin) and nonflavonoid cinnamic acid derivatives (caftaric acid, coutaric acid, and fertaric acid). The nonflavonoid phenols are the primary phenols in white wine. Oxidation followed by polymerization results in the development of a golden brown color.

Young red wines have primarily monomeric anthocyanin pigments, which have a violet red color. As the phenolic compounds are oxidized, anthocyanins polymerize with other flavonoids such as catechins, resulting in a more intense red-brown color. The greatest amount of oxidation and subsequent polymerization of the phenols happens during the first year. As oxidation progresses, the changes in color slows and the color stabilizes.

7.5.1.2   Formation of Acetaldehyde

A by-product of phenol oxidation during maturation is hydrogen peroxide. In a nonenzymatic reaction, hydrogen peroxide will oxidize ethanol in wine to produce acetaldehyde. Other wine components can also be oxidized by the hydrogen peroxide, producing a variety of by-products. Acetaldehyde levels increase as the wine ages.

During barrel storage, anthocyanins and tannins combine in the presence of acetaldehyde and precipitate out of the wine (Zoecklein et al., 1999). Removal of these phenols results in improved wine clarity, color stabilization, and softening of the wine by the reduction of tannic bitterness and astringency.

Acetaldehyde has been found to be an intermediate in the bacterial formation of acetic acid in wines containing lower alcohol levels and more oxygen contact. Acetaldehyde can accumulate, rather than being oxidized into acetic acid, in wine with low oxygen levels and/or alcohol levels greater than 10% (Zoecklein et al., 1999).

When excess acetaldehyde is present, blending might be required to reduce the excess. Additions of SO_2 are also helpful because it binds easily with acetaldehyde.

7.5.2   Microbial Spoilage

Aerobic micro-organisms are responsible for a large percentage of wine spoilage in stored and bottled wines. Exposure to oxygen allows unwanted bacteria and yeasts to thrive, especially in wines containing residual reducing sugars that are not protected with SO_2. The primary culprits are *Acetobacter aceti*, LAB, and *Brettanomyces* yeast. Additional information can be found in Chapter 6.

The control of microbial spoilage is prevention through proper storage containers that reduce oxygen exposure, adequate additions of SO_2 to maintain a molecular level of 0.825 mg/L (Chapter 5), and good sanitation practices. Continual monitoring of the wine's chemistry and plating of the wine will reduce serious spoilage issues by alerting winemakers to potential hazards.

Filtration might be required to eliminate infections upon detection during maturation and/or before bottling. The use of potassium sorbate at a dosage of 200 mg/L is often used to control microbial infections. Not all countries allow the use of sorbate and most will regulate the allowable limits.

More information about microbiological identification and testing can be found in *Bergey's Manual of Systemic Bacteriology* (Williams & Wilkins, 1986) and *Bergey's Manual of Determinative Bacteriology*, 9th ed. (Williams & Wilkins, 1994).

7.5.2.1   Acetic Acid Bacterial Spoilage

Acetic acid bacteria are obligate aerobes that are catalase positive (enzyme that decomposes hydrogen peroxide into water). These bacteria produce acetic acid from the metabolism of glucose and EtOH, which can turn wine into vinegar in a short time. *Acetobacter aceti* (vinegar bacteria) is ethanol tolerant up to approximately 15% v/v and commonly used to produce vinegar commercially. *Acetobacter aceti* is widespread throughout the majority of wineries and poses a real danger to stored wines, not only for the potential detrimental results of increased acetic acid but also acetobacter because produces ethyl acetate as a by-product of metabolism. Acetic acid bacteria have been associated with a flaw called ropiness (increased viscosity) in some wines (Boulton et al., 1999).

Acetic acid bacteria are fairly resistant to molecular SO_2, requiring much larger concentrations to control the growth. They grow best at a pH above the range of most wines, with their lower tolerance limit at a pH of 3.0 (Margalit, 1997).

Close monitoring of acetic acid levels during maturation (see Chapter 6) will help prevent unnecessary damage to the wine. If filtration of the wine is planned, sterile-filtration using a 0.45-μm porosity membrane will remove the bacteria from the wine. Filtration is covered later in this chapter.

7.5.2.2 Lactic Acid Bacterial Spoilage

Chapter 6 outlined the importance of LAB to complete secondary fermentation for the reduction of malic acid to lactic acid and to impart certain flavor components (diacetyl) to the wine, but in a postfermented wine, LAB infections and fermentations can cause severe problems. The overproduction of these bacteria can lead to spoilage problems, especially in wines that contain sugar and have a pH greater than pH 3.3. Increased turbidity, development of haze and sediment, changes in viscosity and color, and sensory changes can occur in wines with LAB spoilage.

Oenococcus oeni and *Pediococcus* are common infections in unprotected wines, mainly because of their ubiquitous availability. Being microaerophilic, these bacteria can exist with very little oxygen and are catalase negative. Some species of lactobacilli can exist in EtOH concentrations in excess of 20% v/v (Boulton et al., 1999). Infections can occur in the bottle or in stored wines that are susceptible, resulting in unwanted malolactic fermentation. One exception is a wine from northern Portugal called Vinhos Verdes that derives its uniqueness from LAB fermentation in the bottle.

Unwanted LAB infections can result in excess diacetyl production and the production of hydrogen sulfide. Diacetyl is a by-product of LAB metabolism, and in lower quantities, such as those found after secondary fermentation, imparts a buttery flavor to the wine, but in excess, it is considered foul.

Hydrogen sulfide (H_2S) develops during fermentation and is the result of the breakdown of amino acids into proteins, which are further broken down into a nitrogen source used by the yeast. Sulfate, sulfite, and elemental sulfur in the must are reduced by the yeast to sulfur. H_2S imparts a rotten egg, sulfur odor to the wine that can be unpleasant at any detection level. Lower levels of H_2S spoilage can be corrected by slight aeration of the wine, but higher concentrations will require the wine to be blended to the point that the H$_2$S is undetectable or chemically treated to correct.

Lactobacilli and pediococci have been associated with a stringy microbial growth in wines called ropiness. Pediococci spoilage imparts an unpleasant vegetative odor to the wine (especially when it occurs in the bottle).

Monitoring sugar levels (Chapter 5), routine organophilic evaluation, and plating of the wine can provide early detection of possible infections. LAB are very sensitive to SO$_2$ and are easily controlled. Sterile-filtration is an additional tool to eliminate the bacterial threat, but only used when necessary. The use of potassium sorbate at the recommended dosage will not affect LAB; they are resistant.

7.5.2.3 Yeast Spoilage

In most wineries, yeasts are indigenous. Wild yeasts come in on the grapes, whereas cultured yeasts are selectively introduced into the juice; both promote fermentation. Oxygen stimulates yeast fermentation (glycolysis), as described in Chapter 6. In an oxygen environment, yeast can exist in a postfermented wine by feeding off the EtOH, residual reducing sugar, wood cellulose, or hemicellulose fragments (Boulton et al., 1999).

Film yeasts exist on the surface of the wine. This film is created by several wild yeasts species such as *Pichia*, *Candida*, *Hansenula*, and other oxidative yeasts. These yeasts can survive on EtOH as their food source in low SO$_2$ wines. Infections can lead to excess production of acetaldehyde (as in the production of flor sherry), acetic acid, and ethyl acetate. Removal of the film, SO$_2$ additions, and proper storage can correct the infection. Oxidative yeasts do not grow well at cellar temperatures less than 12°C (54°F) (Zoecklein et al., 1999).

Wines containing residual reducing sugar and/or low-molecular SO$_2$ can be candidates for wine yeast (Chapter 6) spoilage during storage or in the bottle. *Zygosaccharomyces*, *Saccharomyces*, and *Brettanomyces* contaminations can occur in stored or bottled wines, creating flocculation, granular deposits, off-odors, or increased volatile acidity.

Species of *Zygosaccharomyces* (commonly found in grape juice concentrate) can be resistant to SO$_2$ at the 0.825-mg/L molecular level and will require higher levels to stop their growth. Most contamination occurs in the bottle resulting in a granular deposit.

7.5.2.3.1 Brettanomyces. The yeast creating the most concern in wineries is *Brettanomyces*, the nonsexual, nonsporulating form of *Dekkera*. There are nine different species of *Brettanomyces*, two of which are found in wine: *B. intermidius* and *B. lambiscus*. *Brettanomyces* (Brett) produce the enzyme B-glucosidase, which reacts with sugar cellobiose found in wood to produce the glucose molecules the yeast feed on (Olsen, 1994). New cooperage contains higher levels of cellobiose.

Brett infections in wine will often present with increasing volatile acidity (VA) levels (due to the oxidation of ethanol) and a distinct odor often described as wet dog, horsey, or barnyard (to name a few). Brett metabolism creates the volatile phenols 4-ethyl phenol and 4-ethyl guaiacol, which are known to contribute to this odor. The 4-ethyl phenol is the phenol most associated with the odor and could come from the enzymatic decarboxylation and reduction of hydroxycinnamic acid (Boulton et al., 1999). Brett spoilage of wine comes primarily from the increased VA, but some people regard an excess of the odor as offensive and, therefore, a flaw. Many Bordeaux-styled wines contain higher levels of 4-ethyl phenol, which is felt to add to the wines complexity.

Detection or verification of Brett infections requires microbiological plating of the wine on media prepared with the fungicide cycloheximide and/or determination and quantification of 4-ethyl phenol. Brett is resistant to cycloheximide, making selective plating a good method of identification. They also have a distinct shape and can be identified microscopically (Chapter 6). The growth cycle of Brett peaks at 6–10 months from the onset and then rapidly declines. Plating allows for identification and estimation of colony-forming units (CFU), which, if done early and followed over time, can indicate where the infection is in its growth cycle. Plating cannot tell you the levels of 4-ethyl phenol the yeast has produced and how close to human detection levels it is. Quantification of 4-ethyl phenol is much more involved and requires advanced instrumentation and highly qualified technicians to perform the assay.

Chemical extractions are performed to obtain a small wine sample that is then analyzed via a gas chromatograph–mass spectrometer detector (GC/MSD) and quantified. Wine laboratories most often do not perform extractions because of the odors produced and the safety and health hazards associated with the chemicals. With new technology, these laboratories can now accurately analyze and quantify 4-ethyl phenol to levels less than 10 µg/L (ppb) without extractions.

Stir bar sorptive extraction (SBSE) uses a technique called solid-phase microextraction (SPME). A stir bar is coated with polydimethyl-siloxane (PDMS) sorbant, which develops an equilibrium between the coating and sample, extracting different analytes. This stir bar is placed in the wine sample and stirred for a specific amount of time, removed, rinsed, and then placed in a sampling vial. The vial is introduced into a device where the stir bar is hyperheated and thermally desorbed, extracting the volatile substances. The volatiles are quickly introduced into a subzero chamber (cryotrap) that efficiently collects the volatiles and, in turn, sends them into the gas chromatograph column (Chapter 5), where the analytes are separated and quantified.

Although this is a very expensive piece of equipment, the ability for early detection of a Brett infection and the ability to quantify the level of 4-ethyl phenol can be well worth the investment for larger wineries. The

SBSE/SPME technology can also be applied to the detection of 2,4, 6-trichloroanisole (TCA) found in corked wine, which will be discussed in Chapter 8.

7.5.2.4   Conclusion

Most semidry (off-dry) wines are stored or bottled with higher free SO_2 concentrations to maintain increased molecular SO_2 levels to discourage contamination by these yeasts. Brett is more susceptible to SO_2 and levels of 0.5 mg/L molecular SO_2 can control their growth (Olsen, 1994). Analyses for free SO_2, acetic acid, VA, titratable acidity (TA), pH, and organophilic evaluation should be conducted routinely.

Larger wineries having microbiological evaluation capabilities can choose to routinely plate wine samples to detect any possible contamination or to verify contamination, especially in postbottled wines and topping wines. Unfortunately, current plating techniques take 10–14 days of incubation time before they can be analyzed, which might be problematic for many wineries. New techniques using DNA identification are being tested in the wine industry to allow for faster identification of bacteria and yeasts (see Chapter 8).

Filtration using membranes with a minimum pore size of 0.8–1.1 µm prior to bottling or during maturation will eliminate yeast from the wine (Boulton et al., 1999).

7.6   Maintenance of Maturing Wines

This is the point in the life of a wine when most winemakers sigh with relief. Harvest and fermentation are finished, the wine has been racked and sulfured and placed in good clean cooperage, and now time can work its magic. The wine stored in barrels will begin to soften with reduced bitterness and astringency, the color will intensify and become more stable, and extraction of phenolic and flavor components from the oak will take place. Wine stored in tanks will be waiting for their preparation for bottling while maintaining their characteristics with little to no changes. Things are good.

The average time a wine is held for maturation is 12 months (Boulton et al., 1999). Seasoned winemakers know that this seemingly tranquil time can, in fact, be brewing a load of trouble. Poorly sealed tanks and barrels can lead to oxidation and microbial infections, oxygen can be introduced through wine movements, cellar temperatures can accelerate or slow maturation, reductions in molecular SO_2 levels could allow microbial infections, or any of a dozen more problems can develop.

Winery cellars are constantly busy topping, racking, cleaning, moving, shipping, making additions, and much more. When the cellar is busy, so is the wine laboratory, providing needed information to the winemaking team.

7.6.1 Barrel Storage

During barrel storage, the water and alcohol in wine will evaporate, increasing the concentration, color, flavor, dry extracts, and lowering aroma. According to Margalit (1997), routine barrel storage results in an evaporation rate of 2–5% each year. This equates to a loss of 4.5 liters (1.2 gal) to 11 liters (2.9 gal) in a 225-liter (59 gal) barrel.

The loss of this volume will create a vacuum in a tightly bunged barrel, which can account for the introduction of oxygen and reduction of molecular SO_2. Water will diffuse at a faster rate than EtOH due to the smaller size of the molecule. At a relative humidity level of 74%, the loss of EtOH and water are in equilibrium. Decreases in humidity from equilibrium will result in the loss of more water than EtOH and vice versa. The winemaker might choose to replace the lost volume to reduce headspace; this is called topping.

As the wine matures, primarily red wines, the polymerization and precipitation of phenolic compounds continues, resulting in the development of lees at the bottom of the barrels. This polymerization of phenolic compounds brings about clarity of the wine and color intensity and stabilization. Depending on the varietal, winemakers might choose to remove the wine from the lees, this is called racking.

7.6.1.1 Topping

Topping is a procedure in which wine from the same lot is used to fill the headspace in a barrel. The frequency of topping depends on the wine varietal and the winemaking style, in addition to the evaporation rates found in the individual cellars. Some white varietals are topped monthly, whereas certain red varietals might never be topped.

During a topping procedure, care is taken to use only wine that is free of contamination and flaws. Most winemakers will take this opportunity to adjust the SO_2 levels at the same time, to avoid another invasion of the wine at a later date and to ensure that the wine has been protected after the barrel opening. Analyses for pre-SO_2-addition should include free SO_2, TA, pH, VA or acetic acid, alcohol, and possible plating for microbial infections. Measurements for free SO_2 and pH are performed to determine the amount of SO_2 required to maintain adequate molecular SO_2 levels. The free SO_2 should be reanalyzed and total SO_2 measured after the addition to verify that the correct amount of SO_2 was added. Acidities and plating results are monitored to indicate spoilage. Changes in alcohol levels can indicate significant dilutions of the wine.

Tanks are also topped when they are filled to prevent oxidation. Unlike barrel storage, tanks are not subject to evaporation.

7.6.1.2 Racking

Racking involves moving the wine from the barrel or tank to another container, usually a sanitized tank. Racking off the lees refers to removing the clear wine from the lees that have settled in the bottom of the barrel or tank. The barrels are cleaned and the wine is analyzed and receives any needed adjustments and is returned to the barrel. The terms *rack and return* and *rack and back* are often used. Great care is taken to prevent exposure of the wine to oxygen by using inert gases while transferring, filling barrels, filling headspace, and any other movement that could introduce oxygen. The use of a dissolved oxygen meter is frequently used in the cellar to alert the staff to possible oxygen problems.

Analyses for pre-SO_2-addition should include free SO_2, TA, pH, VA or acetic acid, alcohol, and possible plating for microbial infections. Measurements for free SO_2 and pH are required to determine the amount of SO_2 required to maintain adequate molecular SO_2 levels. Free SO_2 should be reanalyzed and total SO_2 measured after the addition to verify that the correct amount of SO_2 was added. Acidities and plating results are monitored to indicate spoilage. Changes in alcohol levels can indicate significant dilutions of the wine.

Tanks are racked off their lees to other tanks, or racked to barrels.

7.6.1.3   Cellar Temperature

Typical cellar storage temperatures are approximately 13°C (55°F) for white wines and 15°C (59°F) for red wines (Bird, 2000). Low wine temperatures increase the amount of carbon dioxide that goes into solution, which can result in a spritzy or prickly sensation in the mouth as the wine warms and releases the CO_2. Wines that have had an addition of CO_2 as protection from oxygen will lose that barrier as the CO_2 goes into the solution. Oxygen also moves readily into the solution in cold wines. Wines stored to maintain their varietal character are usually stored in temperature-controlled tanks, allowing for lowering of temperatures. The use of these tanks is also beneficial when wine is being cold-stabilized, which is discussed later in this chapter.

As we know, warmer temperatures accelerate chemical reactions and improve microbial growth. Warm cellar conditions will accelerate maturation and can enhance the loss of undesirable flavors (Boulton et al., 1999).

7.7   Organoleptic Evaluation

Laboratory analyses are crucial in winemaking, but just as crucial are the eyes, mouths, and noses of the winemakers and the people who work with the wine. Routine organoleptic evaluation of the wine can alert the winemaking team to the early development of potentially large problems. It is critical that the wine laboratory staff be well trained in the identification of odors, tastes, and mouth feel of wines that could be heading for trouble.

Development of excellent sensory skills is a valuable tool for the winemaker and the millions of people who love wine. Organoleptic evaluations are conducted to identify the good, bad, and the ugly components in wine. This section will only concentrate on those elements that help identify troubled wines. My advice to the reader is to evaluate as many wines as you can and consult some of the hundreds of books on the market that describe organoleptic identifiers and descriptors. A handy tool for the laboratory is a large picture of the aroma wheel developed in 1987 by Ann Noble at the University of California–Davis.

7.7.1   Organoleptic Terminology

Odors are molecular compounds that are volatile. The volatile compounds stimulate the odor receptors in the nose and we perceive the odor. Odors in wines and juices can change when there is a shift in the redox potential. This shift could produce a more volatile form of the compound and change the odor radically. A human's sense of smell and ability to identify odors is far, far greater than their sense of taste.

Wine odors are divided into aroma and bouquet. *Aroma* is the odor derived from the grapes that give the wine their varietal aroma. Odors generated from the processing, maturation, and aging procedures are termed *bouquet*.

The *taste* of wine refers to the tactile response of the tongue to sweetness, acidity, salt, and bitterness (Fig. 7.2). *Astringency* is a feeling sensation in the mouth often accompanied by bitterness and tartness. Astringency causes a puckering, dry feeling felt throughout the mouth. *Flavor* encompasses the tastes, odors, and mouth sensation. *Body* refers to the wine's temperature, taste, and the full or round feeling in the mouth created by the balance of sugar, alcohol, and dry extracts (Margalit, 1997). *Finish* is the final taste left in the mouth after spitting or swallowing.

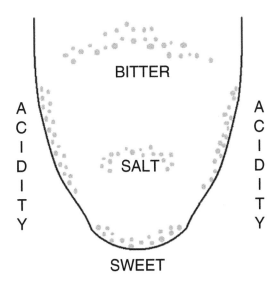

FIGURE 7.2. Tactile diagram of the tongue.

7.7.2   Wine Sensory

Wine odors range from fruity and floral to oxidative and barnyard; tastes range from jam to vinegar. These descriptors are used to help with the organoleptic evaluation of the wine. Microbiological contamination, chemical reaction by-products, oak extracts, and grape varietals all have their distinct flavors in wine. The point at which these flavor elements are perceived by the evaluator (taster) is termed *threshold*. Thresholds are different for every component found in wine and for each tester. One component might produce a strong odor or taste perceptible to the taster at a very low concentration—a low threshold. Conversely, other compounds might require quite high concentrations before an odor or taste can be perceived in the wine—a high threshold.

Monoterpenes, which exist as esters, aldehydes, alcohols, and ketones, are responsible for the floral aroma found in wine and fruit. Pyrazines are primary aromatic compounds responsible for vegetative aroma in wine, fruits, and vegetables.

Phenolic groups contribute greatly to flavor profiles in wine. Flavonols from grapes, such as the catechins and epicatechins, form tannins, which contribute to the bitterness and astringency of the wine (Chapter 5). Nonflavonoid phenols from grapes and those extracted from oak barrels can also contribute to astringency (Margalit, 1997).

Oxidation, microbial contamination, additions, and production processes will add their own elements to the flavor profile—some good, some not so good.

7.7.2.1   Flavor Identification

Table 7.2 list common components in wine associated with wine spoilage and their organoleptic descriptors. Wines should have frequent organoleptic evaluations conducted throughout their evolution. Table 7.3 lists common varietals with their familiar descriptors.

7.8   Fining

Fining is the deliberate addition of a reactive or adsorptive substance to the wine for the removal or reduction of one or more undesirable components, including the removal of unstable colloids (particles too small to remove by filtration), color, certain phenolics, and off-flavors. Fining is performed to enhance the wine's clarity, color, aroma, flavor, and stability (Zoecklein et al., 1999).

Electrical charge interaction, bond formation, and/or absorption and adsorption are actions that take place during fining (Zoecklein et al., 1999). Fining agents will attract components of the opposite charge, forming larger heavier molecules that precipitate out of the wine and are then removed by

TABLE 7.2. Wine component sensory descriptors.

Origin	Descriptor
Bacterial	Vinegar, sauerkraut, sweaty, buttery, acetone, mousey, vegetal
Yeast	Yeasty, mousey, horsey, barnyard, wet dog, rotten eggs, mushroom
Molds	Mildew, musty
Oxidation	Acetaldehyde, sherry, overripe apples
Alcohol (EtOH)	Hot, burning, sweet
CO_2	Spritzy, prickly; threshold 500 mg/L
SO_2	Burnt match, wet wool, skunk, cooked cabbage, sharp, pungent
Sorbate	Soapy, fishy
Acetaldehyde	Oxidized, sherry, nuts, overripe apples; threshold 100–125 mg/L
Acetic acid	Vinegar; threshold 1.2 mg/L in white wine and 0.8 mg/L in red wines
Brettanomyces	Horsey, barnyard, wet wool, tar, tobacco, mousey; threshold 1.0 ng/L (4-ethyl phenol)
Diacetyl	Buttery; threshold 0.1 mg/L
Ethyl acetate	Band-Aid, acetone; threshold 150–200 mg/L
Ethyl mercaptan	Skunk, burnt rubber, rubber tires; Threshold 1 µg/L
H_2S	Rotten eggs, boiled eggs; threshold 50–100 µg/L
Lactic acid	Sauerkraut, sweaty, vinegar esters, milky, yogurt
Lactobacilli	Vegetal
Oenococcus	Buttery
Pediococci	Vegetal, stinky feet
Sorbate and Oenococcus	Geranium
TCA	Mildew, musty; threshold 1.4–4 ng/L

TABLE 7.3. Wine varietal descriptors.

Varietal	Descriptor
Sauvignon blanc	Citrus, mango, melon, cut grass, cat urine, bell pepper
Chardonnay	Peach, citrus, apple, pineapple, banana, spice, vanilla, butter, mineral
Riesling	Floral, green apple, citrus, pineapple, peach, honey, petroleum, pine resin
Pinot noir	Cherry, strawberry, raspberry, oak components, rhubarb, cola, black tea, spice, earthy, leathery
Merlot	Black cherry, cherry, blackberry, weedy, plum, chocolate, oak components
Zinfandel	Jam, blackberry, black cherry, black pepper, spicy, chocolate, alcohol
Syrah	Black cherry, plum, blackberry, black pepper, violets, gamey, raw meat, smoke, tobacco, leather, earthy
Cabernet Sauvignon	Current, cherry, bell pepper, asparagus, cassis, mint, black pepper, oak components, cedar

filtration or racking. Adsorption is the adhesion of a molecular substance to the surface of a liquid or solid, which makes the liquid or solid larger and heavier, allowing it to precipitate out of solution. Absorption is the taking in of one substance by another, again, making it larger and heavier, allowing it to precipitate out of the wine.

Fining agents are processing aids, not ingredients, that act on the surface of the component to be removed. The agents are removed via precipitation with the component or filtered; they do not remain in the wine. Common fining agents are as follows:

- Earths
  Bentonite containing alumino-silicate clay, slight negative charge, adsorptive
- Proteins
  Albumen found in egg whites, positive charge, hydrogen-bonding
  Gelatin found in bones and hides, positive charge, hydrogen-bonding, isoelectric point (PI) pH 4.7
  Isinglass from sturgeon bladders, positive charge, hydrogen-bonding
  Casein found in milk, positive charge, adsorbs, hydrogen-bonding
  Bovine serum albumen banned in the United States and banned in most European countries because of bovine spongiform encephalopathy (mad cow disease)
- Carbon
  Activated carbon (charcoal), adsorbs
- Polysaccharides
  Alginate from marine brown algae, positive charge, bonds to a carrier, facilitates settling
- PVPP
  Synthetic molecule polyvinylpyrrolidone, hydrogen-bond with low-molecular-weight phenolics

Colloidal material carries an electrostatic charge that repels other colloidal material, keeping them well spaced and invisible to the eye. As the colloidal material ages, they lose their charge as a result of proteins being denatured and the molecules rearranged (Bird, 2000). The loss of this electrostatic charge allows the colloidal components to come together, forming visible groups that creates the haze and cloudiness in wine.

Removal of colloidal material such as protein is dependent on the isoelectric point (PI) of the protein. PI is the point where positive and negative forms of the component in question are in equal concentrations, making them electrically neutral (carrying no net charges) and least soluble. Many components, including proteins, have a PI in the pH range of wine (pH 3.0–4.0). Colloids with a PI close to, but above, the wine pH range will have limited solubility and carry an overall positive charge, and those with a PI close to, but lower, than wine pH will have limited solubility and carry an overall negative charge.

Lowering the wine pH prior to fining will increase the amount of positively charged colloidal material. The use of a fining agent with a PI lower than the wine's pH, thus negatively charged, improves the effectiveness of the fining agent to remove the positively charged colloidal material. The pH can be adjusted after fining.

Removal of partially soluble colloids in white wine improves clarity and prevents instability that, under certain conditions, might cause these substances to precipitate out of solution, creating a haze or cloudiness. Colloidal material tends to adhere to tartrate salts and tannin precipitate causing temporary inhibition of tartrate crystal growth in wine. Wines should always be fined before beginning tartrate stabilization. Clarity and stability are achieved by the addition of fining agents, including carbon, but primarily bentonite is used. Gelatin is commonly used to improve white wine clarity, and the addition of isinglass can bring a white wine to brilliance. Tannin or tannic acid can be used with gelatin to enhance clarification (regulated by the Alcohol and Tobacco Tax and Trade Bureau (TTB)).

Tannins in red wines can be removed to reduce the bitterness or astringency, improving wine balance. Tannins carry a negative charge, and positively charged proteinaceous agents such as gelatin, albumen, and isinglass can be used to remove tannins and other phenols. Proteinaceous fining agents have an affinity for polyphenols by hydrogen-bonding between the phenolic hydroxyl and the carbonyl oxygen of the peptide bond (Zoecklein et al., 1999). Monomeric and small polymeric phenols can be reduced with additions of polyvinylpyrrolidone (PVPP). White wines containing tannins may use Kieselsol with gelatin, PVPP, or casein to remove bitterness. Casein is also used to remove excess oak components in white wine. The maximum allowed addition of casein in the United States is 0.2% v/v.

Certain winemaking procedures could require the reduction of color in a wine. Carbon and proteins such as gelatin and casein are the agents of choice. Carbon adsorbs molecules containing benzene rings of weak polarity, such as some phenolic compounds. Carbon has a huge surface area for this purpose. Casein is used in white wines to remove browning and reduce pinking by adsorption and precipitation.

Copper sulfate or carbon is sometimes added to wines for the reduction of off-odors. Isinglass used in white wine can bring out the fruit character without considerable changes in the tannin level. Yeast fining is used to remove herbaceous odors and other off-odors. PVPP has an affinity for catechins and anthocyanins and can be used to reduce browning in young red and white wines.

Casein and yeast fining are employed to reduce copper and iron content.

Regardless of the circumstances, laboratory-controlled fining trials should always be conducted prior to any fining procedure. Various amounts of the fining agent are placed in aliquots of wine and processed according to the instructions for that fining agent. Physical, organoleptic, and laboratory examination of each sample is conducted. The point at which satisfactory fining has taken place without detrimental effects to the wine will indicate the amount of fining agent to add to the entire volume of wine. The volume of fining agent used in the acceptable sample is recalculated for the addition to the larger volume of wine. For example, a laboratory addition of 1.6 mL of a 2% casein solution, made up of 5 g NaC_3 and 20 g/L casein, into 1 liter of wine equates to 3 g/hL (100 liters, 26.4 gal) or 0.25 lb (113.64 g)/1000 gal addition of casein to the larger volume of wine (Zoecklein et al., 1999). Figure 7.3 illustrates two ways to look at the same additions: the amount of 5% w/v bentonite slurry and how that volume equates to pounds per 1000 gal additions.

The most frequently used fining agents in a production winery are bentonite, albumen, gelatin, isinglass, and copper sulfate. Fining agents should be handled carefully and used judiciously. Using the wrong agent or too much of an agent can strip wine flavor and color, generate large volumes of lees, alter the charge of the agent, or leave a residue that might require a second fining.

7.8.1 Bentonite Fining

Bentonite is the most popular fining agent for removal of colloidal materials in wines (primarily white wines) because of its affinity for amino acids. Colloidal material moves in and out of its liquid and solid phases (point of

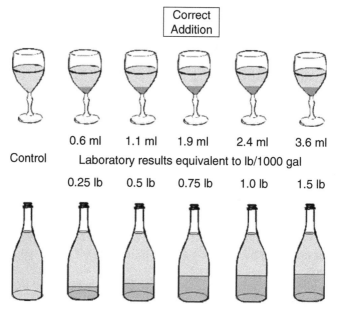

FIGURE 7.3. Fining trial exact results using two different labeling systems. laboratory samples labeled as ml/L addition of 5% bentonite slurry and its equivalent labeled as lbs/1000 gal. The correct addition in this example is 1.9 mL, or 0.75 lb/1000 gal bentonite.

equilibrium, adsorption isotherm). When the colloidal material reaches equilibrium, the resulting concentration will remain in the wine after fining. Negatively charged bentonite clay particles (and ion-exchange resins) have a fixed number of adsorption sites on each particle. As the bentonite particles become saturated with the colloidal material and the adsorption per unit reaches its maximum, the adsorption becomes independent of the colloidal concentration. In wines with a low colloidal concentration, the binding is proportionate to the concentration, but much less adsorption takes place. Longmuir's equation explains this phenomenon (Boulton et al., 1999).

The majority of colloidal material will be removed from the wine in the first few minutes; the wine should not stay in prolonged contact with the bentonite because it could result in a portion of the proteins moving back into the wine. Diatomaceous earth filtration is recommended after fining with bentonite to clarify the wine (discussed later in this chapter).

Sodium bentonite found in Wyoming (United States), has twice the protein exchange of the calcium bentonite found in Europe. Bentonite dust can be harmful and personal protection is advised. Bentonite can cause problems such as sealing percolating ponds and developing poor compaction of sediment.

Most sources recommend making a 5% bentonite slurry prior to addition to the wine. The slurry is made by slowly mixing the measured bentonite with hot water and allowing it to sit and absorb the water. The optimum absorption time is 2 days (Zoecklein et al., 1999) prior to additions to laboratory samples or the main wine volume. The slurry should be slowly added to the wine while stirring constantly. Typical bentonite concentrations added to wine are 0.2–1.5 g/L. Additions of 5% per liter w/v bentonite slurry directly equates to grams per hectoliter (refer to Fig. 7.3).

As mentioned earlier, fining trials should be conducted under laboratory controls. Bentonite fining trials consist of additions of slurry to equal aliquots of the same wine and stirring for a few minutes. It is important to stir the trial samples at a speed similar to the stirring method used in the cellar to stir the larger volume or wine. The 5% slurry preparation for laboratory fining trials is made by adding bentonite to 1 liter of hot water (12 g/L), or 9 g/750 mL, or 4.5 g/375 mL. The wine samples are sterile-filtered and a portion of each sample (including a control) will be heat tested to determine protein stability.

Baseline nephelometric readings are taken using a nephelometer, which employs a strong light beam at a 90° angle that is passed through a glass tube containing the sample. The nephelometer detects and measures the amount of scattered light. A higher percentage of solids to sample volume will scatter light to a greater extent, resulting in higher nephelometer readings (NTU). Following the reading, the samples are placed in an 80°C heat bath for 6 h, after which they are removed and allowed to come to room temperature. Heat testing is the thermal denaturizing of proteins. Prolonged heat with cooling will stimulate the colloidal proteins to combine and fall out of the solution, creating increased haze or cloudiness.

A second nephelometric reading is taken on the room-temperature samples; the difference between the first and second readings will indicate the presence of protein instability. Wineries have different acceptable levels of nephelometric changes, but most agree on a change (Δ) less than 1.5–2.0 NTU in most white wines. In addition, it is important to physically look at the individual wine samples under an intense light source to detect any haze and, if so, to what extent.

The results of the heat test, the organoleptic evaluation of the samples, and the physical appearance of the wine are all taken into consideration to determine the level of bentonite additions required to stabilize the wine. Additional information on fining trials can be found in Chapter 9.

7.8.2   Albumen Fining

Albumen (egg white) fining to remove primarily tannins in red wines and a very few white varietals is used extensively throughout most wineries producing quality red wines. Egg white fining is considered the best method for softening and polishing red wines. Higher polymeric phenols, such as tannins, are negatively charged and attracted to the positive proteinaceous albumen.

Each egg contains 3–4 gs of albumen (Zoecklein et al., 1999). The eggs are cracked and separated and then whisked into water with a small addition of KCl to help reduce the insoluble globular proteins.

Albumen is not regulated by the TTB or Office International de la Vigne et du Vin(OIV), but water is. The TTB limits the amount of water to 0.15 liter/hL or 330 mL per barrel. The OIV allows only the amount of water needed to dissolve dried albumen.

An egg white solution containing 1 gal of water, 1 oz of KCl, and 2 lbs of egg whites is typically made. Additions of this solution cannot exceed 2.5 gal/1000 gal of wine or 0.25 liter/hL per TTB regulations. Barrel additions equate to eight egg whites per 225-liter barrel or 1 g of albumen per 2 gs of tannins (Zoecklein et al., 1999).

Trials are conducted by adding measured portions of a 10% egg white solution to wine samples, mixing gently, and setting aside to allow for settling of the exudates. The wine samples are filtered or decanted and tested. The trial is evaluated by laboratory, physical, and organoleptic results to determine the addition amount. Refer to Chapter 9 for more details.

7.8.3   Gelatin and Isinglass Fining

Gelatin and isinglass are both positively charged proteinaceous fining agents that react with negatively charged tannins and other phenols in wine.

Gelatin binds with larger polymeric phenols and is less effective on color and tannin reduction than other protein agents. The PI of gelatin is in the range 4.7–5.2; therefore, it is positively charged. It is very effective in reducing the harshness of pressed wine. Wines fined with gelatin should be filtered to remove any remaining gelatin. Residual gelatin in the wine can eventually polymerize remaining phenols, creating a haze to develop in bottled wine. Wines that have not been filtered after fining can still pass a heat test because the polymerization takes time to develop.

The gelling power, or the resistance of the gel to deform and its ability to absorb water, is referred to as bloom. Higher bloom numbers indicate a greater adsorbing capability of the gelatin. The recommended bloom range for wine is 80–100. Typical white wine dosage ranges from 20 to 50 mg/L.

Gelatin trials use a 1% or 1.5% w/v solution of high-grade gelatin powder mixed into hot water. Controlled solution additions are made to aliquots of wine, mixed thoroughly, and set aside to allow for settling of the exudates. The wine samples are filtered or decanted and tested. The trial is evaluated by laboratory, physical, and organoleptic results to determine the addition amount. Refer to Chapter 9 for more information. The physical appearance of the acceptable trial sample is clear and bright. Laboratory analyses can include nephelometry, EtOH, TA, and pH.

Isinglass is used primarily to bring out fruit characters in white wines and to bring a brilliance to white wine before bottling, with little affect on tannins. Isinglass is used in méthode champenoise production as a riddling aid. The PI range is pH 5.5–5.8, making isinglass positively charged, and it reacts mainly with monomeric phenols.

Isinglass, from sturgeon collagen, most often comes in two forms: prehydrolyzed and a fibrous form of flocced isinglass. Typically, 10–50 mg/L is used in white wine. Isinglass powder (Drifine) is hydrated with cool water [60°C (15°F)] and acidified to a pH of 2.4–2.9 with citric or tartaric acid. Hot water will partially hydrolyze the isinglass and form smaller molecules (gel), which will render the solution much less effective. The solution is stirred for 24 h. Only high-grade isinglass should be used. Solutions that develop a fishy smell should be discarded.

The fining trial consists of controlled additions of a 0.5% w/v solution (check with manufacturer's recommendations) to aliquots of wine, mixing, and setting aside to allow for settling of the exudates. Physical and organoleptic evaluations are conducted on the samples to determine the level of addition. Refer to Chapter 9 for more details.

Protein-based fining agents should always be made fresh, but they can be kept for several days if an addition of benzoic acid is made.

7.8.4   Copper Sulfate Fining

Copper sulfate ($CuSO_4$) fining aids in the removal of thiols and H_2S, which imparts a rotten egg odor to the wine. Cupric ions (Cu^{2+}) react with the sulfide ions and form a copper sulfide salt (CuS); having limited solubility, it precipitates out of the wine. Additions of $CuSO_4$ range from 0.1 to 1.5 mg/L. Copper content in wine is regulated and the limit in the United States is 0.5 mg/L as copper with a residual level of ≤ 0.2 mg/L and 0.2 mg/L in other countries (Boulton et al., 1999).

Trials are conducted using $CuSO_4$ as Cu^{2+} in milligrams per liter (ppm). Copper sulfate contains approximately 25% Cu^{2+}, and 1 g of $CuSO_4$ contains 0.254 g of Cu^{2+} (63.5$_{MW(molecular\ weight)}$ Cu^{2+} divided by 249.7$_{FW(formula\ weight)}$ $CuSO_4$). Verify the Cu^{2+} content of the chemical you use. To make a 100-mg/L solution as Cu^{2+} requires 0.394 g/L of $CuSO_4$:

$$\frac{(100\,\text{mg/L}) \times (0.254\text{g Cu}^{2+})}{(1\text{g CuSO}_4)} = 0.394\text{g/L}. \qquad (1)$$

Aliquots of wine receive controlled additions ranging from 0.1 up to 0.5 mg/L. The samples are covered to await organoleptic evaluation to determine the level of addition to the body of wine. Refer to Chapter 9 for more details.

7.9   Wine Stability

Unstable elements in wine will eventually polymerize and precipitate, leaving a residue, sediment, and/or crystal deposit in the storage container or bottle. This is a natural occurrence as wine matures and changes chemically. The majority of wine consumers prefer clean, sediment-free, and crystal-free wines. It is strictly cosmetic and has no bearing on the quality of the wine. Fining and stabilization are methods to rid wines of insoluble material that might create these cosmetic flaws. This section will be concerned with tartrate instability and the stabilization of wine to prevent crystal formation in bottled wines.

Stabilization of wines for cosmetic purposes requires intervention that can affect the quality of the wine. Stabilization is accepted in white wine production much more than red wine. Many quality winemaking professionals around the world refuse to alter a perfectly good wine (and possible strip away flavors) just to make it more presentable. The majority of white wine production strives for brilliantly clear wines in the bottle, making stabilization a necessity.

Tartaric acid (H_2T), potassium, and calcium are all found at various quantities in juice and wine (Chapter 5). It is the salts created by these constituents during the maturation process that lead to the development of crystals. The ionized forms of H_2T found in wine are potassium bitartrate (KHT) and calcium tartrate (CaT). The concentration of these tartaric forms is pH dependent. The pK value for the transition of H_2T to bitartrate (HT^-) is pH 2.98 and the pK value for the transition of HT^- to tartrate (T^{2-}) is pH 4.34 (Zoecklein et al., 1999). At wine pH 3.7, the majority of H_2T exists as HT^-.

Alcohol concentration and temperature directly affect the solubility of the tartrates. As the alcohol content of the juice or wine increases, the solubility decreases and more tartrates precipitate out of the wine. Lowering the wine temperature also decreases solubility. Polyphenols in red wines and proteins in white wines can impede the formation of tartrates by combining with H_2T. Wines must always be fined before they are stabilized. The use of metatartaric acid inhibits the formation of KHT, but its use is not permitted in the United States.

There are two stages to crystal formation: nuclei formation and crystallization. It begins with a tiny particle that acts as a nucleus upon which small tartrate crystals begin to form and build upon one another until the crystal is large and heavy enough to precipitate out of the wine. Deliberate additions of KHT to wine act as nuclei and promote crystal formation; this is called seeding.

The method of choice for tartrate stabilization in wines is cold stabilization with or without seeding (contact process). The wines are placed in refrigerated tanks where the temperature is reduced to near their freezing point. Once the wine is chilled, it is seeded with KHT and stirred constantly to keep the KHT nuclei in contact with the wine, enhancing crystal formation and the removal of tartrates. Within the first 3 h, there is a rapid reduction of tartrates in most wines. After several hours, the wine is cold-filtered to remove the crystals. Any increase in temperature will allow a portion of the crystals to go back into solution. Wines with higher alcohols might require colder temperatures. According to Zoecklein et al. (1999), the formula for determining the temperature for optimum cold stabilization is calculated as

$$\text{Temperature } (-\,^{\circ}\text{C}) = \tfrac{1}{2}\,\text{EtOH \% v/v} - 1 \tag{1}$$

After cold stabilization and the precipitation and subsequent loss of KHT, there is a shift of the pH and TA due to the generation or removal of protons. Wines with less than a pH of 3.65 generate one free proton per molecule of KHT precipitated, showing a reduction in pH of 0.2 units and TA of up to 2 g/L. Wines with greater than pH 3.65 have an increase in pH while lowering TA, due to the removal of one proton per tartrate anion (Zoecklein et al., 1999).

Ion exchange is another method used to remove tartrates. The wine is first refrigerated and sent through an ion exchanger, where the cation K^+ is exchanged for an H^+ or Na^+. The removal of the K^+ reduces the solubility coefficient, reducing the possibility of KHT formation at low temperatures. This method is not normally used in premium wine production.

Laboratory analysis can confirm tartrate stability. Currently, there are two methods commonly used: the freeze test and cold conductivity. It is prudent to state that the validity of these tests rests on cellar practices, removal of colloidal material and polymeric phenols, and the quality of the sample obtained. Samples received in the laboratory should be at the same temperature as the wine and processed immediately.

7.9.1   Freeze Test

The freeze test is a visual test for the detection of crystal formation. Many winemaking professionals find this test a bit primitive, whereas others swear by it. Laboratories sometimes conduct a freeze test in addition to other verification methods, such as cold conductivity analysis.

The theory behind the test is based on the concentration of tartrates. Freezing the water in the sample increases the relative concentration of tartrates, which enhances nuclei and crystal formation (Zoecklein et al., 1999). Freezing alters the coagulation of colloidal components in the wine, which can interfere with nucleation and crystal growth, leading to false-positive results.

After stabilization and before filtration, a cold wine sample is brought to the laboratory, where it is sterile-filtered, placed in a glass tube, and frozen solid. The wine could be seeded with KHT depending on the reason for testing (confirmation of stability or trial). Seeding can obscure the visual interpretation of the test with a noted haze. The freezing time required for the sample is dependent on the alcohol content of the wine. Typically, wines are placed in the freezer for 4–16 h.

The sample is removed from the freezer and allowed to thaw at room temperature. At the point that the wine is liquefied, the sample is inspected under an intense light for indications of crystal formation. If crystals do appear, the sample is reinspected when the sample reaches room temperature to verify if the crystal have gone back into solution. The absence of crystals upon thawing or at room temperature indicates a stable wine.

Stability of CaT is most often practiced in sparkling and fortified wine production. The concentration of CaT is pH dependent and alcohol affects the solubility, as mentioned earlier. Reduction of temperature has no effect on CaT solubility; therefore, the freeze test is eliminated as an analytical tool for verification of stability. Testing for decreases in tartaric acid and calcium concentrations would be indicated. Tartaric acid can be tested using high-performance liquid chromatography, spectrophotometry, and paper chromatography. Measurements of calcium can be obtained with ion selective electrode (ISE).

7.9.2   Cold Conductivity

During KHT cold stabilization, there is a loss of free K^+, resulting in a change in the electrical conductance of the wine. Changes in pH can alter conductance by altering the concentration of potassium. Hydrogen ions affect conductance to a degree because they are 10 times more conductive than K^+. Potassium ions exist in larger quantities at wine pH, but as pH levels become more acidic (pH 3.0), H^+ increases, thus having more influence on electrical conductance.

If KHT is added to a cold wine sample that is stable, the conductivity will increase because of the addition of free K^+. If the conductance falls, it would indicate precipitation of tartrates and the loss of free K^+; the wine is unstable. A constant reading indicates the wine is at the correct temperature where it is just stable.

Cold conductivity testing involves a cold bath that can maintain a constant accurate temperature of 0°C (32°F) and below, a stirring mechanism, and a calibrated conductivity meter. A chilled wine sample is brought to the laboratory, immediately sterile-filtered, and degassed. The sample is placed in the cold bath at a temperature near freezing. White wines are tested at 0°C (32°F) and red wines are tested at 5°C (41°F) (Zoecklein et al., 1999), with aged red wines being tested at temperature up to 10°C (50°F) (Boulton et al., 1999). The conductivity probe is inserted into the sample and the conductivity is read when the sample reaches testing temperature. An aliquot of KHT is added to the chilled sample and stirred for a few minutes. Another conductivity reading is taken when the temperature of the wine stabilizes. The change in conductivity from the initial reading (C1) to the post KHT reading (C2) will indicate the level of stability. A change of < 3–5% indicates a stable wine:

$$\frac{C1 - C2}{C1} \times 100 = \% \, \Delta \, \text{Conductivity} \tag{1}$$

The wine sample can be filtered again and tested for changes in pH and TA. More information can be found in Chapter 9.

7.10 Filtration

Removal of solids found in wine is carried out to improve wine clarity, reduce microbes, and remove unwanted substances in the wine. To filter or not to filter depends on the varietal, winemaking style, and the necessity for the filtration. Quality wine production frowns on filtration because it can strip flavors from the wine, but at times it might be necessary. Wines are filtered for a variety of reasons:

- Removal of solid particles: grape pulp and so forth, postfermentation
- Removal of yeast: postfermentation, prevent fermentation (wine containing residual reducing sugar)
- Removal of bacteria: infected wines, prevent infections
- Removal of colloidal material: clarity, postfining
- Removal of tartrates: poststabilization

Deleterious effects can be avoided if the correct method and procedures are chosen. Depth and surface filtration are the two methods used in wineries. Depth filtration filters by trapping particles throughout the material and surface filtration acts as a sieve having a specific surface pore size that removes particles larger than the pore size. Determination of which method to use is based on the material to be removed and the particle size of that material. Determination of the filter material to use depends on the particle sizes required:

- Coarse filtration: particles larger than 36.0 μm
- Moderate filtration: particles larger than 14.0 μm

- Guard filtration: particles larger than 2 µm
- Membrane filtration: particles larger than 0.8 µm (all yeast)
- Ultra membrane filtration (sterile filtration): particles larger than 0.45 µm (most wine bacteria)
- True sterile filtration: particles larger than 0.2 µm (all bacteria)

Coarse filtration is accomplished by depth filtration using diatomaceous earth (DE), Kieselguhr, diatomite, or pads (sheets). Diatomaceous earth is the fossilized shells of a unicellular aquatic plant called a diatom, which are ground into various particle sizes. The smaller the particle sizes, the smaller the particles it filters out of the wine. Coarse filtration using larger particles of DE (rough filtration) can remove larger particles; using a fine particle size of DE (pink powder) can filter down to 0.8 µm (polished filtration). Filtration to the lower micrometer level is not recommended because the method is not absolute and yeast can get through.

Pad, or sheet, filtration employs a sheet of cellulose material ranging in thickness. Particles are trapped within the material. The thicker the pads, the finer the particles trapped. This method is commonly called plate and frame filtration, which describes the equipment used to hold the pads in place and the mode of distribution of the wine. Filtration to the lower micrometer level is not recommended because the method is not absolute and yeast and bacteria can get through the pad filters if enough pressure builds in the system.

Moderate filtration usually follows coarse filtration and uses primarily depth filtration with pad filters. Again, filtration to the lower micrometer level is not recommended because the method is not absolute.

Guard filtration uses the depth filtration method, but the filter is placed in a cartridge and used as a prefilter for fairly clean wines during bottling. These filters are nominally rated for particle size down to around 2 µm and are normally placed at the front (upstream) of a series of finer membrane surface filters.

Membrane surface filters come in cartridge form and are placed after (downstream) a guard filter. Membrane filters are manufactured from polycarbonate, polysulfone, or polypropylene, with different pore sizes down to 0.8 µm, which will remove all material including yeasts.

Ultra membrane surface filtration is often referred to as sterile filtration in the wine industry, but it is not truly a sterile filtration. The filters are in cartridge form and placed downstream of the membrane filter. Ultra filtration is the final filtration prior to the wine being bottled (only those requiring sterile filtration). Ultra filtration removes particles down to 0.45 µm, which includes all yeasts and wine bacteria.

Sterile membrane surface filtration is indeed sterile filtration and removes all particles down to 0.2 µm, which includes all yeasts and all bacteria. This very small pore size not only removes all the yeast and bacteria but it will also remove important flavor components and body characteristics in the wine. Sterile-filtration is not used in the wine industry.

Sulfured wines must be fully protected prior to any filtration process. The risk of oxygen exposure can be high during these processes. Following any filtration process, wine should be analyzed to determine O_2, alcohol, TA, pH, and possibly free SO_2. These are indicators of unintentional dilutions or oxygen exposure during the movement of the wine.

7.11   Blending

Blending of wines is a complex and personal endeavor. Blending is strictly based on organoleptic preferences. A blend is put together by the winemakers according the flavor profiles of the individual wines (components) available to them. A good blend will achieve a well-balanced wine that maintains its varietal characteristics, has complexity, and is void of deficiencies. Wines can be blended at any point of development.

Prior to blending, it is helpful to have current accurate analyses for EtOH, TA, pH, VA or acetic acid, LMA, RS, and current microbiological information if available. Typically, there are a few rules of thumb used to begin a trial blend: Acidic or tannic wines can be softened by added EtOH levels; reduction of sugar perception can be achieved by added EtOH levels; and more acidic wine will increase the sugar perception.

Pearson's square is a blending tool used for two-component blends and is adaptable for larger blends. The square utilizes a desired value and the ratio of the same component values of two wines. This ratio gives the breakdown of each component required to reach the desired value. Figure 7.4 illustrates the use of Pearson's square to reach a desired value using EtOH as an example as described by Boulton et al. (1999). Very sophisticated computer programs exist to aid complicated blending of many components and are used throughout the wine industry.

Blends are treated like a new wine and require an entirely new set of analyses that covers free and total SO_2, TA, pH, VA or acetic acid, LMA, RS, stability tests, and microbiological analyses if needed. Wines that have been fined or stabilized prior to blending can develop instabilities as the chemistries of the wines blend and might require repeating the tests and the process.

Small-scale trial blends are made for organoleptic evaluation and analyses, including stability testing. It is important that trials be conducted under the same conditions (such as temperature) as the production blend.

Blend approval leads to the actual blending of the bodies of wine. Blending is based on volume rather than weight.

Blended wine is typically prepared for bottling if the blend is formed after maturation and stabilization. Blends made prior to maturation can be placed in barrels or containers and allowed to mature prior to bottling preparation.

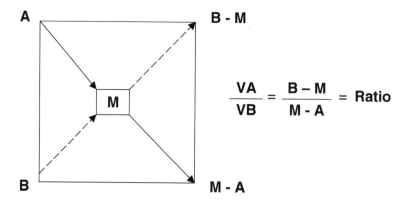

$$\frac{VA}{VB} = \frac{B - M}{M - A} = \text{Ratio}$$

M = Desired Value
A = Component Value
B = Component Value
(One component value must be lower than the desired value)
V = Volume
If the desired value is 12% v/v, component A is 11%, and B is 14%
the ratio would be:

14 – 12/12 – 11 = 2 / 1 ratio
or
1/3 VB and 2/3 VA = Blend Volume resulting in 12% v/v ETOH

FIGURE 7.4. Pearson's square as a blending tool using ethanol as an example.

7.11.1 Fractional Blending

The simplest, but unusual, blending technique is the solara system of fractional blending developed in Portugal and Spain. The solara system consists of barrels of sherry or port systematically blended year to year. Each new vintage is placed on top of the barrels from the previous year. The oldest wines in barrels are at the bottom of the stacks and are a homogeneous blend of all the previous years, giving the wine consistency from year to year. There are very few true vintage sherries or ports.

Each year, the oldest barrels have a certain volume removed for bottling. That volume is replaced from the barrels of wine located directly above it in the next row and that row's volume is replaced by the earlier year's barrels above it and so on.

8

Bottling Basics

8.1 Introduction

Years of work, worry, and winemaking approaches the finale—bottling day. This is the moment of truth when winemaker's and winery's reputations will be tested as the wine is bottled and subsequently released to the public. The day of bottling is usually highly charged with anticipation and fear. The smallest breach in bottling protocol can spell disaster for a great wine.

The bottling process seems very straightforward but there are numerous areas fraught with dangers resulting in microbial contamination and oxidation. Successful bottling relies on properly prepared wine, identification of problematic areas, continual scrutiny throughout the bottling process, and tight quality control.

8.2 Packaging

Preparation for bottling begins long before the big day. Labels are designed and sent to government agencies for approval far in advance of the bottling date. Bottles, corks, and capsules or wax to cover the corks are purchased and delivered a few weeks prior to bottling.

Estimations on the final volume of wine to be bottled are made in advance and the number of bottles, corks, and labels required for the bottling are ordered (check with manufacturers for exact counts):

750-mL bottles packed 12 to a case = 56 cases = 1 pallet
Estimated number of pallets of 750-mL bottles = No. of gallons/131
350-mL bottles packed 12 to a case = 100 cases = 1 pallet
Estimated number of pallets of 375-mL bottles = No. of gallons/117

Corks and bottles should be received at the winery in time to test glass sterility and cork contamination, but not too far in advance to risk contamination by lengthy storage. Labels should arrive as early as possible to allow time for any changes that could result in possible resubmission for approval.

Worldwide, there are several methods of packaging wine other than by bottle and cork closures:

- Polyvinyl chloride (PVC): wine for daily consumption; very short shelf life
- Aluminum cans: 250-mL cans; limited shelf life due to the acids in the wine
- Polyethylene terephthalate: 187.5-mL containers used chiefly by airlines; short shelf life
- Bag in a box: 2-, 3-, 5-, 10-, and 20-liter volumes; longer shelf life; layered bag containing a light oxygen barrier of polyvinyl alcohol (PVA) or tin foil sandwiched between layers of high-density polyethylene
- Screw-on bottle caps: closure for wines that will not benefit from micro-oxidation
- Plastic bottle corks: closure for wines that will not benefit from long-term bottle aging

The varietal and winemaking style will dictate the optimum type of packaging required to present the wine to the marketplace.

8.2.1   Labels

In the United States, the proposed wine label with an application are sent to the Alcohol and Tobacco Tax and Trade Bureau (TTB) for approval. Included on the TTB F 5100.31 application is information regarding the vintage, appellation, bottling location, net contents, sulfite warning (if SO_2 is greater than 10 mg/L), and alcohol percentage. Most of this information will appear on the label and must be accurate.

The United States taxes wine according to the percent of alcohol and carbon dioxide content (CO_2 in still wines should not exceed 3.92 g/L; greater CO_2 content increases the taxes). Actual alcohol content cannot vary more than 1.5% v/v in wines classified as table wines (7–14.0% v/v or lower) from the alcohol level stated on the label. Fortified wines (14.1–24% v/v) cannot vary actual alcohol contents more than 1.0% v/v from the concentration stated on the label. Many California wines contain alcohol levels very close to the 14.0% tax cutoff. Making a commitment to an alcohol concentration for label approval months before the product is ready for bottling can be a bit tricky. If label approval was granted by the TTB for 13.5% v/v as a table wine and the actual alcohol level of the finished wine is 14.06% v/v, a new label-approval process must take place and the wine will be taxed as a fortified wine.

Wine label regulations can be located in the code of federal regulations (CFR) as follows:

- Alcohol content: 27 CFR 4.36
- Appellations of origin: 27 CFR 4.25a
- Declaration of sulfites: 27 CFR 4.32e

- Estate bottled: 27 CFR 4.26
- Health warning statement: 27 CFR Part 16
- Name and address: 27 CFR 4.35a
- Net contents: 27 CFR 4.37
- Varietal designations: 27 CFR 4.23, 4.28, 4.91
- Vintage date: 27 CFR 4.27
- Viticultural areas: 27 CFR Part 9

Wines that are slated for bottle aging at the winery are not required to be labeled until they are ready for sale.

8.2.2   Bottles

There are several different styles and colors of wine bottle used in the industry. The bottle styles are derived from traditional French designs representing Burgundy, Bordeaux, and Alsace regions. Bottles differ in shape, having different slopes at the shoulders of the bottle (see Fig. 8.1) and glass color is dependent on the wine varietal and the need to protect it from light. Bottle aged wines require significant protection from light and are either deep green or brown, whereas more fruity wines are bottled in clear to slightly colored glass.

Bottle necks have been standardized to accommodate general cork size, as illustrated in Figure 8.1. Note the smaller opening compared to the neck. This design allows the inserted cork to expand further in the neck, assuring a secure closure.

In 1979, United States Customs set the standard bottle size for imported wines at 750 mL. The European Union (EU) standardized their bottle sizes as well and adopted 750 mL as the standard bottle volume to reduce problems with exportation of wine to the United States. Regardless of bottle style, the volumes will be consistent in each category, as shown in Table 8.1.

8.2.3   Cork

Corks make a nearly perfect closure for wine bottles. They have compressibility and elasticity, adhere to the glass walls, insulate, are immune to rot and decay, and are impermeable to liquids. Corks are replaced every 30–40 years in long-term bottle-aged premium wines.

Western European oak trees *Quercus suber* and *Quercus occidentalis* are the primary sources of cork in today's market. Portugal is the number 1 producer of cork for the world market, followed by Spain, France, Italy, Tunisia, Morocco, and Algeria.

Quercus suber tree bark is used most often. Tree bark is harvested on an average of every 9 years beginning when the tree is 25–30 years old. The bark is stripped off the trees, cut into planks, and left outside to cure. After approximately 6 months to 1 year, the bark planks are boiled, dried, graded,

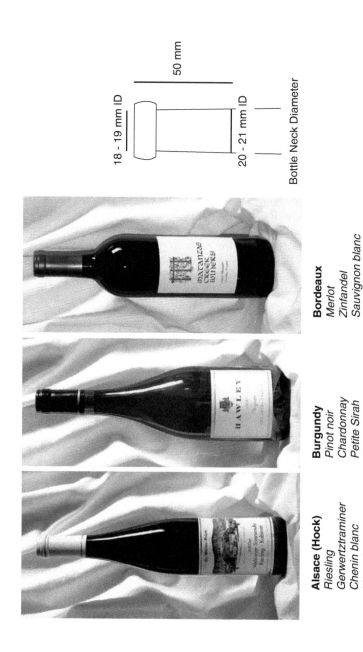

Alsace (Hock)
Riesling
Gerwertztraminer
Chenin blanc
Muscat
Brown - Rhine
Green - Moselle

Burgundy
Pinot noir
Chardonnay
Petite Sirah
Gammay
Green - Red Wine
Clear or Light Green
- White Wine

Bordeaux
Merlot
Zinfandel
Sauvignon blanc
Cabernet Sauvignon

18 - 19 mm ID

50 mm

20 - 21 mm ID

Bottle Neck Diameter

FIGURE 8.1. Traditional bottle shapes and colors. Varietals are associated with bottle shape. A diagram of a standard bottle neck for all types of bottle is also shown.

TABLE 8.1. Traditional wine bottle volumes.

375 mL = half bottle
750 mL = typical size bottle
1.5 liters = magnum
2.25 liters = Marie-Jean
3 liters = double magnum
3 liters = Jeroboam (Champagne, Burgundy)
4.5 liters = Jeroboam (Bordeaux, Cabernet Sauvignon)
4.5 liters = Reboboam (Champagne, Burgundy)
6 liters = Imperial (Bordeaux, Cabernet Sauvignon)
6 liters = Methuselah (Champagne, Burgundy)
9 liters = Salmanazar (Champagne, Burgundy)
12 liters = Balthazar (Champagne, Burgundy)
15 liters = Nebuchadnezzar (Champagne, Burgundy)

punched, polished, bleached, dried, and hand sorted. There are five grades of cork according to the Cork Quality Council (CQC) in the United States:

- Grade A: top quality, excellent surfaces, no major visual flaws
- Grade B: good quality, good visual appearance, no major flaws
- Grade C: average quality, average appearance, one or more major cosmetic flaws
- Grade D: poor quality, poor appearance, many cosmetic flaws
- Grade E: rejected from other grades

Corks should contain approximately 6–8% moisture and have a compressibility of 85%. The standard diameter is 24 mm and the lengths can vary between 40 and 50 mm, the later being used for sparkling wine bottles. Longer cork lengths are used for wines that will be bottle aged. Wine bottles are standardized to accept general cork sizes (Fig. 8.1).

Colmated corks are lesser quality corks that are coated and sealed with an approved (US Food and Drug Administration) colmating material (glue), making them more usable and economical. Agglomerate corks are lesser quality corks coated and sealed with Teflon. Conglomerate corks are made from cork layers or pieces stuck together with glue or paraffin and are much less expensive. These corks are intended for wines that do not require extensive bottle aging and will be consumed within a few years.

The final step to cork production is packaging and sterilization. Most corks are placed in polyethylene bags, which are then filled with SO_2, which can reduce mold populations by 80–100% (minimum SO_2 levels of 200 mg/L). Gamma-irradiation is also being used to reduce microbial growth.

8.2.3.1 Cork Quality Control

Cork quality control (QC) can be a full-time job for many larger wine laboratories. QC examinations including grading, sensory, moisture content, dimensions, and sterility are required for corks prior to bottling. Frequently, wineries will have their names branded on the surface of the cork and this requires two separate QC examinations: an examination before and after the branding. One winery might reject a lot of cork, whereas another winery might find those same corks appropriate for their purposes. If the corks have been branded, the cork producers cannot sell the corks to another buyer if a winery rejects them.

The first examination is to approve the quality of the corks to be purchased and branded. A representative sampling of corks from each lot (approximately 0.1% of the total) is evaluated and rejected or approved according to lot, box, and possibly down to individual bags. The level of examination is usually discussed with the cork suppliers and agreements made on the type of examination criteria to be used and the procedures for rejected corks.

The corks are graded on their visual appearance and major flaws such as cracks or fissures. The dimensions are measured to assure correct cork size. The compressibility of the cork is dependent on the moisture content and should be between 6% and 8%. If the moisture is too low, the cork can slip too far down the neck of the bottle or crumble, making a poor seal. A cork with excess moisture might be difficult to insert into the bottle or contain mold. If the moisture content is too low, the corks will be sent back to the supplier for rehydration; if too high, the corks can be retested in a few days. Two commonly used meters for moisture content measurements are the Aquaboy and the DC 2011 (Moisture Register Products).

Sensory testing consists of putting cork samples into a volume of sterile water or wine and letting them soak for 1 or 2 days at room temperature. The samples are inspected for cork dust and smelled to detect moldy odors or off-odors. Most wineries use a 10% v/v alcohol white wine for their cork soak. The volume of wine used should be as little as possible but enough to float the cork and interface with all cork surfaces: one cork per 100 mL wine, or 250 mL wine for four corks in one container.

Corks contain volatile compounds, including phenolics, fatty acid esters, and aldehydes. Organoleptic detection of a moldy odor can indicate the presence of 2,4,6-trichloranisol (TCA), which is a by-product of fungal activity. Chemical odors can indicate poorly rinsed corks after the bleaching process. Detection of TCA and/or off-odors should be reported to the supplier and the corks rejected because these odors can transfer to the wine.

Cork dust is generated from cork processing and is usually eliminated by the rinsing process prior to packaging. Corks of lesser quality or low moisture content tend to breakdown during packaging and shipping, creating higher levels of dust. Excess cork dust can dirty a wine, making it less desirable to the consumer.

If the corks pass all of the above preliminary examinations, they are approved for purchase and branded. When the branded corks arrive at the winery, a random representative sampling of the lot (approximately 0.1% of the total) begins the second stage of cork approval. The more corks used in the sample, the better the chance of detecting problems.

Stage 2 involves a second sensory and moisture evaluation, determination of sterility, SO_2 measurement, and capillary test (wicking). Sulfur dioxide levels in sealed bags should be approximately 200–300 mg/L. Low SO_2 levels indicate inadequate SO_2 additions prior to shipment; these corks should be returned to the supplier. Excessively high SO_2 levels should be held and retested until the SO_2 levels come closer in line to the acceptable range. Corks purchased without SO_2 protection are subject to a short wash with a solution containing 1000 mg/L SO_2 prior to bottling (Margalit, 1997).

The capillary or wicking test involves standing 10 corks in 5 Petri dishes, each containing 25 mL of a colored solution or red wine for 24 h (Cork Quality Council, 1994). Migration of the liquid up into the cork is measured. Samples showing moderate wicking or greater will make a poor seal and are rejected. Minimal wicking is acceptable.

Microbiological plating is required to evaluate the sterility of the corks. Wineries with moderate to large laboratories will incubate and examine these plates (Petri dishes containing nutrient agar for growth), whereas smaller wineries will need to take their cork samples to a certified laboratory for evaluation. The plating procedure involves placing corks in sterile saline solution, soaking for a few minutes, and filtering the saline solution through 0.45-μm filters. The filters are aseptically placed on pH-adjusted WL agar (Wallerstein Laboratory) and incubated for 7 days. Growth could show in 4–5 days, but the full 7 days might be required for slower-growing organisms. The plates are read and might require microscopic evaluation to determine the type of growth (yeast, mold, or bacterial) (see Fig. 8.2, Color Plate 2). Mold colonies will appear as fuzzy, or hairy, and can range in color from white to dark brown. Bacterial colonies have a smooth surface and typically are yellowish in color. Yeast colonies tend to be smaller, milky in color, with a smooth rounded surface. Cork contamination comes from *Bacillus* sp. and various other molds, including *Penicillium rogueforti*.

Cork samples with positive growth should be replated to confirm contamination before contacting the supplier.

8.2.3.2   TCA

Earlier in this chapter and in Chapter 7, TCA has been mentioned in conjunction with moldy odors in wines that originate from molds in the cork, hence the term "cork taint" or "corkiness." Levels of TCA can be detected at very low thresholds, usually 1.4–10 ng/L (parts per trillion). TCA in its pure form is very potent and it has been said "one-half tablespoon could ruin all the wine in France."

It is believe that chlorine used in the processing of cork plays a significant role in the development of TCA (Boulton et al., 1999; Margalit, 1997; Zoecklein et al., 1999). Cork phenols exposed to chlorine form chlorophenols, which are biodegraded by fungal activity into 2,4, 6-trichloroanisole. Many cork producers have now switched to hydrogen peroxide as a bleaching agent and use nonchlorinated water during the processing.

Although this will most likely solve the majority of TCA problems, there could be other sources of TCA or TCA-like chemicals produced from oak products. Also found in cork is 2-methoxyphenol, or guaiacol, produced from the degradation of lignin (Margalit, 1997). Guaiacol produces a smoky medicinal odor, which is frequently associated with the presence of TCA.

Sulfur dioxide additions used for microbial control could eliminate the mold on the corks, but does not rid the cork of existing TCA. It is imperative to test the corks before they are used.

Most wineries use the sensory test to detect the presence of TCA. Although this has been a method used for many years, it is fallible, inconsistent, and the validity rest on the person conducting the organoleptic examination. Every person has a different detection level for most odors. One person might detect TCA at a 3-ng/L level and another might not detect it at a 20-ng/L level. Sensitivity to TCA and other odors is essential for the success of organoleptic evaluations of wine.

As mentioned in Chapter 7 (Section 7.5.2), the current method for detection of TCA is SBSE and thermal desorption using a GC/MS device. TCA has been accurately quantified to below 1 ng/L. The development of this technology has drastically improved the ability of large wineries, wine laboratories, and cork producers to detect TCA, improving cork QC. Commercial wine laboratories can easily provide this service to smaller wineries.

With this technology, there arises questions such as how much TCA is passable? Can TCA increase with time from lower nondetected levels to detectable levels? Should specified procedures be implemented to ensure consistency in testing throughout the industry? How will the increase in cork rejection affect the price of cork? Many questions remain.

8.3   Bottling Lines

There are all shapes and sizes of bottling lines: from a manual one bottle at a time, to computerized lines handling hundreds of bottles a minute. Regardless of the type of bottling line, there are procedures and precautions practiced by all. A common layout of a bottling line is shown in Figure 8.3.

The object is to get the wine into the bottle cleanly with no residual oxygen or contamination. Several processes are used throughout the world, such as tunnel pasteurization, thermotic bottling, flash pasteurization, and cold sterile filtration. Flash pasteurization and cold sterile filtration are used most often for wine.

Flash pasteurization utilizes a heat exchanger that quickly raises the temperature of the wine to 95°C(230°F) for a few seconds and then cools it down quickly before it enters the filler bowl. Heating the wine eliminates yeasts and bacteria.

The cold filtration process sends chilled wine through a series of cartridge filters graduated down to 0.45-μm pore size just before the wine enters the filler

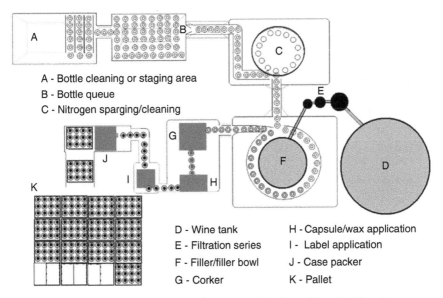

A - Bottle cleaning or staging area
B - Bottle queue
C - Nitrogen sparging/cleaning

D - Wine tank	H - Capsule/wax application
E - Filtration series	I - Label application
F - Filler/filler bowl	J - Case packer
G - Corker	K - Pallet

FIGURE 8.3. Typical bottling line configuration for cold sterile filtration.

bowl (see Fig. 8.3). It is imperative when using this method that the bottles and closure be free of contamination. Tanks, hoses, and all bottling equipment must be sterilized before bottling. Personnel should all be aware and trained in aseptic technique as well as procedures for stopping the line or removing the product if there is even a remote possibility of contamination or oxidation. The sterile filters must be watched carefully and monitored for pressure buildup, which can indicate a dirty filter and lead to contamination. The corker should be in close proximity to the filler, corking the bottles as soon as is possible.

8.4 Preparing Wine for Bottling

Wines ready to be bottled have been carefully blended, stabilized, filtered, and analyzed and the packaging supplies received, tested, and inspected.

Final filtration decisions have been made per the winemaking style and varietal. Wines that have completed primary and secondary fermentation and contain no residual reducing sugar (RS) or L-malic acid (LMA), usually do not need sterile filtration bottling because of the lack of oxygen and nutrients and the alcohol level of the wine. Other wines will require serial membrane filtration just prior to bottle filling.

Preparation of the wine for the chosen bottling method commences with moving the wine to the bottling tanks and adjusting the temperature to the appropriate bottling temperature.

8.4.1 Required Analyses

Prior to bottling, the winemaker will review all current analyses, including the following:

Free SO_2	Heat stability (i.e., white wine)	O_2
Total SO_2	Cold stability (i.e., white wine)	4-Ethyl phenol
EtOH	Cold conductivity (i.e., white wine)	TCA
Titratable acid and pH	Filterability	Specific gravity
Volatile acid	Density	Temperature
RS	CO_2	
LMA	All plating results (i.e., red wine and corks)	

Evaluation of the blend components and general chemistry will confirm the blend specifications and any need for adjustment. Verification of wine stability will assure a clean, clear wine after bottling.

Final filtration decisions will be dependant on the winemaking style, wine filterability, sugar and LMA content, and the red wine plating results.

Cork and bottle approval is determined by the plating, organoleptic, and TCA results.

8.4.1.1 Sulfur Dioxide

Prebottling sulfur dioxide levels should be high enough to maintain adequate molecular SO_2 after the bottling process. During bottling, there might be exposure and pickup of oxygen at the filling and corking stages in the range 0.5–1.5 mg/L. It is estimated that 1 mg/L of oxygen reacts with 8 mg/L of total SO_2, producing acetaldehyde until the oxygen is utilized (Boulton et al., 1999). Knowing the bottling equipment and the typical amount of oxygen pickup will allow for proper additions of SO_2. Sulfur dioxide levels are analyzed before, during, and after bottling.

Total SO_2 levels in bottled wines are regulated worldwide. The United States and most countries allow up to 350 mg/L of total SO_2 (100 mg/L for US organic wine) in bottled wine across the board but the European Union and Australia have broken down the allowable concentration of total SO_2 to include the following (Bird, 2000):

- Red – dry = 160 mg/L
- Red – RS ≥ 5 g/L = 210 mg/L
- White – dry = 210 mg/L
- White – RS ≥ 5 g/L = 260 mg/L
- Spätlese, White Bordeaux = 300 mg/L
- Auslese = 350 mg/L

- Trokenbeerenauslese, Beerenauslese, Ausbruch, Sauterns, Graves Supérieurs = 400 mg/L
- Australian – dry = 250 mg/L
- Australian – sweet = 300 mg/L

Approved methods for measurement of total SO_2 in the United States by the TTB is the aeration–oxidation method (see Chapter 9).

8.4.1.2   EtOH

It is recommended that two approved methods be used by wineries to confirm the EtOH content of the wine prior to bottling to verify the concentration in % v/v represented on the label. Organisation International de la Vigne et du Vin (OIV), Association of Official Analytical Chemist International (AOAC), and TTB-approved laboratories and others would only require one method. Ebulliometry, distillation, densitometer, and gas chromatography (GC) are acceptable TTB methods (see Chapter 9).

Sampling should be representative of the entire volume and samples mixed prior to analysis. Obtaining two confirming values (especially in wines near a tax demarcation or very close to the allowable variance) demonstrates the winery's desire to comply with all set regulations.

Checking the EtOH levels after the wine has been transferred to the bottling tanks and directly after bottling can detect any dilution problems.

8.4.1.3   Acidity

Volatile acidity levels in bottled wines are regulated by most wine-producing countries requiring analysis by approved methods. The approved TTB methods for VA are Cash still (Chapter 9) and segmented flow. The general VA limits (expressed as acetic acid exclusive of sulfur dioxide) are as follows:

- United States
  red = 1.4 g/L; white = 1.2 g/L
- California (US)
  red = 1.2 g/L; white = 1.1 g/L
- France
  red and white = 1.1 g/L
- Germany
  red = 1.6 g/L; white = 1.2 g/L
- Australia
  red and white = 1.5 g/L
- South Africa
  red and white = 1.2 g/L

Other acids are also regulated in the United States, such as citric acid with additions not to exceed 0.7 g/L, fumaric acid with additions not to exceed 3.0 g/L, and sorbic acid not to exceed 300 mg/L. Chemical additions to reduce acidity, such as potassium bicarbonate, calcium carbonate, and calcium sulfate, are also limited and regulated (a listing of regulated sustances may be obtained from TTB or their website www.TTB.gov). Acidity levels in wines prepared for bottling should be well under the specified limits. If acid adjustments are necessary for any reason, stabilization testing should be repeated with the possibility of wine restabilization.

8.4.1.4   Microbial Growth

The presence of RS, LMA, or Brett can present the danger of microbial growth in the bottled wine.

Residual reducing sugar provides food for any yeast or bacteria that might be present in the bottled wine. Most wines with RS are sterile-filtered during the bottling process, thus eliminating the food source and danger of contamination.

L-malic acid in the wine can result in a secondary fermentation in the bottled wine and sterile filtration is recommended.

Indication of Brett (*Brettanomyces*) from plating, sensory, or 4-ethyl phenol results might also require sterile filtration of the wine to assure elimination of the yeast and prevent its growth and formation of 4-ethyl phenol in the bottle.

Analyses for RS and LMA are also used before during and after bottling to determine possible dilution because their levels are very low and any dilution can be readily detected.

Postbottled wine is routinely plated to determine microbial contamination and will be discussed later in this chapter.

8.4.1.5   Carbon Dioxide

Carbon dioxide is added to finished wine prior to bottling to accentuate tactile perception of acidity, odor, fruitiness, and freshness. Fruity wines will have larger additions of CO_2 than many red wines. Still wines will contain less than 3.92 g/L CO_2.

An added bonus of CO_2 additions prior to bottling is the ability of CO_2 to displace oxygen, thus assisting in reducing oxidation and loss of molecular SO_2.

Excess amounts of CO_2 can increase the acidity of the wine because it exists as carbonic acid at wine pH. The saturation level of CO_2 in wine is 1.5 g/L at 20°C and can contribute up to 2.0 g/L (as tartaric acid) to wine titratable acidity (Boulton et al., 1999). In this case, wines can be sparged with N_2 to eliminate the excess CO_2.

Carbon dioxide analysis is performed before, during, and after bottling to document proper levels. Typical CO_2 analysis is conducted on a sample of wine at bottling temperature using a piece of equipment called a Carbodoseur. This very simple device is a graduated glass cylinder with a screw top containing a long glass open-ended impinger tube. The cylinder is filled to a graduated mark with the wine sample, the top is screwed on, a thumb is placed over the open impinger tube, and the sample is vigorously agitated. The Carbodoseur is held vertical and the thumb is carefully removed, allowing the CO_2 gas to escape without losing any wine volume. This is repeated several times until no gas is generated. The volume of remaining wine is determined by the graduation lines; the temperature of the sample is obtained and noted. Subtraction of the remaining wine volume from the initial volume equates to the loss of CO_2 gas and is calculated into milligrams per liter according to the temperature. More information can be found in Chapter 9.

8.4.1.6   Oxygen and Wine Temperature

Oxygen levels are closely monitored as the wine is prepared for bottling and throughout the bottling process. As discussed in several sections of this book, oxidation must be avoided in the majority of cases. Oxygen is much more soluble in cold wines and great care must be taken to avoid any air contact when using the cold sterilization method. Oxygen levels are kept below 1.0 mg/L.

Dissolved oxygen meters are used to determine the oxygen content in wine, must, and juice. This technology employs a gold cathode, silver anode, KCl electrolyte solution, and a membrane. An electric potential develops across the membrane and oxygen molecules are transferred at the gold cathode when voltage is applied through the silver anode. The rate of oxygen diffusion across the membrane is electronically measured.

It is important when using such a meter that it is in good working order, calibrated properly, and temperature compensated and a constant flow of sample surrounds the sensing cathode. Dissolved oxygen analysis is conducted with each wine movement and throughout the bottling process.

The set wine temperature will be determined by the method of bottling. Wine temperature affects the solubility of gases in the wine and microbial growth. Cold sterilization bottling requires the wine to maintain a very low temperature through the filters, into the filler bowl, and into the bottle. Flash pasteurization requires the wine to be heated to a constant 95°C (203°F) for a few minutes before entering the filler and bottle. Both methods remove microbes.

8.4.1.7   Wine Density

Wine density is measured to determine the volume of wine in a bottle, using the weight of a known volume of wine. Density equals the weight of the substance in grams at 20°C (68°F) divided by the volume. For example,

$$\text{Density} = 100 \text{ mL Weight (in g) of } 100 \text{ mL of wine at } 20\,^\circ\text{C} \qquad (1)$$

$$= 100 \text{ mL } X \text{ g}$$

$$= 100 \, X \text{ g}/100 \text{ mL}$$

$$= X \text{ g/mL}$$

$$= X \text{ g/mL} \times 750 \text{ mL (standard bottle volume)}$$

$$= X \text{ g}/750\text{-mL bottle}$$

Density analysis is used to monitor the volume of wine in the bottles during the bottling process to ensure proper filling.

8.5   Prebottling Activities

Prior to starting the bottling process, quality control procedures are implemented to assure a successful bottling. Final analyses of free SO_2, CO_2, dissolved O_2, and temperature is conducted. Blend and tank identification are reconfirmed. The hose lines are traced to the bottling line and the connections are re-examined.

The entire bottling line is sanitized and sterilized, and the filters are tested to confirm proper functioning.

8.5.1   Bottling Line Sanitation

There are three distinct steps in preparing the line for bottling: cleaning, sanitizing, and sterilization. Procedures might vary at different line sections such as the filler bowl, filters, hoses and lines, and corker, as shown in Table 8.2. The areas that require sterilization are everything downstream from and including the membrane filters, filler bowl, vacuum lines, and corker. All other areas such as the bottle sparger, hoses, transfer lines, and tanks should be heavily sanitized, at minimum.

Chlorine compounds, citric acid, 70% v/v EtOH, TSP(sodium triphosphate), and alkaline cleaners are used to clean and sanitize the equipment. The primary methods used to attempt sterilization of the bottling lines are hot water, dry heat (thermal conductivity), and ozone. Dry-heat sterilization consists of heating the bottling system's metal parts via thermal conductivity to a temperature of 82°C (180°F) for approximately 30 min. The ozone chemical sterilization method uses ozone gas as an oxidizer to eliminate microbes throughout the bottling lines by injecting a volume of the gas into the system followed by a 20-min water flushing. Because ozone is a toxic gas, it can cause severe irritation to the skin and eyes. Many wineries prefer to use the standard hot-water method of sterilization despite the safety hazards associated with extremely hot water or steam.

Hot-water sterilization consists of heating water to above 82°C (180°F) and flushing it through a clean bottling system for a minimum of 30 min, maintaining 82°C throughout the system. Steam can also be used.

It is important when timing sterilization to begin the timing sequence when the water temperature is reached and can be maintained for the specified period. Monitor temperature at the coldest point in the system. Manufacturer's recommendations for bottling line preparation should be followed to prevent damage to parts, seals, gaskets, and filters.

Sanitation and sterilization quality control is monitored by frequent sampling (swabbing with plating) from various locations on the bottling line. Key

TABLE 8.2. Bottling line sanitation and common procedures.

Equipment	Cleaning	Sanitizing	Sterilization
Filter	1. Hot-water rinse	1. Flush with caustic cleaner (alkaline, TSP*) for 15 min	Hot water 82°C (180°F) for 30 min
		2. Citric acid rinse for 10 min	
		3. Water rinse for 15 min	
Filler bowl	1. Hot-water rinse	1. Hypochlorite wash for 15 min	Hot water 82°C (180°F) for 40 min
		2. Hot-water rinse for 15 min	
		3. Citric acid rinse for 15 min	
		4. Hot-water rinse for 15 min	
Lines and hoses	1. TSP	1. Hypochlorite wash for 15 min	Hot water 82°C (180°F) for 90 min
		2. Citric acid rinse for 15 min	
	2. Hot-water rinse	3. Hot-water rinse for 15 min	
Corker (jaws: disassemble for sanitizing and sterilization)	70% EtOH	1. Hypochlorite wash for 15 min	Hot water 82°C (180°F) for 20 min, or steam for 45 min, or iodine soak
		2. Citric acid rinse for 15 min	
		3. Hot-water rinse for 15 min	
Bottle sparger (nitrogen purge)	1. Vacuum		70% EtOH soak for 10 min
	2. 70% EtOH		

*TSP (Sodium triphosphate)

spots to swab are corker jaws, filler spouts, centering bell, bottle sparger, filler bowl, areas around seals, and cork bins (hopper).

A newer method called bioluminescence is being used in many wineries to monitor the cleanliness of the bottling lines. Bioluminescence utilizes the light-reflecting properties of adenosine triphosphate (ATP), which is a chemical produced by the microbes. Concentrations of ATP are proportional to the biomass present, resulting in changes in light reflectance. Measurement of these changes can alert bottling line personnel to potential contamination problems immediately rather than waiting for plating results.

8.5.2   Membrane Filter Integrity Test

Membrane filters are placed just before the filler bowl and are the last physical adjustment to the wine before it enters the bottle. Prebottle filtration is accomplished using a series of membrane filters usually beginning with a pore size of 2.0 μm or 1.2 μm. Subsequent filters will have a gradual reduction in pore size until the proper level of filtration is reached, typically 1.2 μm → 0.8 μm → 0.65 μm → 0.45 μm.

The membrane filter integrity test is commonly referred to as the "Bubble Point Test." The test consists of passing N_2 gas through the wet filter cartridge and reading the pressure at the point N_2 bubbles are noted on the outlet side of the filter cartridge.

Nitrogen is introduced into the filter system containing the wine to be bottled to a pressure of approximately 80% of the working filter pressure. The pressure is increased by 1 psi/min until the bubbles are detected. As the filters become dirty, the pressure will begin to increase to a point when the filter can no longer maintain its integrity and fails, requiring replacement (see manufacturer's filter specifications). It is prudent to assess imminent filter failure well before it occurs to prevent contamination of the wine.

A low bubble point can indicate flawed membranes or improper seating of the cartridge. Bubble point tests should be conducted throughout the bottling, primarily after periods of inactivity such as lunch and breaks.

8.6   Bottling

Bottling lines vary in size, output, mechanics, and function. Each line is different and strict attention must be paid to the manufacturer's recommendations for successful bottling. This section attempts to provide the reader with a generalized overview of the bottling process. Review Figure 8.3 to understand the progression of bottles as they move along a typical bottling line.

8.6.1   Bottle Queue

Larger wineries today purchase their empty wine bottles presterilized, which reduces the chance of contamination plus reduces labor costs resulting from hand washing of bottles. Smaller wineries might wash and sterilize their bottles prior to bottling. Solutions containing SO_2 (500 mg/L) or peroxide producing peracetate is used to sterilize the bottles, which are then, in turn, rinsed with filtered watered. ozone followed by a N_2 gas sparging can also be used to sterilize bottles.

Dry sterile bottles are placed neck up at the beginning of the bottling line. A canopy cover placed over the open bottles will protect them from dust or debris. If the bottles' finish (top) is accidentally touched or contaminated, a 70% EtOH solution can be sprayed over the top of the bottles to reduce any contamination.

The bottles proceed on a conveyor belt, or track, to a bottle sparger using vacuum and N_2 to remove any dust, dirt, or box particles from the bottle. Sparging with N_2 also temporarily reduces the amount of oxygen in the bottle if placed just before the filling station. A good N_2 flush can reduce the oxygen content to less than 1.0 mg/L (Boulton et al., 1999).

8.6.2 Membrane Filter

The wine is moved from the holding tank to the filtration system. Normal working pressure of the filtration system can be obtained from the manufacturer. Typical working pressures are less than 15 psi. The working pressure will increase as the filter becomes dirty; at approximately 30–40 psi the filters should be replaced (refer to manufacturer's information). Filter system pressure is monitored throughout bottling.

Before starting the line, it is prudent to circulate wine through the filters for approximately 10 min to remove any remaining water and to wet the filters with the bottling wine. This wine can be decanted or circulated through the filler bowl and spouts into sterile empty bottles to rinse any water from the system.

8.6.3 Filler

The size of the bottling line determines the number of filler spouts on a filler bowl. Smaller lines have 6–12 spouts, whereas larger continuous-flow lines can have 40–120 spouts. Filler units use different methods to fill the bottles: gravity fill, counterpressure, and vacuum with counterpressure.

Gravity fill utilizes the pressure difference between the elevated bowl and the empty bottles. Counterpressure fillers introduce pressurized wine into the bottle and slowly vent the gas as the bottle fills, creating a counterpressure. This method can inadvertently introduce oxygen into the wine. Vacuum and counterpressure units draw a vacuum in the bottle to fill and then switches to counterpressure venting of gas to slow the fill.

The filling spouts work by engaging the top of the bottle, which activates a valve to open. The length of the fill can be decided by the weight of the bottle on a pressure-sensitive plate that will disengage the bottle and close the valve when the proper weight (fill) has been reached, or simply by timing the valve closure.

Each of these spouts and the filler bowl are sterile and great care must be taken in maintaining the sterility. Many fillers and spouts are disassembled, scrubbed, and reassembled prior to the sterilization procedure. Wine should be rinsed (once or twice) through the filter, filler, and spouts prior to bottling to remove any remaining water used during the sterilization procedures. Contamination of a spout during bottling (such as touching the spout) can be minimized by spraying the surfaces with a 70% EtOH solution. The filler room itself must be equipped with a dust-free, positive-pressure filtered-air system. All personnel entering the room should wear appropriate clothing and head covers. Another source of contamination is created from condensation when bottling cold wine. As condensation develops on the bowl, it might drip down the bell, contaminating the wine and the bottle finish.

Periods of delayed bottling (1–2 h depending on the wine's susceptibility to contamination) might require the bowl be emptied of wine and the line resterilized. Smaller filler bowls are recommended in cases where there is contamination of the bowl and the wine must be discarded. If the line stops for any reason, always stop the filler bowl first to reduce waste. Filled bottles not yet corked when the line stops should be decanted.

The fill level of the bottles is critical by legal, winemaking, and economic standards. The volume of wine in the bottle is indicated on the approved label and care must be taken to assure that the volume is correct. Fill levels are monitored regularly throughout the bottling process. Bottles that are over-filled not only cost the winery in lost revenues but also can result in bottle leakage. As the temperature rises and the wine expands, the EtOH, CO_2, and water vapor pressure increases, resulting in a reduction of headspace in the bottle. Up to 6 mL of headspace can be lost due to this expansion (Boulton et al., 1999). Increasing pressure can push a cork up, allowing oxygen to enter the wine and wine to leak out. Bottles filled too low contain too much head-space, increasing the amount of air (oxygen) present in the bottle. Fill heights should leave 1–2 cm of space between the top of the wine and the bottom of the cork (closure). Adjustments of volume and cork insertion depths can be made.

Filling the bowl with wine should be as gentle a process as possible. Excessive splashing or allowing the wine level to drop too low can introduce oxygen.

8.6.4 Corking Machine

Corking machines are supplied with corks that have been placed in a sterilized hopper. Bags of corks and the bag opener are sprayed with a 70% EtOH solution before the bags are opened. The corks are dumped into the hopper by a properly attired worker. The cork is fed into as set of sterile "jaws" that grabs the cork, radially compresses it, and inserts it into the bottle neck at a pressure of 5–6 atm (Margalit, 1996). The cork springs back to fit snuggly in the neck of the bottle and remains in place by a high friction constant. Corks compressed to more than 33% of their original size could cause internal structural damage of the cork and is not recommended (Zoecklein et al., 1999). Corks from each head should be inspected for flaws, grease, creasing, and dust prior to the initiation of bottling.

Corkers might employ a vacuum system. Insertion of the cork into a bottle in a nonvacuum system can increase the air pressure in the bottle head-space up to 2 atm (Boulton et al., 1999). A vacuum system will reduce the pressure below 1 atm (0.3–0.6 atm per Boulton et al. (1999). Bottles corked under vacuum will exhibit a short period of slight foaming of the wine as gases come out of solution to equalize the headspace pressure. Atmospheric pressure during bottling should be noted and regarded with care. Adjustments to the vacuum level and fill heights along with storage considerations might be required.

Cork insertion depth is monitored throughout the bottling. Corks should never extend beyond the bottle finish or inserted too deep (1–2 cm from top of wine to bottom of cork).

The jaws and hopper of the corking machine creates a primary contamination problem, necessitating great care to maintain sterility. The jaws and hopper should be periodically spayed with 70% EtOH throughout the bottling, in addition to the procedures for filling the hopper and direct contamination. A technique of heating the jaws to 80°C or 90°C helps reduce contamination on higher-quality natural cork but can cause problems with corks treated with paraffin or silicon.

8.6.5   Quality Control

Laboratory QC is imperative up to this point in the bottling process. Postbottling samples for analyses are pulled immediately after the line begins. Samples are removed after the corker and always identified by date, time, and pallet number. Identification of the samples will be crucial if problems develop. Wine bottled during a problematic period can be isolated, held for retesting, decanted, or discarded.

Bottles are inspected for particulate matter, fill levels, and cork depth. The samples are analyzed for vacuum levels, dissolved oxygen, wine temperature, wine chemistry (including CO_2, density, SO_2, TA, pH, VA, RS, EtOH, and LMA), and sterile plated for microbiological monitoring. The key analyses indicating problems requiring immediate attention (such as dilution, oxidation, and equipment settings) must be performed quickly and include CO_2, SO_2, O_2, EtOH, fill heights, cork depth, temperature, and vacuum levels. In the event of problems, adjustments are made and sampling continued until the problem is solved. Samples are again collected after a short period of bottling, with the analyses repeated.

Additional sampling is periodically performed throughout the bottling at scheduled intervals and in the event of possible problems. The length of these intervals will depend on the speed of the bottling line being used. The object is to detect problems early to lessen the amount of wine affected.

Bottled wine requiring decanting should be analyzed for SO_2, EtOH, and, possibly, RS. The entire process from pulling corks to returning the wine to the bottling tank should be as sanitary as possible, especially in those wines that will not be sterile-filtered.

Performing QC requires properly working equipment. Always calibrate and check the function of all instruments prior to bottling.

8.6.6   Capsules, Wax, Labeling, and Storage

Most wineries feel that the corked top of the bottle should be covered to protect the cork and improve the appearance of the bottle. Wax has been a traditional method of sealing the bottle and is still extensively used today. Melted wax can be applied as a small cap atop the cork, or the entire neck of the bottle can be hand dipped. The most common method to finish off the bottle is using capsules. Capsules were previously made of lead, which was pliable and could be molded easily to the top of the bottle. Today, lead capsules are prohibited by many countries and have been replaced by materials such as tinfoil, aluminum, and plastic. Capsules can be twisted on by hand or shrunk onto the top of the bottle.

Bottling lines can incorporate wax or capsulation machines in the production line. Cosmetic appearance is the concern and bottles should be inspected periodically throughout the bottling for potential problems and flaws in the application.

Labeling can be a frustrating function of the bottling line. Neck labels, front labels, and back labels are applied to the majority of wine bottles. Labels are imperative to brand identification and require strict packaging quality control.

A labeling machine will press or roll the label onto the bottle and can utilize a non-water-based glue to apply the labels or utilize self-adhesive labels, which are becoming more popular. Labels that are crooked, have excess glue, or marred in any way should be removed and the wine bottle relabeled. The glues or adhesives used on the labels might be difficult to remove and require the bottles to be soaked in water and the labels manually scrapped off. This is very labor-intensive and the number of bottles requiring rework should be minimized with strict QC practices.

After the labeling process, the bottles are placed in cases (neck up or down depending on the winemaker and packaging) and they, in turn, are placed on a pallet or cart. The wine is stored in a cool, low-humidity environment until it is shipped. Wines are often retained at the winery because of a phenomenon referred to as bottle sickness or bottle shock, which occurs in some wines in the first month directly after bottling. Bottle sickness is a temporary condition resulting in a decrease in flavor. It is thought that acetaldehyde forms in the bottle because of the exposure of oxygen introduced during the bottling process. Maintaining low oxygen levels and the use of adequate amounts of free SO_2 to react with the oxygen can reduce or eliminate bottle sickness. Typically, Chenin blanc, Riesling, Gewürztraminer, Muscat, French Colombard, and Blanc de noir are shipped quickly. Chardonnay, Sauvignon blanc, and Semillon can be held 1–2 years before release and red wines can be held up to 3 years before release (Margalit, 1996).

Wines that are to be bottle-aged at the winery, such as sparkling wines, are placed in a cool, low-humidity environment until they are ready to be labeled and shipped. Excess humidity can increase the chance of mold developing around the cork, and temperature fluctuations can lead to leaking and oxidation. A constant temperature is imperative when storing wine. The heating and cooling of the wine and the related pressure changes will work against the cork, pushing it out of the bottle over time. The suggested temperature to store bottled wine is 13°C (55°F).

Sparkling wine can be bottled-aged on yeast up to 5 years, but most are aged 6 months to 1 year. Champagne in France is aged for a minimum of 1–3 years and Spanish Cava is aged a minimum of 9 months.

8.7   Microbiological Monitoring

The plating of wine to detect microbial contamination has been discussed in several sections of this book. It is important to mention a few key suggestions and elements for this procedure.

Successful plating and general identification of microbes requires personnel trained in aseptic technique and microbe identification, plus several critical pieces of equipment:

- Steam autoclave: capable of producing temperature greater than 120°C (250°F) for a minimum of 15 min to sterilize filtration setups, tools, water, and plated media: (Sterile equipment and media may be purchased.)
- Incubator: temperature control to 30°C (86°F)
- Microscope: 10× and 40× dry lenses plus a 100× oil emersion lens; phase contrast

Smaller wineries with limited bottling and/or red wine production might find it more economical to send their wine samples to a credited laboratory for plating, but larger wineries will greatly benefit from a microbiology section in the laboratory. A 70% v/v EtOH solution is used throughout the plating procedure to maintain aseptic conditions.

8.7.1   Plating Media

Typical micro-organisms of primary interest to wineries are yeasts, molds, and bacteria (fungi and bacteria). Yeasts such as *Saccharomyces cerevisiae*, *Brettanomyces, Kloeckera*, and film yeasts create the largest problems in wineries. Spoilage bacteria such as *Lactobacillus*, *Acetobacter*, *Oenococcus, Pediococcus*, and *Penicillium rogueforti* are of greater or equal importance.

Yeasts or bacteria detected in sterile-filtered bottled wines indicate a breech in the sterility of the bottling process. This puts the off-dry to sweet wines at risk of refermentation in bottle or wine spoilage. Wines bottled without sterile filtration are assessed for the amount of contamination indicted by the plating results. The pore size will dictate what type of contamination can be expected. Wines that were filtered to the 0.8-µm level should be void of yeasts but could grow some bacteria; larger pore sizes will allow all yeasts and bacteria to enter the wine and will subsequently show up on the microplating. Additional information about yeasts and bacteria taxonomy and identification can be found in Chapter 6.

A medium is a sterile nutrient system for the cultivation of organism. Media can be made in the laboratory, poured into small glass or heat-resistant plastic plates (Petri dishes), and autoclaved for sterility. Laboratories can purchase prepared sterile plates ready for plating. The type of medium used depends on the organisms to be detected. Additions of certain chemicals can isolate a species of yeast or bacteria resistant to the chemical by killing organisms that are not resistant. This isolation technique allows for easier identification of targeted microbes (selective plating).

Typical prepared or purchased media for yeasts and some bacteria propagation are Wallerstein Laboratory Nutrient medium (WLN), Wallerstein Laboratory Differential medium (WLD), malt extract medium, and wort (grain or malt) medium. Apple Rogosa nutrient agar media with cycloheximide can be used to propagate bacteria cultures.

Isolation of Brett is achieved using one of the above media with the addition of cycloheximide ($C_{15}H_2 3NO_4$) also known as Actidione. Brett is one of a very few wine yeasts resistant to cycloheximide that allows isolation of this particular yeast. Prepared medium levels of 10 mg/L cycloheximide are sufficient to reduce all other yeasts. Preparation of media requires the addition of 20 mg/L of cycloheximide to compensate for the half-life of cycloheximide when the media is autoclaved for sterility. The addition of thiamine (Tryptone ™) to malt media at a rate of 2 g/L will encourage Brett growth (Boulton et al., 1999).

Lactobacillus sp. *Acetobacter* sp., and selective *Saccharomyces* sp. can be isolated with chemical additions. Additional information concerning identification and propagation of bacteria can be found in *Bergey's Manual of Determinative Bacteriology*, 9th edn., (Williams & Wilkins, 1994).

8.7.2   *Plate Inoculation, Incubation, and Identification*

Plate inoculation can be achieved by using a sterile loop, probe, or swab to transfer the sample to the sterile media plate. The transfer methods are useful for small samples, isolating a colony from another plate, and swabs taken from equipment. Using an aseptic technique, a portion of the sample is lightly streaked on a small section of the media plate in parallel lines. Repeat twice using two additional sample portions on the same plate with each portion perpendicular to the other. Liquids can be sterile-filtered through a 47-mm, 0.45-μm, or 0.2-μm filter with grids placing the filter on the sterile media plate. Filtration is the best method to plate wine samples.

Filtration requires a sterile filter holder, sterile filter, careful sterilization of the bottle finish, and sterile instruments to transfer the filter. Filter holders are sterilized by autoclave or a laboratory might wish to purchase sterile disposable units if they do not have an autoclave or method of sterilization.

Optimally, the samples are filtered and plated in a laminar flow hood to reduce unwanted plate and sample contamination resulting from airborne microbes. The laminar flow hood takes up room air, filters it, and blows it forward across the work area, keeping unclean air from the work area. Laboratories without laminar flow hoods should filter and plate their samples in a closed area free of drafts.

The entire work area (including the vacuum manifold, or lines used to assist filtration) should be clean, sprayed with 70% v/v EtOH, and allowed to sit for 10 min before using. The sterile filter holders (filter funnel) are carefully attached to the vacuum system manifold or to a Buchner flask attached to a vacuum source (one sterile setup per sample). Tweezers tips or forceps are dipped into alcohol and placed over an open flame (flamed) to sterilize. A sterile filter is picked up by the sterile forceps and place on the filter holder. The bottle finish is spayed with 70% EtOH and allowed to sit a minute or two; the excess alcohol is shaken off and the bottle flamed. The cork is pulled halfway out and the process is repeated until the cork is removed completely (Boulton et al., 1999). If the sheer quantity of samples prohibits the use of the above cork-removal method (or the bottles do not have a cork), the bottle finish is sprayed with 70% EtOH before and after the cork is removed. Any part of the corkscrew device that touches the bottle finish or cork (screw and foot) should be sterilized with 70% EtOH and flamed or autoclaved before use.

The amount of sample to be filtered is dependent on the expected level of contamination. Plating samples that are too contaminated (> 50 colonies per filter) will make it difficult to accurately count, isolate, and identify the colonies. Wines that are turbid or unfiltered or where the expected colony count is unknown will require dilution of a smaller wine sample. For these types of sample, it is recommended that two samples be prepared at different dilution levels in the hope that one will be within the counting range of 10–200 CFU (colony-forming units). If the counts of the two samples are in range, each should be counted separately and the mean reported. Counts above 200 CFU are reported as TNTC (too numerous to count) and the test should be repeated using a new sample. Sterile-filtered wine has a very low expected colony count, so an entire bottle is filtered.

Vacuum is applied to the filtration setup and the bottled wine is poured into the filtration cup and allowed to pass through the filter (a sterile water control is filtered at the same time for quality control evaluation of the plating procedure). Immediately turn off the vacuum when the sample has passed through the filter, and using a sterile technique, transfer the filter to the sterile media plate; the used wine is discarded. You can use gravity to drain the wine through the filter only if the wine is very clean or has been sterile-bottled. The gravity method is not recommended because it takes time to allow the wine to drain, thus increasing the chance of sample contamination.

Using a sterile technique, place the filter on the plate, grid side up, cover, identify, and place upside down in the incubator. Yeast plates are incubated at temperatures between 20°C and 30°C (68–86°F) and will likely show some growth in 2–3 days *(Saccharomyces)*. Incubation of the plates for 7 days is recommended to compensate for any slow-growing organisms (*Brettanomyces* will show growth in 5–7 days). Bacteria plates are incubated at temperatures not to exceed 25°C (77°F) and will likely show growth in 10 days. Single plates that are used for both yeast and bacterial growth are incubated between 20°C and 25°C for a minimum of 10 days.

At the end of the incubation period, a visual examination should include the following:

- Colony count for each type of organism found (yeast, bacteria, or mold); calculate the number of organisms per milliliter of sample used (taking into account dilution factors)
- Colony size for each
- Description of the colony for each, including form, shape, elevation, texture, and color
- Color changes of any colony grown on certain media (WL agar), which would indicate acidic changes
- Note any abnormal odors detected from the colonies (*Do not* directly sniff the plate)

Figure 8.2 shows a typical plating result of a contaminated bottled wine. Earlier in this chapter (Section 8.2.3.1) and in Chapter 6, the identification of yeasts and bacterium has been discussed. Identification of organisms should be conducted with aseptic techniques and protocol. The wine laboratory technician should not touch the cultures with bare hands or sniff the cultures which could inadvertently infect the technician. Winemaker's might require additional plating of positive growth samples to identify the organism and isolate the origin of the contamination. Further identification might require microscopic examination, replating a colony on selective chemically treated differential media, Gram's colorant test, and/or DNA identification.

Microscopic evaluation includes the following:

- Cell morphology (shape and size)
- Budding characteristics (yeast) or the presence of pairs or chains (bacteria), including the number of cells in a chain
- Presence of spores, flagella, or other characteristics

Bacterial identification for Gram-positive or Gram-negative attributes can be performed easily. The sample is placed on a stick and dried by the flame from a Bunsen burner. The sample stick is dipped into violet colorant, then dipped into an organic solvent (alcohol/acetone solution), and, finally, dipped in a rose colorant. Gram-positive bacteria will not be affected by the organic solvent and will remain violet colored; therefore, Gram-negative bacteria will react with the organic solvent and will be rose colored. Further bacterial identification information can be found in *Bergey's Manual of Determinative Bacteriology*, 9th ed., (Williams & Wilkins, 1994).

The technological advancements in microbial DNA identification have enabled laboratories to identify and quantify microbial contamination without plating. In order to perform these tests, wineries must hire qualified and trained personnel as well as investing in highly technical equipment. The major impact on the wine industry is the ability to quickly identify wine contamination from spoilage organisms before they can be organoleptically detected and the immediate release of wine for shipping. For large wineries using "just-in-time" bottling (wines bottled and shipped according to the demand), the technology allows immediate shipment of certain wines without waiting for the plating results, totally eliminating the warehousing procedure. Currently in wineries, the most economical method of microbial DNA identification is polymerase chain reaction (PCR). The method is based on the amplification of the genome portions of an organism with the use of genetic markers or primers (oligonucleotides), which hybridize with the DNA matrix. An electrophoretic profile of the amplification obtained allows different levels of classification with a particular genus or species of organism to be achieved.

Other identification methods include genomic analysis and hybridization. Genomic analysis compares the DNA of one organism to the known DNA of another. Electrophoretic profiles of the cytoplasmic soluble proteins of the cell are used for comparison to known electrophoretic profiles and ribosome sedimentation speed (Svedberg value). Hybridization utilizes the ability of a single strand of DNA to separate under certain conditions and reassociate when the conditions are reversed. Two single strands of DNA, one a known control with a DNA probe and the other the unknown, are separated and reassociated. The reassociation results in one hybrid strand due to the complementary DNA sequences. If the resulting homology is 70% or more, the unknown organism is considered the same species.

In the future, general identification of fungi and bacteria will utilize one or more of these methods, eliminating the need for plating. Look for technologies such as RFLP (restriction fragment length polymorphism), TAFE (traverse alternating field electrophoresis), FIGE (field-inversion gel electrophoresis) and others to become the norm.

Ideally, bottled wines that have been sterile-filtered will have clean (no growth) plates. If growth exists and the control plate is clean, contamination of the bottling process has occurred. If the control plate is also contaminated, the entire new group of samples will need to be replated. Most often, resampling involves randomly pulling 12 bottles of wine per pallet. Keeping the samples well identified throughout the process will eliminate unnecessary plating of samples and allow the clean portion of the bottling to be shipped.

8.8   Required Analyses for International Trade

Bottled wines shipped to foreign countries require documentation and approval from the TTB in the United States before shipping. Receiving countries require certain analyses documentation from the shipper before receiving the wine through customs. Completed documents, which include analyses from a TTB-certified laboratory, are sent to TTB. When approved and stamped, the documents are returned to the shipper, copies made, and the original is then sent to the person or company receiving the wine (consignee), allowing them to retrieve the product from customs in their country.

The EU, which includes Austria, The Netherlands, Luxembourg, Germany, Denmark, Belgium, France, Italy, Greece, Spain, Portugal, Finland, Sweden, Ireland, and the United Kingdom, accepts the TTB documentation Form VI-1. This form has a short and a long version. The long version is filled out by shippers that are not currently registered with the TTB, are not the producing winery, or are shipping bulk wine. Wineries can submit Form IV to the TTB, which states that the winery is the producer of the wine to be exported and the wine has been made according to the United States rules and regulation of production. If approved, the winery will be regarded as Self-certified, which means they can fill out their own short version of the VI-1 Form, including the analyses obtained from the certified TTB laboratory.

Japan requires a certificate of analyses to include EtOH, free and total SO_2, RS, sorbic acid, total dry extract, and volume confirmation. Switzerland (not being a part of the EU) does not require a processed VI-1 Form but does require a certificate of analyses. The Swiss will accept the analyses on the VI-1 Form signed by the winery and there is no need to get approval or notify the TTB. The Pacific Rim requires analyses to include EtOH and methanol, free and total SO_2, RS, sorbic acid, total dry extract, and volume confirmation. Other countries might have their own importation regulations that will need to be learned and complied with prior to exporting the wine.

The information required for analyses and the VI-1 Form is as follows:

- Two labeled and finished bottles of wine sent for analyses
- Precise label information of each wine
- Quantity of wine being shipped
- Name and address of the consignee
- Noncertified analyses such as free SO_2, density, RS, pH, and their tolerances
- The amount of any additives to the wine during processing
- Information regarding the production of the wine (i.e., stabilization, filtration, acid adjustments)
- Microbiological data
- Varietal, vintage, appellation, and composition information

Certified analyses required for the VI-1 Long Form is EtOH, alcohol (total), TA, VA, citric acid, total dry extract, and total SO_2 to be documented and signed by the certified laboratory. RS analysis is often included and red wines might include hybrids qualitative analyses. The VI-1 Short Form requires EtOH, TA, and total SO_2 analyses documented and signed by the certified laboratory. Japan requires EtOH, extract, sorbate, and total SO_2 analyses documented and signed by the certified laboratory.

With global commerce increasing and the strong desire of countries to reach agreement on the analyses requirements for worldwide export, wineries will be required to provide much more analysis documentation in the future. Health concerns related to pesticides, fungicides, genetically modified ingredients, and allergy-related additives are just a few of the topics currently being discussed that someday might require wineries to provide analytical documentation relating to these area.

8.9   Bottle Aging

The ideal temperature to keep bottled wine for aging is 13°C (55°F) at a lower humidity. All attempts should be made to maintain a constant temperature in the wine storage area. Heating and cooling of the wine increases and decreases the bottle pressure, which can slowly move a cork up the neck of the bottle, allowing air into the bottle and possible leakage of wine.

Many fruity and sherry wines are intended to be consumed shortly after bottling and require little to no aging. Red wines and a few white wines will benefit from the development of dimethyl sulfide and acetaldehyde, which occurs during bottle aging and is part of the phenomena termed *bottle bouquet*.

Aged red wine loses its fruitiness over time and becomes more red-brown in color, and the tannins soften, resulting in greater complexity and softer mouth feel. Chapter 7 discusses maturation and aging in more detail.

Typical cellar time for bottle aged wines is as follows:

- Dry white wine: 0–4 years (5 years maximum)
- Dry red wine: 0–10 years
- Port wine: 0–10+ years
- Bottle fermented sparkling wine: 1–3+ years

There are many red wines that are cellared for many more years. Wines cellared for long periods of time should be recorked approximately every 20 years.

9

Analytical Procedures

9.1   Introduction

Wine laboratories vary extensively in size and sophistication. The methods and procedures outlined in this chapter are those that are performed in most wine laboratories. The choice of procedure or method will vary based on the analyte of interest, available equipment, cost, experience of the analyst, degree of accuracy required, sample load, time restrictions, and staffing. Additional methods and/or alternate procedures can be found in reference texts, such as *Wine Analysis and Production* and *Methods for Analysis of Must and Wines* in the Bibliography.

To understand the analysis results, you must understand the theory behind it. To achieve confidence in your analyses, you must understand and practice excellent laboratory techniques and skills, quality control, and quality assurance.

This chapter will cover analytical procedures. Chapter 10 will cover reagents and standardization of reagents. Each procedure contains a summary (or a text reference) explaining the concept behind the analysis, method type, typical equipment used, reagents required for the analysis, required accuracy of the measurements, application notes, and a listing of the most frequent analytical errors.

Just a few reminders before you begin:

- Please read Chapters 2 and 3 before you attempt any laboratory analysis.
- Temperature is critical when measuring liquids; always check the temperature of the samples.
- Unless a container or measuring device is clean and dry, always rinse the container used for measurement or analysis with the final wine sample before use.
- Practice and perfect your pipetting technique, which is paramount for accurate measurements.
- Know how to read a meniscus correctly.
- Recheck all readings and notations for errors.
- Read the manufacturer's operations manuals.
- Keep all equipment clean and in excellent working condition.
- Safety should be foremost in your mind in preparation and performance of any analyses.

9.2   Procedures

9.2.1   *Spectrometric Color and Phenolic Measurement*

Concept: Measurement of phenolic compounds and pigments: SO_2 (bleaching effect) used to bind with monomeric anthocyanins; HCl used to lower pH to increase flavylium anthocyanins (red color form) from the colorless water base in red wine; acetaldehyde used to bind any SO_2 that could bleach anthocyanins; distilled water dilution to breakdown copigmentation of monomeric anthocyanins. Refer to Chapter 5, Section 5.3.8.

Method Type: Spectrometric with selective chemical analyses

Equipment Required:

- Spectrometer single beam – absorbance (A) wavelength settings of 280 nm, 320 nm, 420 nm, and 520 nm; adjusted (manually or automatically) for dark current at 0% transmission (T); light at $100\%T$ using a distilled water blank to set instrument's normal absorbance; zero instrument (some instruments may zero to air); $A = \log (100/\%T)$; consult manufacturer's operation manual. Calculate $\Delta A = $ (sample $A_1 - A_2$) – (blank $A_1 - A_2$).
- Methacrylate cells (cuvette): 10 mm (use of only 10-mm cells requires dilution of red wine in most tests; absorbance changes are not linear to dilution)
- Quartz cells (cuvette): 1 mm and holders (preferred unless otherwise indicated)
- Micropipette (accurate range 0–100 µL): manual or digital with appropriate tips
- Serological glass pipettes: 0–10 mL (or autopipette, with appropriate tips)
- Test tubes: 15-mL capacity with closure
- Class A volumetric flask, 10 mL
- Pasteur pipettes: used to transfer samples to cuvettes
- Laboratory wipes

Reagents Required:

- Copigmentation buffer
- 20% w/v SO_2 (sulfur dioxide) solution
- 10% C_2H_4O (acetaldehyde) solution
- $1N$ HCl (hydrochloric acid)
- Distilled water (dilutions if required and cell blank)

Measurement Accuracy: SO_2, C_2H_4O, and buffer approximate; HCl, distilled water dilutions, and wine samples accurate.

Note: Absorbance readings accurate in the range of 0.1–2.0 absorbance units (AU); $A < 0.001$ will be distorted by electronic noise; $A > 2.0$ is too dense for accurate measurement, dilute except where noted; dilute samples with distilled water by factors of 10 (10×, 20×, 100×, etc.). All results must be corrected to the reportable 10-mm cell light path length cell (i.e., use of a 1-mm, 2-mm, or 5-mm cuvette cell would result in multiplying A by 10, 5, or 2, respectively). A_{280} measures all phenols; A_{320} measures hydroxycinnamates in white wine; A_{420} measures brown pigments; A_{520} measures red pigments (anthocyanins). SO_2 and sugars could interfere with absorbance readings with the exception of A_{280} readings. There are no set tables or scales to compare results. Results are utilized for historic evaluation of the wine/juice defined by varietal and used in combination with other chemical and sensory evaluations.

Possible Errors: Incorrect dilutions; pipetting errors; spectrometer not zeroed prior to reading; wrong wavelength set; improper mixing; calculation errors; lamp failures; corrections not made for cell size or dilution; absorbance range > 2.0.

Procedure:

- *Visible Color, Brown Pigments, Hydroxycinnamates, and Phenols*

  White wine: filter sample with 0.45-μm filter; place in test tube
  Red wine: place red wine sample in test tube
  Using the Pasteur pipette, transfer wine sample to 1-mm quartz cuvette.

  Set spectrometer to 520 nm; zero; read the absorbance; do not dilute sample. Set spectrometer to 420 nm; zero; read the absorbance; dilute as needed. White wines only: set spectrometer to 320 nm; zero; read the absorbance; no dilution should be required. Set spectrometer to 280 nm; zero; read the absorbance; dilute as needed.
  Dilutions: Place appropriate aliquot of wine sample in test tube; add appropriate volume of distilled water; cap and mix thoroughly; transfer diluted sample to 1-mm cuvette; repeat the readings. Correct all readings for 10-mm cuvette; correct red wines for any dilution (not indicated in formulas). Note results:

$$\text{Visible color} = A_{520}, \qquad (1)$$

where $A_{520} = \text{AU} \times 10$;

$$\text{Brown pigments} = A_{420}, \qquad (2)$$

where $A_{420} = \text{AU} \times 10$;

$$\text{Hydroxycinnamates (white wine)} = A_{320} - 1.4, \qquad (3)$$

where $A_{320} = \text{AU} \times 10$; $1.4 = \text{AU}$ nonphenolic material

$$\text{Phenols} = A_{280} - 4, \qquad (4)$$

where A_{280} = AU × 10; 4 = AU nonphenolic material.

- *Total Red Pigment and Total Phenols (No Copigmentation, Maximum Red Pigment)*

Add to 10-mL volumetric flask:
    100-μL wine sample
    Bring to volume with 1N HCl

Cap and mix thoroughly. Incubate at 20°C for 3 hours. Transfer incubated sample to 10-mm cuvette. Set spectrometer to 520 nm; zero; read the absorbance. Set spectrometer to 280 nm; zero; read the absorbance. Dilute as needed; correct for dilution and note:

$$\text{Total red pigment} = A_{520}^{HCl}, \qquad (1)$$

where A_{520}^{HCl} = AU × 100;

$$\text{Total phenols} = A_{280}^{HCl} - 4 \qquad (2)$$

where A_{280}^{HCl} = AU × 100; 4 = AU nonphenolic material
or

$$\text{Vineyard tracking of phenols} = A_{280} - 4. \qquad (3)$$

- *Polymeric Anthocyanins, SO_2 Resistant Pigments*

Add to test tube:
    2-mL wine sample
    40 μL of 20% w/v SO_2 solution

Cap and mix thoroughly. Incubate for 1 min. Transfer incubated sample to 1-mm quartz cuvette. Set spectrometer to 520 nm; zero; read the absorbance. Correct for 10-mm cuvette and note:

$$A_{520}^{SO_2} = \text{AU} \times 10. \qquad (1)$$

- *SO_2 Bleached Pigments*

Add to test tube:
    2-mL wine sample
    20 μL of 10% acetaldehyde solution

Cap and mix thoroughly. Incubate for 45 min. Transfer incubated sample to 1-mm quartz cuvette. Set spectrometer to 520 nm; zero; read the absorbance. Correct for 10-mm cuvette and note:

$$A_{520}^{ACT} = \text{AU} \times 10.$$

- *Polymeric Anthocyanins*

Add to test tube:
    100-μL wine sample
    1.9 mL Co-pigmentation buffer

Cap and mix thoroughly. Incubate a few minutes. Transfer incubated sample to 10-mm cuvette. Set spectrometer to 520 nm; zero; read the absorbance. Correct for dilution and note:

$$A_{520}^{BUF} = AU \times 20. \tag{1}$$

- Discard samples and thoroughly clean and dry equipment.

Additional determinations using the above results:

Total monomeric anthocyanins (mg/L) $= 20 \times [A_{520}^{HCl} - (5/3)(A_{520}^{SO_2})]$ (1)

Color Density $= A_{420} + A_{520}$ (for Organisation International de la Vigne et du Vin (OIV), add A_{620}) (2)

Color hue $= A_{420} / A_{520}$ (3)

Total flavonoids (white wine) $= (A_{280} - 4) - [0.66(A_{320} - 1.4)]$ (4)

Copigmentation color $= A_{520} - A_{520}^{BUF}$ (5)

Visible anthocyanins (flavylium form) $= 20(A_{520} - A_{520}^{SO_2})$ (6)

Percent anthocyanins (flavylium form) =

$$\left\{ \left(A_{520} - A_{520}^{SO_2} \right) \left[A_{520}^{HCL} - (5/3)(A_{520}^{SO_2}) \right] \right\} \times 100 \tag{7}$$

Chemical age factor color proportion (polymeric anthocyanins) $= A_{520}^{SO_2} / A_{520}^{HCl}$ (8)

SO_2 bleaching (loss of visible color) $= A_{520}^{ACE} - A_{520}$ (9)

9.2.2 Density Measurement, Direct Method

Concept: The mass of a substance *divided* by the *volume* of the substance.

$$\text{Density (g/cc)} = \text{Substance mass/Substance volume.} \tag{1}$$

Refer to Chapter 1, Chapter 5 (Section 5.3.2), and Chapter 8 (Section 8.4.1). **Method Type:** Direct measurements, accepted by the Alcohol and Tobacco Tax and Trade Bureau (TTB) (4–2004) for bottle fill.
Sample Volume Required: 150 mL
Equipment Required:

- Analytical balance, calibrated
- Class A volumetric flask, 100 mL: Class A volumetric flask ranging from 10 to 100 mL can be used in lieu of a pyncometer (Howe, 2000a)
- Water bath, 20°C (68°F)
- Thermometer: accurate; clean and dry with each use
- Pasteur pipette
- Lint-free cloth

Reagents Required: None

Measurement Accuracy: Temperature of liquid solution must be accurate.

Note: The formula for the calculation and conversion of the density of wine in grams per liter to pounds per gallon

$$\text{Density (lb/gal) } (20°C) = 453.59 \text{ g/gal } [(X \text{ g}/100 \text{ mL}) \times 10] \times 3.7854 \quad (1)$$

or

$$\text{Density (lb/gal) } (20°C) = (X \text{ g}) (0.0834542) \quad\quad\quad (2)$$

where 10 converts g/100 mL to g/L; 3.7854 converts g/L to g/gal, 453.59 g/gal is the weight of 1 gal of water and converts grams to pounds.

Possible Errors: Inaccurate balance; neglecting to tare the balance; incorrect measuring temperature; incorrect meniscus reading resulting in incorrect volume; wrong type of flask; calculation error; reporting error.

Procedure:

Place a clean, dry, volumetric flask on the balance and tare (zero).

Weigh the flask and note the weight.

Add the wine sample to just above the volume mark.

Obtain the wine temperature; if not 20°C, cover the flask and place in a 20°C water bath to bring the wine to temperature.

At temperature, or slightly above, remove the flask, dry with a lint-free towel, remove the cover, tare the balance, and place the flask on the balance.

Quickly insert the thermometer, obtain the temperature reading, and note.

Bring the flask to volume by using the Pasteur pipette to remove the excess volume (read bottom of the meniscus).

Read the weight of the filled flask and note.

Subtract the weight of the empty flask from the filled flask. Report the density to three or more decimal points and the temperature of the measurement [i.e. Density (20)].

Discard sample; clean and dry all equipment thoroughly.

9.2.3 Specific Gravity Measurement

Concept: The *density* of a solution *compared* to the *density* of water (no units):

$$\text{Specific gravity} = \frac{\text{Density of the same volume of water}}{\text{Density of a volume of solution}}$$

Refer to Chapter 1, Chapter 5 (Section 5.3.2), and Chapter 6 (Section 6.4.3).

Method Type: Hydrometry, accepted by the AOAC.

Sample Volume Required: 375 mL

Equipment Required:

- Hydrometer: certified and calibrated
- Graduated cylinder: 250-mL glass or plastic

- Thermometer: accurate; clean and dry with each use
- Water bath, 20°C
- Laboratory wipes

Reagents Required: Surfactant agent such as antifoam

Measurement Accuracy: Temperature and hydrometer readings must be accurate; wine sample volume can be approximate.

Note: Any change in temperature of the sample above or below the hydrometer calibration temperature (commonly 20°C; will be noted on hydrometer or certification) must be taken into consideration.

Temperature correction: each degree above 20°C, add (+) 0.0002,   (1)

each degree below 20°C, subtract (−) 0.0002.   (2)

For accuracy, read the specific gravity (SG) at the hydrometer calibration temperature rather than using temperature correction. Report the measurement temperature of the sample (20°C) over the density of water at 20°C (20/20), which equals 1.000. No units.

Possible Errors: Using an inaccurate hydrometer or thermometer; misreading the meniscus or thermometer; wrong temperature correction or no temperature correction where indicated; leaving sample exposed to air too long [can lose up to 0.1 g/L/h alcohol in wine samples, which will change the readings (Howe, 2000a)]; hydrometer stem not wiped clean before reading; excess gas in sample; bubbles around hydrometer; hydrometer touching side of cylinder; dirty hydrometer (oil, grease, detergent, or dirt; keep clean and do not touch bulb of hydrometer); solid material in sample.

Procedure:

Obtain the wine/juice temperature. If the sample is not 20°C, place the sample bottle in a 20°C water bath until it reaches temperature.

Thoroughly rinse a clean, graduated cylinder with wine/juice sample; discard sample.

Fill the cylinder to approximately the 200-mL range.

Observe the sample and the content of gas (CO_2 in the sample will result in erroneously high readings by increasing the buoyancy of the hydrometer). Juice samples containing excess CO_2 require degassing (do not degas wine samples as loss of alcohol will effect reading). Degas sample by pouring the contents of the cylinder into a clean, juice-rinsed container; pour the contents of the container back into the cylinder and repeat until there is minimal foam or gas production (driving off CO_2). Excess foam might require the addition of a drop or two of anti-foam.

Place a clean, dry hydrometer in the sample by the stem. Gently spin the hydrometer to release any adhering air bubbles.

Ensure the hydrometer is free floating; touching the sidewall of the cylinder will adversely affect the reading.

Carefully lift the hydrometer and wipe the stem (fluid on the stem will weigh it down, adversely affecting the reading).

Gently release the hydrometer and allow it to equilibrate.

At eye level, read the SG at the closest graduation to the bottom of the meniscus, and note.

Insert the thermometer into the cylinder; obtain the sample temperature and note.

Make temperature correction to the SG reading as necessary.

Report the SG to four places with measuring temperature over 20°C [i.e. SG (20/20)].

Discard sample; clean and dry all equipment thoroughly.

9.2.4   Soluble Solids Measurement via Hydrometer

Concept: The measurement of the soluble solids, which equates to sugar content in juice and is measured as °Brix, °Baumé, or Oechslé. This is a measurement of the *weight* of a solution *compared* to the *weight* of a solution containing sucrose (i.e. 1 °Brix = 1 g of sucrose per 100 g of solution). °Brix will be measured in this procedure. Refer to Chapter 5 (Section 5.3.2) and Chapter 6 (Section 6.4.3 and Fig. 6.9). Monitoring of fermentation dictates daily analysis of °Brix until two to three consecutive negative readings are obtained.

Method Type: Hydrometry
Sample Volume Required: 300 mL minimum
Equipment Required:

- Hydrometer: certified, calibrated, and of the correct range (0–35 °Brix, −5 to +5 °Brix)
- Graduated cylinder: 250 mL, plastic
- Thermometer: accurate; clean and dry with each use
- Strainer: to remove solid material from juice
- Containers: 3 liters, plastic
- Bucket, plastic (or cart)
- Towels, or paper towels

Reagents Required:

- 70% v/v EtOH (ethanol) solution in a plastic spray bottle
- Surfactant agent such as antifoam

Measurement Accuracy: Temperature and hydrometer readings must be accurate; wine sample volume can be approximate. Because the majority of wineries take measurements in the cellar, conditions are not ideal but every attempt should be made to eliminate variables.

Note: Any change in temperature of the sample above or below the hydrometer calibration temperature (commonly 20°C and will be noted on hydrometer or certification) must be taken into consideration.

Temperature correction: each degree above 20°C, add (+) 0.06,   (1)

each degree below 20°C, subtract (−) 0.06.   (2)

The factor for °Baumé is ± 0.05. As the alcohol rises during fermentation, the buoyancy of the hydrometer will decrease, resulting in negative readings at the end of fermentation. Samples are most often obtained and measured in the cellar from tanks and barrels (bbl). Safety precautions (Chapters 2 and 6) regarding CO_2 exposure and safety in the cellar must be taken. Never use a hose to siphon juice or wine from a barrel (potential high levels of CO_2 or SO_2). Protect equipment when working in the cellar.

Possible Errors: Using inaccurate hydrometer or thermometer; misreading the meniscus or thermometer; using incorrect range of hydrometer; mistaken temperature correction or no temperature correction where indicated; hydrometer stem not wiped clean before reading; excess gas in sample; bubbles around hydrometer; hydrometer touching side of cylinder; dirty hydrometer (oil, grease, detergent, or dirt; keep clean and do not touch bulb of hydrometer); solid material in sample; not flushing the sampling valve on tank; inadequate barrels sampled.

Procedure:

Place all required equipment in bucket or on a cart.
Cross-reference wine lot identification.
Obtain a 300-mL clean sample in a 3-L container:

- *Tank*: Place a 3-L container under valve opening, carefully and slowly open the sample valve or racking valve, allow the container to fill, discard and repeat (flushes the valve area to obtain best sample; juice can also be returned to tank upon completion of test), and fill the container to approximately the 300-mL mark. Strain a small amount of the juice through a sieve placed on top of a second 3-L container, rinse the container and discard the juice, and repeat but place the juice in the cylinder, rinse thoroughly, and discard the juice. Strain the remaining sample into the second container. Place the cylinder inside the first container (to contain overflow) and sit on a level surface; fill the cylinder with sample to approximately the 200-mL mark. For tanks with no valves or with bins, dip a 3-L container down into the top middle portion of the tank or bin or use a tank sampler with a chain to obtain the sample and continue as stated above.
- *Barrels:* Sample at least two barrels. Brush any debris from around bung hole and remove fermentation bung, placing it on a clean surface. Using a clean wine thief or pipette, obtain a sample of juice, rinse thief or pipette with the sample and expel into a 3-L container, rinse the container, and discard. Obtain an equal volume of sample from the middle of each barrel to be tested and place in the 3-L container, replace bung

(never leave a barrel without the proper bung), and pour a small amount of juice into the cylinder, rinse, and discard (barrel fermented juice usually has very little solid material and straining is not required). Place the cylinder inside the second container (to contain overflow) and sit on a level surface; fill the cylinder with sample to approximately the 200-mL mark.

Observe the sample and the content of gas (CO_2 in the sample will result in erroneously high readings by increasing the buoyancy of the hydrometer). Samples containing excess CO_2 require degassing by pouring the contents of the cylinder into the first container; then pouring the contents of the container back into the cylinder, repeat until foaming and gas generation is minimal (driving off CO_2).

Place a clean dry hydrometer in the sample by the stem. Gently spin the hydrometer to release any adhering air bubbles. Excess foam might require the addition of a drop or two of antifoam to properly read the meniscus.

Ensure the hydrometer is free floating; touching the sidewall of the cylinder will adversely affect the reading.

Carefully lift the hydrometer and wipe the stem (fluid on the stem will weigh it down, resulting in an erroneously low reading).

Gently release the hydrometer and allow it to equilibrate.

At eye level, read the specific gravity at the closest graduation to the bottom of the meniscus, and note.

Insert the thermometer into the cylinder; obtain the sample temperature and note.

Make temperature correction to the °Brix reading as necessary.

Report the °Brix and measurement temperature.

Discard sample; clean all equipment thoroughly and shake off excess water. Wash down tank sampling sites using a bucket of water or a hose (unwashed valves attract insects). Spray sampling valves with 70% EtOH solution. Wash spills or excess juice with water to nearest drain. Wash down barrel areas if needed.

9.2.5 Density, Specific Gravity, and °Brix Measurements via Density Meter

Concept: Change in oscillation frequency of a U-shaped tube proportionate to the mass of a liquid. Refer to Chapter 1, Chapter 5 (Section 5.3.2 and Fig. 5.2), and Chapter 6 (Section 6.4.3).

Method Type: Densitometry, accepted by TTB and AOAC.

Sample Volume Required: 50 mL

Equipment Required:

- Density meter (such as the Anton Paar DMA 35n®): clean and set for the appropriate measurement and measuring temperature; calibration verified

with wine standard and noted; recleaned (consult manufacturer's operation manual for details)
- Laboratory wipes
- Small containers: glass or plastic
- Thermometer: accurate; clean and dry with each use
- Water bath set at measuring temperature (if required)

Reagents Required:

- Deionized water
- All-purpose cleaner such as 2% RBS® or similar laboratory glass cleaner

Measurement Accuracy: Sample temperature and volume accurate

Note: It is imperative to keep the internal U-tube and sampling section of the instrument clean and free of any contamination. Instruments should be protected to prevent breakage of the internal U-tube. These devices are sensitive to gas or air bubbles in the sample and will result in erroneous readings. Most instruments will be factory set for density and specific gravity automatic calculations, °Brix may be offered by only a few companies and, to date, only Anton Paar offers negative °Brix instrument programming. The procedure is the same for all parameters, only the instrument setting is different and the type of sample tested (juice or wine).

Possible Errors: Air or gas bubbles in U-tube; dirty U-tube or sampling tube; leak in sampling tube; improper meter setting; sample not at measuring temperature; uncalibrated instrument; improper use of instrument.

Procedure:

Turn on meter and note if the meter setting is correct. Follow this procedure to obtain a known wine standard measurement. A standard's measurement within acceptable parameters (i.e., ± 0.0005 for specific gravity) indicates a calibrated instrument. Analyze a known standard periodically for quality control. A 2 °Brix solution can be used to assure calibration when using the instrument for fermentation monitoring.

Clean sampling tube with a wipe and insert into a small volume of cleaner placed in a container.

Aspirate cleaner; expel to discard.

Clean sampling tube with a wipe and insert into a small volume of clean deionized water placed in a second container. Aspirate water, expel to discard, and repeat many times to thoroughly rinse.

Obtain sample temperature. If the sample temperature is not within measuring range (flashing temperature reading on meter), allow the sample to come to temperature in the meter (flashing will cease when measuring temperature is reached) before taking the reading. Samples can also be placed in a water bath at the optimum specified measuring range until it reaches temperature.

Clean sampling tube with a wipe and insert into the sample. Aspirate sample, expel to discard, and repeat two more times to rinse. Slowly aspirate sample.

Observe the U-tube for the presence of bubbles. For gaseous samples, apply slight back-pressure on the U-tube volume (can be accomplished by slightly depressing the plunger on certain instruments, which will add pressure to the contents in the U-tube), or holding a hand-held unit on its side to allow the gas and pressure to escape from the U-tube measuring chamber. Consult manufacturer for suggestions. Fermenting juice samples will require agitation and gas release to drive off the CO_2, or degassing before aspirating the sample.

Hold the hand-held units in the proper reading position. Note the sample temperature and the measurement. Make temperature corrections to the measurement if required. Report the measurement to four places, with the measurement temperature and standard temperature [i.e. SG (20/20) no units; Density (20) g/cc]. The DMA 35n allows adjustments of the measurement standard temperature.

Clean immediately by repeating the above cleaning steps.

Turn off the instrument; place in a clean, dry, and safe location.

Discard sample. Clean equipment thoroughly.

9.2.6   Soluble Solids Measurement via Refractometer

Concept: The ability of light to refract (bend) as it passes through a liquid. The degree of refraction is dependent on the density of the liquid. Refer to Chapter 5 (Section 5.3.2 and Fig. 5.3).

Method Type: Refractive index

Sample Volume Required: 5 mL, juice only

Equipment Required:

- Refractometer: bench style or hand-held
- Laboratory wipes or soft cloth
- Light source: natural or artificial
- Dropper
- Thermometer: accurate; clean and dry with each use

Reagents Required:

- Distilled water
- 70% v/v EtOH (ethanol) solution
- 20% sucrose solution (20 °Brix)

Measurement Accuracy: Approximate

Note: Determine if the instrument is temperature compensated. Most lenses are susceptible to scratching; handle with care. Keep lens clean at all times. Temperature corrections are required on non-temperature-compensated instruments.

Temperature correction: each degree above 20°C, add (+) 0.07,   (1)

each degree below 20°C, subtract (−) 0.07.   (2)

Possible Errors: Uncalibrated instrument; dirty or scratched lens; unsuitable sample volume; too many solids; misreading scale; measurement not temperature corrected.

Procedure:

Hand-Held instruments:

Lift the protective cover and clean lens and cover with 70% EtOH solution; rinse with distilled water; dry with a laboratory wipe or soft cloth. Inspect the lens for scratches.

Hold the instrument horizontally with the lens in the up position. Use a clean dropper and place one or two drops of distilled water or a known °Brix solution on the lens; close cover.

Face a light source; keep the instrument horizontal and look through the eyepiece. Adjust the eyepiece until the scale comes into focus.

Read the measurement on the scale at the point where the light and dark sections meet (see Fig. 9.1) (water will read 0 and the 20 °Brix solution will

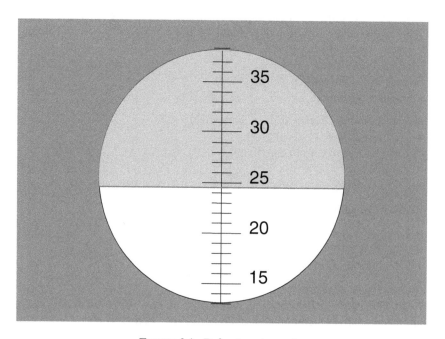

FIGURE 9.1. Refractometer scale.

read 20). If these reading are not correct, consult the manufacturer's manual for calibration method and recalibrate the instrument (instruments are recalibrated with a 20 °Brix solution just before harvest field work commences each year).

Clean lens and cover with the EtOH solution; dry with laboratory wipe or soft cloth.

Obtain the temperature of the juice sample.

Use a clean dropper, aspirate a juice sample, and expel and discard. Aspirate a second juice sample.

Hold the refractometer horizontally, lens up, and place one to two drops of juice on the lens; close cover.

Face a light source; keep the instrument horizontal and look through the eyepiece. Adjust the eyepiece until the scale comes into focus. Read the measurement and correct for temperature if required and note.

Carefully rinse the lens and cover with distilled water, clean both with the EtOH solution, rinse both with distilled water, and dry with a wipe of soft cloth.

Discard juice sample. Clean equipment thoroughly.

Bench models:

Most bench instruments are temperature compensated with digital readout.

Clean lens with the EtOH solution, rinse with distilled water, and dry with a laboratory wipe or soft cloth.

Use a clean dropper, place one or two drops of a water or °Brix solution on the lens and; note. If these reading are not correct, consult the manufacturer's manual for the calibration method and recalibrate the instrument (instruments should be recalibrated just before field work and harvest commences yearly).

Clean lens with the EtOH solution, rinse with distilled water, and dry with wipe or soft cloth.

Use a clean dropper, aspirate a juice sample, and expel and discard. Aspirate a second juice sample and place one or two drops on the lens. Note the measurement. If the instrument is not temperature compensated, make temperature corrections by obtaining the temperature of the juice sample.

Immediately clean lens with the EtOH solution, rinse with distilled water, and dry with wipe or soft cloth.

Discard sample. Clean all equipment thoroughly.

9.2.7   pH Measurement

Concept: See Chapter 5 (Sections 5.3.6 and 5.3.7).
Method Type: pH meter, accepted by AOAC
Sample Volume Required: 125 mL

Equipment Required:

- pH/Ion meter: temperature compensated; equipped with reference electrode and pH electrode (or combination); many meters and electrode models available; consult manufacturer's operation manual for details
- Magnetic stir plate: with stir bars (clean and dry)
- Chemical-resistant graduated beakers: 150 mL (clean and dry)
- Container or beaker: 300 mL
- Laboratory wipes
- Thermometer: accurate; clean and dry with each use
- Water bath: at measuring temperature if required
- Laboratory wash bottles

Reagents Required:

- pH buffer 4
- pH buffer 7
- Deionized water

Measurement Accuracy: All measurements accurate

Note: Read all instrument instructions and prepare the meter and electrodes for operation. Calibrate the meter daily or more frequently if required. Temperature-sensitive assay; ensure all calibration buffers and samples are within the proper measuring temperature for the equipment used. The fill hole of the electrode should be open when in use and closed when not in use.

Possible Errors: Failure to calibrate the instrument; reagent and sample measurements inaccurate; bubbles around the electrode membrane; electrode filling solution too low; crystal development in electrode; degraded storage solution; damaged electrode membrane; reference electrode faulty; faulty meter; expired reagents; sample temperature out of measurement range; improper pipetting technique; fill hole closed; improper stirring.

Procedure:

Turn instrument on and set to the pH mode. Continue setup procedure per manufacturer's manual.

Obtain temperatures of all liquids to ensure proper measurement and calibration temperature.

Clean electrode; rinse thoroughly with deionized water; pat dry *do not wipe* electrode tip to remove excess water (wiping an electrode can create static charges, which will interfere with the ability of the electrode to function properly and can also damage the electrode membrane).

Allow the electrode to sit in pH 4 buffer to acclimate after cleaning.

Calibrate the instrument using pH 4 and 7 buffer.

Consult manufacturer's operation manual for details (usually a good slope is close to 100%).

Analyze a known standard after calibration and periodically for quality control.

To a 150 ml graduated beaker, add approximately 100 mL sample (enough sample to cover the tips of the electrodes)

Place beaker on the stir plate; add stir bar; and adjust to a slow speed.

Slowly insert dry electrode into the sample to a level just above the stir bar and placing them at a slight angle to prevent the formation of air bubbles on the membrane. Begin measurement.

Allow the reading to stabilize; note the reading.

Carefully remove the electrode from the sample. Place the 300-mL beaker around, but not touching, the electrode. Using laboratory wash bottle containing deionized water, thoroughly rinse off electrode followed by a wash with pH 4 buffer.

Discard sample and thoroughly clean and dry equipment.

Store electrode in the proper storage solution; turn off meter; return all chemicals to the proper storage area.

9.2.8 Titratable Acids Assay, Manual

Concept: See Chapter 5 (Sections 5.3.6 and 5.3.7).
Method Type: Titration
Sample Volume Required: 15 mL, degassed, clarified if turbid
Equipment Required:

- pH/Ion meter: temperature compensated; equipped with reference electrode and pH electrode (or combination); many meters and electrode models available; consult manufacturer's operation manual for details
- Magnetic stir plate: with stir bars (clean and dry)
- Chemical-resistant graduated beaker: 250 mL (clean and dry)
- Container or beaker: 300 mL
- Volumetric pipettes: 25 mL and 5 mL
- Titration burette: glass, bottle top, or automated
- Laboratory wipes
- Thermometer: accurate; clean and dry with each use
- Water bath: at measuring temperature if required
- Laboratory wash bottles

Reagents Required:

- pH buffer 4
- pH buffer 7
- Deionized water
- Red wine
- 0.1N NaOH (sodium hydroxide)

Measurement Accuracy: All measurements accurate
Note: Read all instrument instructions and prepare the meter and electrodes for operation. Calibrate the meter daily or more frequently if required. Temperature-sensitive assay; ensure all calibration buffers and samples are

within the proper measuring temperature for the equipment used. The fill hole of the electrode should be open when in use and closed when not in use. Never allow reagent volume to become too low in the container. Standardize when refilling reagents.

Possible Errors: Failure to calibrate the instrument; reagent and sample measurements inaccurate; bubbles around the electrode membrane; electrode filling solution too low; crystal development in electrode; degraded storage solution; damaged electrode membrane; reference electrode faulty; faulty meter; expired reagents; sample temperature out of measurement range; improper pipetting technique; fill hole closed; improper stirring.

Procedure:

Turn instrument on and set to the pH mode. Continue setup procedure per manufacturer's manual.

Obtain temperatures of all liquids to ensure proper measurement and calibration temperature.

Clean electrode; rinse thoroughly with deionized water; pat dry *do not wipe* electrode tip to remove excess water (wiping an electrode can create static charges which will interfere with the ability of the electrode to function properly and can also damage the electrode membrane).

Allow the electrodes to sit in pH 4 buffer to acclimate after cleaning.

Calibrate the instrument using pH 4 and 7 buffer.

Consult manufacturer's operation manual for details (usually a good slope is close to 100%).

Wine buffer system:

To a 250-mL graduated beaker, add 25 mL of deionized water and 25 mL of red wine.

Place beaker on the stir plate; add stir bar; and adjust to a slow speed.

Slowly insert dry electrode into the solution to a level just above the stir bar and placing them at a slight angle to prevent the formation of air bubbles on the membrane.

Fill the burette with $0.1N$ NaOH (glass burette will be filled to a specific level) and place the delivery tip level to the top of the beaker and to the opposite side of the pH electrodes.

Watching the pH meter, slowly begin to add (titrate) the NaOH solution; stop titration when a pH of 8.2 has been reached and is maintained for 1 min. The amount of NaOH added is called the titer. This wine buffer is the beginning point (pH 8.2); each sample will be added to the wine buffer lowering the pH (more acidic); additional NaOH will be added to return to pH 8.2; and the volume used noted. The wine buffer is used until the liquid level reaches approximately the 200-mL mark or covers the electrode by 2 in. A known standard should be analyzed at the beginning of analysis and periodically for quality control.

Sample analysis:

Add to the wine buffer: 5 mL of sample.

Reset burette (or note starting volume) and titrate with NaOH until pH 8.2; note volume.

Calculate the TA with the formula:

$$TA \text{ (g/L as } H_2T) = \text{Titer volume (mL)} \times 1.5 \qquad (1)$$

or

$$TA \text{ (g/L as } H_2T) = 75 \times \text{NaOH molarity} \left(\frac{\text{titer volume (mL)}}{\text{Sample volume (mL)}} \right), \qquad (2)$$

where 1.5 relates to the concentration of the NaOH, number of moles titrated, and the volume of sample used. For a 10-mL sample this value would be 0.75 (Iland et al., 2000).

Upon completion of all analyses, carefully remove the electrode from the buffer. Place the 300 ml beaker around, but not touching, the electrode. Using laboratory wash bottle containing deionized water, thoroughly rinse off electrode followed by a wash with pH 4 buffer.

Discard sample and thoroughly clean and dry equipment.

Store electrode in the proper storage solution; turn off meter; return all chemicals to the proper storage area.

9.2.9   Titratable Acid and pH Assay via Automated Titration

Concept: The concept is the same as described in the pH and titratable acid procedure.

Method Type: Automated titration, accepted by TTB and AOAC.

Sample Volume Required: Various, degassed, clarified if turbid

Equipment Required:

- Automated system with titration; pH meter; sample changer; stirrer; printer
- Sample vials or containers to fit system (clean and dry)
- Volumetric pipettes: various
- Thermometer: accurate; clean and dry with each use
- Water bath: at measuring temperature if required
- Laboratory wash bottles
- Laboratory wipes

Reagents Required:

- $0.1N$ NaOH (sodium hydroxide)
- pH buffer 4
- pH buffer 7
- Deionized water

Measurement Accuracy: Accurate

Note: There are various systems available. Some systems are equipped with nitrogen degassing units to eliminate the predegassing step. Consult the manufacturer's operational manual for all details. Proper sample preparation and measurement are required. The analysis is performed in the same manner as stated above, except that each individual sample is titrated to pH 8.2; analysis time is reduced; and calculations are automatic. Calibration of the system is identical to that of any pH meter. Most units are temperature compensated. Sample size varies depending on the individual systems. Standards should be used for quality control at the beginning of each sample load and periodically throughout large groups of samples to alert the technician to possible problems. Temperature-sensitive assay; ensure all calibration buffers and samples are within the proper measuring temperature for the equipment used. The fill hole of the electrode should be open when in use and closed when not in use. Never allow reagent volume to become too low in the container. Standardize when refilling reagents.

Possible Errors: Failure to calibrate the instrument; bubbles around the electrode membrane; air bubbles in the reagent titration lines; electrode filling solution too low; crystal development in electrode; degraded storage solution; damaged electrode membrane; reference electrode or temperature probe faulty; faulty meter; expired reagents; sample temperature out of measurement range; improper pipetting technique; fill hole closed; improper stirring; improper instrument settings; degassed sample in queue too long, reuptake or gas production (fermenting juice).

Procedure:

Turn instrument on and continue setup procedure per manufacturer's manual.

Obtain temperatures of all liquids to ensure proper measurement and calibration temperature.

Clean electrode; rinse thoroughly with deionized water; pat dry *do not wipe* electrode tip to remove excess water (wiping an electrode can create static charges, which will interfere with the ability of the electrode to function properly and can also damage the electrode membrane).

Allow the electrodes to sit in pH 4 buffer to acclimate after cleaning.

Calibrate the instrument using pH 4 and 7 buffer.

Consult manufacturer's operation manual for details (usually a good slope is close to 100%).

Fill reagent reservoir and check lines for air bubbles. A known standard should be analyzed at the beginning of analysis and periodically for quality control.

Pipette prepared samples into clean, dry sample cups. If the system does not contain an automatic degassing sequence, do not load more samples into the sample changer than can be analyzed within 0.5 h, or sooner for juices (point at which samples will need to be degassed).

Initiate measurement sequence. Note the measurement of the known standard. If the measurements are out of specification, stop the sequence; troubleshoot (might require recalibration of the instrument). If the known standard is within specification, continue to monitor the instrument's function; check all known standard samples results as they become available.

Discard sample and thoroughly clean and dry equipment.

Store electrode in the proper storage solution; turn off meter; return all chemicals to the proper storage area.

9.2.10   Fluoride (F) Assay via ISE

Concept: See Chapter 5 (Section 5.3.7), which refers to ISE measurements.
Method Type: Ion selective electrode
Sample Volume Required: 75 mL
Equipment Required:

- pH/Ion meter: temperature compensated; equipped with reference electrode and fluoride-ion-specific electrode; many meters and electrode models available, consult manufacturer's operation manual for details
- Magnetic stir plate: with stir bars (clean and dry)
- Chemical-resistant graduated beakers: 150 mL (clean and dry)
- Container or beaker: 300 mL
- Volumetric pipettes: 1 mL, 10 mL, 50 mL
- Laboratory wipes
- Thermometer: accurate; clean and dry with each use
- Water bath: at measuring temperature if required
- Laboratory wash bottle

Reagents Required:

- Fluoride standard: 100 mg/L (ppm)
- TISAB II (ISA, ionic strength adjustor)
- Deionized water

Measurement Accuracy: All measurements accurate
Note: Read all instrument instructions and prepare the meter and electrodes for operation. Clear all previous fluoride standard measurements from meter and calibrate the instrument to determine the electrode slope (usually a good slope is close to 100%).
Possible Errors: Failure to calibrate the instrument; reagent and sample measurements inaccurate; bubbles around the electrode membrane; electrode filling solution too low; crystal development inside electrode; damaged electrode membrane; reference electrode faulty; faulty meter; expired reagents; sample temperature out of measurement range; improper pipetting technique; improper stirring.

Procedure:

Turn instrument on and set to the ion mode. Continue setup procedure per manufacturer's manual for known addition method.

Obtain temperatures of all liquids to ensure proper measurement and calibration temperature.

Calibrate meter for electrode slope.

Suggested calibration:

Add to 150 mL beaker using appropriate volumetric pipettes; 1 mL of 100 mg/l (ppm) fluoride standard, 50 mL of deionized water, and 50 mL of TSAB II.

Place beaker on the stir plate; add stir bar; and adjust to a moderate speed.

Slowly insert dry electrodes into the solution to a level just above the stir bar and placing them at a slight angle to prevent the formation of air bubbles on the membrane.

Begin measurement of the standard. When the reading is stable, input the first standard reading as 1.0 mg/L (ppm) into the meter.

Carefully remove the electrodes from the solution.

Place the 300-mL beaker around, but not touching, the electrodes. Using laboratory wash bottle containing deionized water, thoroughly rinse off electrodes. Pat dry (*do not wipe*) electrode tip to remove excess water (wiping an electrode can create static charges which will interfere with the ability of the electrode to function properly and can also damage the electrode membrane).

Discard sample.

In a second 150-mL beaker, add the following: 10 mL of 100 mg/L (ppm) fluoride standard, 50 mL of deionized water, and 50 mL of TSAB II.

Repeat the above procedure.

Begin measurement of the standard. When the reading is stable, input the second known addition standard reading as 10.0 mg/L (ppm) into the meter.

Discard sample.

Repeat the above electrode cleaning procedure.

Note and record electrode slope if within acceptable limits; if not, check for possible errors and repeat the above.

Thoroughly clean all equipment. To continue analysis of samples you may reuse the TSAB and the 10-mL volumetric pipette used for the fluoride standard and clean when finished testing.

The baseline and known addition standards are now defined. Proper quality control would indicate analyzing a standard with known fluoride content and noting the results on a continuing log (see Chapter 3).

Sample analysis:

To a 150-mL graduated beaker, add 50 mL of sample or control sample and 50 mL of TSAB II.

Place beaker on the stir plate; add stir bar; and adjust to a moderate speed.

Slowly insert dry electrodes into the solution to a level just above the stir bar and placing them at a slight angle to prevent the formation of air bubbles on the membrane.

Begin measurement of the sample for the first reading; note reading.

Add to the solution, 10 mL of 100 mg/L (ppm) fluoride standard.

Begin measurement of the sample for the second reading; note reading.

Data entry into the meter for the calculation is as follows

Sample volume   50 mL

TISAB II volume (ISA)   50 mL

Standard volume and concentration   10 mL, 100 mg/L (ppm)

The meter will make the calculations based on the calibration parameters and display the results.

Note the results and the units (i.e., mg/L or ppm).

Discard sample and thoroughly clean and dry equipment.

Store electrodes in their proper storage solution; turn off meter; return all chemicals to the proper storage area.

9.2.11 Ammonia (NH$_3$) Assay via ISE

Concept: See Chapter 5 (Sections 5.3.7 and 5.4.1), which refers to ISE measurements.

Method Type: Ion selective electrode

Sample Volume Required: 75 mL (clarified)

Equipment Required:

- pH/Ion meter: temperature compensated; equipped with reference electrode and ammonia ion specific electrode; many meters and electrode models available, consult manufacturer's operation manual for details
- Magnetic stir plate: with stir bars (clean and dry)
- Chemical-resistant graduated beakers: 150 mL (clean and dry)
- Container or beaker: 300 mL
- Volumetric pipettes: 10 mL, 50 mL
- Laboratory wipes
- Thermometer: accurate; clean and dry with each use
- Water bath: at measuring temperature if required
- Laboratory wash bottle

Reagents Required:

- Ammonia standard: 100 mg/L (ppm)
- Ammonia standard: 50 mg/L (ppm)
- Ammonia ISA (ionic strength adjustor) or similar
- Deionized water

Measurement Accuracy: All measurements accurate

Note: Read all instrument instructions and prepare the meter and electrodes for operation. Adding the ISA alters the pH of the sample and should not be added until ready to begin measuring. The ammonia level will peak and then begin to drop as the ammonia begins to dissipate from the sample; do not leave unattended; note the highest reading.

Possible Errors: Failure to calibrate the instrument; reagent and sample measurements inaccurate; bubbles around the electrode membrane; electrode filling solution too low; crystal development in electrode; degraded storage solution; damaged electrode membrane; reference electrode faulty; faulty meter; expired reagents; sample temperature out of measurement range; improper pipetting technique; failure to read the ammonia reading at the highest point; improper stirring.

Procedure:

Turn instrument on and set to the ion mode. Continue setup procedure per manufacturer's manual.

Obtain temperatures of all liquids to ensure proper measurement temperature.

Determine the quality of the current meter calibration by measuring the 50- and 100-mg/L (ppm) ammonia standards; measurements should be within ± 5% of the standard solution. Recalibrate the instrument if necessary; consult the manufacturer's operation manual for details. To a 150-mL graduated beaker, add 50 mL of sample or control sample, and 10 mL of ammonia ISA.

Immediately place beaker on the stir plate; add stir bar; and adjust to a slow speed.

Slowly insert dry electrodes into the sample solution to a level just above the stir bar and placing them at a slight angle to prevent the formation of air bubbles on the membrane.

Place beaker on the stir plate; add stir bar; and adjust to a moderate speed.

Begin measurement. Continually watch the meter reading until the values begin to drop. Note the highest value and the units (i.e., mg/L or ppm).

Carefully remove the electrodes from the solution. Place the 300-mL beaker around, but not touching, the electrodes. Using laboratory wash bottle containing deionized water, thoroughly rinse off electrodes. Pat dry (*do not wipe*) electrode tip to remove excess water (wiping an electrode can create static charges which will interfere with the ability of the electrode to function properly and can also damage the electrode membrane).

Discard sample and thoroughly clean and dry equipment.

Store electrodes in their proper storage solution (i.e., 1000 mg/L ammonia solution); turn off meter; return all chemicals to their proper storage area.

9.2.12   Free and Total Sulfur Dioxide (SO_2) Assay via Ripper

Concept: Refer to Chapter 5 (Section 5.4.3). Based on the redox reaction (Margalit, 1996)

$$SO_2 + 2H_2O + I_2 \rightarrow H_2SO_4 + 2HI.$$

Total SO_2 analysis requires hydrolysis of the acetaldehyde–sulfurous bonds by adding a strong base to the sample and incubating before the acid addition (Ough et al., 1988).

Method Type: Ripper, titrametric
Sample Volume Required: 75 mL; juice may require clarification; *Do not* degas sample.
Equipment Required:

- Graduated cylinder: 25 mL
- Volumetric pipettes: 25 mL, 5 mL (or bottle top dispensers)
- Erlenmeyer flask: 250 mL, wide mouth
- Flask stoppers
- Burette: glass, bottle top, or automatic
- Light box: with reference light gels for color-matching red/white wine and juice
- Timer
- Thermometer: accurate; clean and dry with each use
- Laboratory wipes

Reagents Required:

- 0.02N I_2 (iodine)
- 1N NaOH (sodium hydroxide)
- 1% Starch Indicator Solution
- 1 + 3H_2SO_4 (25% v/v sulfuric acid)

Measurement Accuracy: Sample and iodine are accurate; NaOH, H_2SO_4, and starch are approximate.
Note: Standardize NaOH and iodine. Iodine must be protected from light exposure. Samples require rapid titration with gentle mixing immediately after the acid is added (SO_2 liberated). Free SO_2 results tend to be inaccurately high due to SO_2 being pulled from the dissociation of anthocyanins and bisulfite, and in wines containing ascorbic acid (Zoecklein et al., 1999). Nonsulfite iodine-reducing substances are present in wines and contribute to the inaccuracy of the assay (Ough et al., 1988). Light gels corresponding to the end-point color desired for red (deep blue/red) and white (moderate tone of lavender) wine placed on the light box, as examples will assist the technicians in reaching consistent end points. The use of a stir plate and stir bar set at a low speed can prevent overagitation of sample. Never allow reagent volume to become too low in the container. Standardize when refilling reagents.
Possible Errors: Inaccurate sample volume; inaccurate sample measurement temperature; inaccurate reagent concentration; inaccurate burette; oxidized iodine; failure to standardize reagents; inaccurate measurement of reagents; failure to hydrolyze and incubate total SO_2 sample; end-point color not reached or maintained for the specified time (overtitration or undertitration); failure to clear digital burette's previous results or note manual burette beginning volume before the next titration; too vigorous sample agitation; mathematical errors.

Procedure:
Free SO$_2$

Turn on light box; note the desired end point color to reach; clear burette reading or note the volume level on glass burette; note sample and reagent temperatures.

Add to Erlenmeyer flask, 25 mL of wine or juice sample, 5 mL of 1% Starch Indicator Solution, and 5 mL of $1 + 3H_2SO_4$.

Place the flask over the light source.

Immediately titrate with 0.02N iodine; SO$_2$ has been liberated and the titration should be as fast as can accurately be performed while gently swirling flask to mix. As the end point nears, slow the titration until the point is reached; a few drops of titrate may be added to allow for a persistent end point for 30 s. Do not continually keep adding titrate to the sample (iodine uptake will continue resulting in inaccurately high results).

Note titer volume (mL); calculate the SO$_2$ to the nearest whole number:

$$SO_2 \text{ mg/L (ppm)} = \frac{\text{Sample volume (in mL)}}{[(I_2 \text{ titer in mL})(N \text{ of } I_2)(32)(1000)]}, \qquad (1)$$

or for a 25-mL sample volume and 0.02N I_2, the above formula can be abbreviated:

$$SO_2 \text{ mg/L (ppm)} = I_2 \text{ titer (in mL)} \times 25.6, \qquad (2)$$

or via I_2 standardization factor (refer to Chapter 10, Section 10.3.3),

$$SO_2 \text{ mg/L (ppm)} = I_2 \text{ titer (in mL)} \times \text{Standardization factor.} \qquad (3)$$

Discard sample and thoroughly clean and dry equipment; return all chemicals to their proper storage area.

Total SO$_2$:

Add to the Erlenmeyer flask, 25 mL of wine or juice sample and 25 mL of 1N NaOH.

Place stopper on flask; incubate at room temperature for 10 min.

Refer to the above free SO$_2$ procedure (the hydrolyzed sample replaces the wine or juice sample).

9.2.13   Free and Total Sulfur Dioxide (SO$_2$) Assay via Aeration–Oxidation

Concept: Refer to Chapter 5 (Section 5.4.3 and Fig. 5.9) for details and equipment setup.

Method Type: Aeration–oxidation–distillation–titration, accepted by TTB and AOAC

Sample Volume Required: 50 mL; juice sample may require clarification; *do not degas sample*

Equipment Required:

- Aeration setup: including vacuum system (aspiration) or nitrogen source (blow by); Graham condenser and heat source (if performing total or bound SO_2 assay); ice bath for free SO_2 assay; all appropriate glassware as described in Chapter 5 (Section 5.4.3)
- Volumetric pipettes: 20 mL, 10 mL (or bottle top dispensers set at 10 mL)
- Thermometer: accurate; clean and dry with each use
- Heat source: Bunsen burner, microburner, heating mantle
- Timer
- Laboratory wash bottle
- Burette: glass, bottle top, or automatic
- Dropper bottles
- Laboratory wipes

Reagents Required:

- 0.3% H_2O_2 (hydrogen peroxide)
- Color indicator solution
- one + $3H_3PO_4$ (25% v/v phosphoric acid)
- $0.01N$ NaOH (sodium hydroxide)
- Dilute $0.01N$ NaOH
- Deionized water
- Anti-foaming agent

Measurement Accuracy: Indicator solution, H_2O_2, and H_3PO_4 approximate; sample and NaOH titrate accurate.

Note: Distillation of volatile acids from high-volatile-acidity (VA) wines and carbon dioxide can affect the total SO_2 result. Reagents should be standardized frequently. Never allow reagent volume to become too low in the container. Standardize when refilling reagents. All liquids should be within the measurement range temperature.

Possible Errors: Inaccurate sample volume; inaccurate sample measurement temperature; inaccurate reagent concentration; inaccurate burette; failure to standardize reagents; inaccurate measurement of reagents; incorrect coolant temperature in condenser (results in poor distillation of volatile acids); incorrect ice-bath temperature (increased temperature will increase the dissociation of bound SO_2); failure to zero the burette or note beginning volume in glass burette; incorrect airflow or nitrogen flow; bubble tube too short (tip should be at the bottom of the flask or tube to allow for adequate bubbling); allowing sample to sit after acid added; loose connection; dirty glassware; end-point color not reached; mathematical errors.

Procedure:
Free SO_2

Turn on vacuum and set flow rate in the range from 800 to 1200 cc/min.
To pear shape flask (H_2O_2 receptacle), add 10 mL of 0.3% H_2O_2 and 1 to 3 drops of Color indicator solution.

Slowly titrate $0.01N$ NaOH into the H_2O_2 flask until the purple/lavender color turns an olive green or gray-green color; this color will be the end point to achieve further in the test.

Dry the stem of the vacuum connecting tube with a wipe and attach the H_2O_2 flask. Note the bubbling and adjust vacuum to prevent loss of H_2O_2.

To round-bottom flask (sample receptacle), add 20 mL of wine/juice sample and 10 mL of $1 + 3H_3PO_4$.

Immediately connect the sample flask to the condenser, or to the connecter tube going to the vacuum connecting tube and pear flask assembly.

Place an ice water bath around the sample flask; maintain temperature; keep the bath-water level above sample volume level in the flask.

Adjust the Pasteur pipette on the air inlet side of the sample flask so the tip sits close to the bottom of the flask for maximum aeration of the sample; note bubbling and adjust the vacuum flow rate if necessary for vigorous bubbling without losing sample.

Set timer for 10 min of aeration.

Check all connections for leaks.

Disconnect sample flask; allow the lines to clear with air; turn off vacuum; remove the H_2O_2 flask half way; rinse the stem with a small amount of deionized water and collect in the H_2O_2 flask.

Zero burette or note the volume on glass burette; titrate $0.01N$ NaOH into the H_2O_2 until it reaches the predetermined end point. Note the volume of titer used (mL).

Calculate the SO_2:

$$SO_2 \text{ mg/L (ppm)} = \frac{[(\text{NaOH titer (in mL)}) (N \text{ of NaOH}) (32) (1000)]}{\text{Sample volume (in mL)}}, \quad (1)$$

or for a 25-mL sample volume and $0.02N$ I_2, the above formula can be abbreviated,

$$SO_2 \text{ mg/L (ppm)} = \text{NaOH titer (in mL)} \times 16, \quad (2)$$

or via NaOH standardization factor (see Chapter 10, Section 10.3.2),

$$SO_2 \text{ mg/L (ppm)} = \text{NaOH titer (in mL)} \times \text{Standardization factor.} \quad (3)$$

Discard sample unless performing total SO_2 analysis on the same sample (see the following); discard H_2O_2; thoroughly clean and dry equipment.

Return all chemicals to their proper storage area.

Procedure:

Total SO$_2$

To pear shape flask (H_2O_2), add 10 mL of 0.3% H_2O_2 and 1 to 3 drops of Color Indicator Solution.

Slowly titrate $0.01N$ NaOH into the H_2O_2 flask until the purple/lavender color turns an olive green or gray-green color; this color will be the end point to achieve further in the test.

Dry the stem of the vacuum connecting tube with a wipe and attach the H_2O_2 flask. Dry and attach the sample flask used in the above analysis (for bound SO_2) to the condenser, or pipette a new sample into a clean round bottom flask; add 10 mL of $1 + 3$ H_3PO_4 (for total SO_2) and attach to condenser.

Turn on water, or coolant circulator, to condenser.

Place heat source under sample flask; let the sample come to boiling.

Turn on vacuum; set timer and aerate for 15 min.

Turn off heat source and remove from under sample flask; remove H_2O_2 flask; titrate and calculate SO_2 as in the above procedure.

Disconnect sample flask; turn of vacuum; turn off water, or coolant circulator, to condenser.

Discard sample and H_2O_2; thoroughly clean and dry equipment; return all chemicals to their proper storage area.

$$\text{Total } SO_2 \text{ mg/L (ppm)} = \text{Bound } SO_2 + \text{Free } SO_2. \tag{1}$$

9.2.14 Reducing Sugar and Residual Reducing Sugar (RS) Enzymatic Assay

Concept: Refer to Chapter 5 (Section 5.3.3).
Method Type: UV Spectrometric/enzymatic
Sample Volume Required: 200 µL, sterile-filtered to 0.45 µm
Equipment Required:

- Spectrometer, single beam: absorbance (A) wavelength settings of 340 nm, adjusted (manually or automatically) for dark current at 0% transmission (T); light at 100%T using a water blank $A = \log(100/\%T)$; zero instrument to air; consult manufacturer's operation manual
- Micro pipetters: accurate/certified 2000 µL, 1000 µL, 100 µL with corresponding ranges that include 10, 20, 200, 900, and 1900 µL; manual or digital with appropriate tips
- Methacrylate cells (cuvette): 10 mm
- Cuvette tray
- Water bath set at measuring temperature
- Timer
- 10-mL sample cups (clean and dry) for mixing
- Laboratory wrapping film (i.e., Parafilm M™)

Reagents Required:

- D-glucose/D-fructose enzyme kit such as Roche Chemical manufacturers or use bulk chemicals as listed:

Triethanolamine hydrochloride buffer (pH 7.6) containing NADP (nicotinamide–adenine dinucleotide phosphate) and ATP (adenosine-5'-triphosphate) $MgSO_4$ (magnesium sulfate)

HK/G6P DH (hexokinase/glucose 6–phosphate dehydrogenase)
PGI (phosphoglucose isomerase)
RS standard (glucose)

- Distilled water (pure)

Measurement Accuracy: Accurate

Note: Enzyme kits are more economical in medium to small wineries. When using enzyme kits, follow manufacturers' instructions. Larger wineries will use their own chemicals, but the procedure is similar to that of the kits. Absorbance readings accurate in the range of 0.1–2.0 absorbance units (AU); $A < 0.001$ will be distorted by electronic noise; $A > 2.0$ is too dense for accurate measurement.

Possible Errors: Poor pipetting technique; inaccurate sample or reagent volume; instrument improperly set; instrument uncalibrated; inaccurate pipette; expired reagents; contaminated reagents; incorrect measuring temperature; loss of sample volume during mixing; fingerprints on cuvette; inaccurate incubation time; mathematical errors for manual computation; instrument error; absorbance range > 2.0.

Procedure:

General procedure for standard wine dilution of **20X**; sugar concentration should be in the range of 0.08 to 0.5 g/l; dilute as necessary (i.e., 10X, 40X, etc.) (refer to kit manufacturer's instructions).

Turn on spectrometer; set at 340 nm; zero.

Label one cup and one cuvette for each sample plus one blank and one standard to be tested. The quantity of samples tested in each session should not exceed what can comfortably and accurately be analyzed within the specified time frame.

Obtain temperatures of reagents and samples to ensure proper measurement.

In sample mixing cup, add and mix 100 μL (0.1 mL) of wine/juice sample and 1900 μL (1.9 mL) of distilled water.

Per manufacturer's instructions, pipette an aliquot of the above diluted sample or undiluted standard to the 10-mm cuvette using a clean tip with each sample (remember to pipette onto the cuvette wall; do not touch pipette to wall). Add the same amount of distilled water to the blank.

Pipette to all cuvettes, including the blank, using separate tips for each reagent: aliquot of buffer and aliquot of distilled water.

With clean hands, carefully pick up cuvette without touching the side surfaces; place a small piece of film over the top; gently invert five or six times until well mixed.

Incubate for three minutes. For multiple samples start the timer after mixing the first cuvette.

Place the blank in the first position or read and record the absorbance (some instruments will store the absorbencies and automatically calculate the results)

Read or place the standard in position followed by each sample (A_1). Note the location of each sample when using a multi sample spectrometer.

Place samples back into the cuvette rack.

Pipette to all cuvettes, including, the blank using separate tips for each reagent: Aliquot of HK/G6P-PH and aliquot of PGI.

With clean hands, carefully pick up cuvette without touching the side surfaces; place a small piece of film over the top; gently invert five or six times until well mixed.

Incubate for 12 min. For multiple samples, start the timer after mixing the first cuvette.

Read the cuvettes in the same order as mentioned above (A_2).

Calculate the results as glucose read at 340 nm with a 20X sample dilution according to the following formula (or as instructed by the kit manufacturer):

$$RS\ (g/L) = \left(\frac{V \times MW}{c \times d \times v \times 1000} \right) \times \Delta A \times D$$

$$= \left(\frac{3.03\ (\text{avg.}) \times 180.16}{6.3 \times 1.0 \times 0.1 \times 1000} \right) \times \Delta A \times 20 \tag{1}$$

or a general calculation,

$$RS\ (g/L) = (\Delta A \times 0.8665) \times 20, \tag{2}$$

where V is the final cuvette volume (mL), MW is the molecular weight of analyte, c is the NADPH coefficient, d is the light path (cm), v is the sample volume (mL); $\Delta A = $ (sample $A_2 - A_1$) − (blank $A_2 - A_1$), and D is the dilution factor.

Note calculations for the standard. If the results are not within an acceptable range (close to 100%), do not report results; repeat the analysis.

Discard all cuvettes, sample cups, pipette tips, and clean all equipment thoroughly. Return all chemicals to their proper storage area.

9.2.15 L-Malic Acid (LMA) Enzymatic Assay

Concept: Refer to Chapter 6 (Section 6.5.5).
Method Type: UV Spectrometric/enzymatic
Sample Volume Required: 200 µL, sterile-filtered to 0.45 µm
Equipment Required:

- Spectrometer, single beam: absorbance (A) wavelength settings of 340 nm, adjusted (manually or automatically) for dark current at 0% transmission (T); light at 100%T using a water blank, $A = \log (100/\%T)$ zero instrument to air; consult manufacturer's operation manual
- Micro pipetters: accurate/certified 2000 µL, 1000 µL, 100 µL with corresponding ranges that include 10, 20, 200, 900, and 1900 µL; manual or digital with appropriate tips
- Methacrylate cells (cuvette): 10 mm

- Cuvette tray
- Water bath set at measuring temperature
- Timer
- 10-mL sample cups (clean and dry) for mixing
- Laboratory wrapping film (i.e., Parafilm M™)

Reagents Required:

- L-Malic acid enzyme test kit (manufacturers such as Roche Chemical) or use bulk chemicals as listed:

  Glycylglycine buffer (pH 10.0) with $C_5H_9NO_4$ (L-glutamic acid)
  NAD (nicotinamide–adenine dinucleotide)
  GOT (glutamate–oxaloacetate transaminase)
  L-MDH (L-malate dehydrogenase)
  L-Malic acid standard

- Distilled water (pure)

Measurement Accuracy: Accurate

Note: Enzyme kits are more economical in medium to small wineries. When using enzyme kits, follow manufacturer's instructions. Larger wineries will use their own chemicals, but the procedure is similar to that of the kits. Absorbance readings accurate in the range of 0.1–2.0 absorbance units (AU); $A < 0.001$ will be distorted by electronic noise; $A > 2.0$ is too dense for accurate measurement.

Possible Errors: Poor pipetting technique; inaccurate sample or reagent volume; instrument improperly set; instrument uncalibrated; inaccurate pipette; expired reagents; contaminated reagents; incorrect measuring temperature; loss of sample volume during mixing; fingerprints on cuvette; inaccurate incubation time; mathematical errors for manual computation; instrument error; absorbance range > 2.0.

Procedure:

General procedure for standard wine dilution of 20X; LMA concentration should be in the range of 0.04–0.2 g/L; dilute as necessary (i.e., 10X, 40X, etc.) (refer to kit manufacturer's instructions).

Turn on spectrometer; set at 340 nm; zero.

Label one cup and one cuvette for each sample plus one blank and one standard to be tested. Do not attempt to test more than eight samples in each session.

Obtain temperatures of reagents and samples to ensure proper measurement. In sample mixing cup, add and mix 100 μL (0.1 mL) of wine/juice sample and 1900 μL (1.9 mL) of distilled water.

Per manufacturer's instructions, pipette an aliquot of the above diluted sample or undiluted standard to the 10-mm cuvette using a clean tip with each sample (remember to pipette onto the cuvette wall; do not touch pipette to wall). Add the same amount of distilled water to the blank.

Pipette to all cuvettes, including the blank, using separate tips for each reagent: aliquot of buffer solution, aliquot of NAD, aliquot of GOT, and aliquot of distilled water.

With clean hands, carefully pick up cuvette without touching the side surfaces; place a small piece of film over the top; gently invert five or six times until well mixed. Incubate for 3 min. For multiple samples, start the timer after mixing the first cuvette.

Place the blank in the first position or read and record the absorbance (some instruments will store the absorbencies and automatically calculate the results).

Read or place the standard in position followed by each sample (A_1). Note the location of each sample when using a multisample spectrometer.

Place samples back into the cuvette rack.

Pipette to all cuvettes, including the blank, using separate tips for each reagent: aliquot of L-MDH.

With clean hands, carefully pick up cuvette without touching the side surfaces; place a small piece of film over the top; gently invert five or six times until well mixed. Incubate for 7 mins. For multiple samples, start the timer after mixing the first cuvette.

Immediately read the cuvettes in the same order as mentioned above (A_2).

Calculate the results read at 340 nm with a 20X sample dilution according to the following formula (or as instructed by the kit manufacturer):

$$\text{LMA (g/L)} = \left(\frac{(V \times MW)}{c \times d \times v \times 1000} \right) \times \Delta A \times D \qquad (1)$$

$$= \left(\frac{2.22 \times 134.09}{6.3 \times 1.0 \times 0.1 \times 1000} \right) \times \Delta A \times 20,$$

or a general calculation,

$$\text{LMA (g/L)} = (\Delta A \times 0.4725) \times 20, \qquad (2)$$

where V is the final cuvette volume (mL), MW is the molecular weight of analyte, c is the NADH coefficient, d is the light path (cm), v is the sample volume (mL), ΔA = sample (A_2–A_1) – blank (A_2–A_1), and D is the dilution factor.

Note calculations for the standard. If the results are not within an acceptable range (close to 100%), do not report results; repeat the analysis.

Discard all cuvettes, sample cups, pipette tips, and clean all equipment thoroughly. Return all chemicals to their proper storage area.

9.2.16 Acetic Acid (AC) Enzymatic Assay

Concept: Refer to Chapter 5 (Section 5.3.3) and Chapter 6 (Section 6.3.1). Acetic acid reaction in this method:

$$\text{AC + ATP + CoA} \xrightarrow{\text{ACS}} \text{Acetyl-CoA + AMP + Pyrophosphate}, \qquad (1)$$

$$\text{Acetyl-CoA + Oxaloacetate + H}_2\text{O} \xrightarrow{\text{CS}} \text{Citrate + CoA}, \qquad (2)$$

$$\text{L-Malate} + \text{NAD} \xleftarrow{\quad\text{MDH}\quad} \text{Oxaloacetate} + \text{NADH} + \text{H}. \qquad (3)$$

NADH is measured. The NADH formed in the reaction is not linear to the concentration of acetic acid and the absorbance measurements must be corrected (note the calculations at the end of this section).

Method Type: UV Spectrometric/enzymatic

Sample Volume Required: 200 μL, sterile-filtered to 0.45 μm

Equipment Required:

- Spectrometer, single beam: absorbance (A) wavelength settings of 340 nm, adjusted (manually or automatically) for dark current at 0% transmission (T); light at 100%T using a water blank, $A = \log (100/\%T)$; zero instrument to air; consult manufacturer's operation manual
- Micropipetters: accurate/certified 2000 μL, 1000 μL, 100 μL with corresponding ranges that include 10, 20, 200, 900, and 1900 μL; manual or digital with appropriate tips
- Methacrylate cells (cuvette): 10 mm
- Cuvette tray
- Water bath set at measuring temperature
- Timer
- 10-mL sample cups (clean and dry) for mixing
- Laboratory wrapping film (i.e., Parafilm M™)

Reagents Required:

- *Acetic acid enzyme test kit (manufacturers such as Roche Chemical) or use bulk chemicals as listed:*

Triethanolamine buffer pH 8.4 containing $MgCl_2$ (magnesium chloride) and L-malic acid
ATP (Adenosine-5′-triphosphate)
CoA (coenzyme A)
NAD (nicotinamide–adenine dinucleotide)
L-MDH (L-malate dehydrogenase)
CS (citrate synthase)
ACS (acetyl-CoA-synthetase)
Acetic acid standard

- *Distilled water (pure)*

Measurement Accuracy: Accurate

Note: Enzyme kits are more economical in medium to small wineries. When using enzyme kits, follow manufacturer's instructions. Larger wineries will use their own chemicals, but the procedure is the similar to that of the kits. Absorbance readings accurate in the range of 0.1–2.0 absorbance units (AU); $A < 0.001$ will be distorted by electronic noise; $A > 2.0$ is too dense for accurate measurement.

Possible Errors: Poor pipetting technique; inaccurate sample or reagent volume; instrument improperly set; instrument uncalibrated; inaccurate pipette;

expired reagents; contaminated reagents; incorrect measuring temperature; loss of sample volume during mixing; fingerprints on cuvette; inaccurate incubation time; mathematical errors for manual computation; instrument error; absorbance range > 2.0.

Procedure:

General procedure for standard wine dilution of 10X; acetic acid concentration should be in the range of 0.03–0.15 g/L; dilute as necessary (i.e., 20X, 40X, etc.) (refer to kit manufacturer's instructions).

Turn on spectrometer; set at 340 nm; zero.

Label one cup and one cuvette for each sample plus one blank and one standard to be tested. Do not attempt to test more than eight samples in each session.

Obtain temperatures of reagents and samples to ensure proper measurement.

In sample mixing cup, add and mix, 100 µL (0.1 mL) of wine/juice sample and 900 µL (0.9 mL) of distilled water.

Per manufacturer's instructions, pipette an aliquot of the above diluted sample or undiluted standard to the 10-mm cuvette using a clean tip with each sample (remember to pipette onto the cuvette wall; do not touch pipette to wall). Add the same amount of distilled water to the blank.

Pipette to all cuvettes, including the blank, using separate tips for each reagent: aliquot of buffer solution, aliquot of NAD, aliquot of ATP, aliquot of CoA, and aliquot of distilled water.

With clean hands, carefully pick up cuvette without touching the side surfaces; place a small piece of film over the top; gently invert five or six times until well mixed.

Place the blank in the first position or read and record the absorbance (some instruments will store the absorbencies and automatically calculate the results)

Read or place the standard in position followed by each sample (A_1). Note the location of each sample when using a multisample spectrometer.

Place samples back into the cuvette rack.

Pipette to all cuvettes, including the blank, using separate tips for each reagent: aliquot of L-MDH and aliquot of CS.

With clean hands, carefully pick up cuvette without touching the side surfaces; place a small piece of film over the top; gently invert five or six times until well mixed.

Incubate for 3 mins. For multiple samples start the timer after mixing the first cuvette.

Read the cuvettes in the same order as mentioned above (A_2).

Pipette to all cuvettes, including the blank, using separate tips for each reagent: aliquot of ACS.

With clean hands, carefully pick up cuvette without touching the side surfaces; place a small piece of film over the top; gently invert five or six times until well mixed.

Incubate for 15 min until the reaction has stopped. For multiple samples, start the timer after mixing the first cuvette.

Read the cuvettes in the same order as mentioned above (A_3).

Calculate the results read at 340 nm with a 10X sample dilution according to the following formula (or as instructed by the kit manufacturer):

Calculation for ΔA

$$[s(A_3-A_1)-b(A_3-A_1)]-\left[\frac{s(A_2-A_1)^2-b(A_2-A_1)^2}{s[A_3-A_1)-b(A_3-A_1)}\right], \tag{1}$$

$$\text{Calculation for AC (g/L)} = \left(\frac{V\times MW}{c\times d\times v\times 1000}\right)\times\Delta A\times D \tag{2}$$

$$= \left(\frac{3.23\times 60.05}{6.3\times 1.0\times 0.1\times 1000}\right)\times\Delta A\times 10$$

or a general calculation

$$\text{AC (g/L)} = (\Delta A\times 0.3079)\times 10, \tag{3}$$

where s is the sample, b is the blank, V is the final cuvette volume (mL), MW is the molecular weight of analyte, c is the NADH coefficient, d is the light path (cm), v is the sample volume (mL), and D is the dilution factor.

Note calculations for the standard. If the results are not within an acceptable range (close to 100%), do not report results; repeat the analysis.

Discard all cuvettes, sample cups, pipette tips, and clean all equipment thoroughly. Return all chemicals to their proper storage area.

9.2.17   Volatile Acidity Assay via Distillation

Concept: Refer to Chapter 6 (Section 6.3.1 and Fig. 6.1) for details and description of distillation equipment.

Method Type: Distillation/titrametric

Sample Volume Required: 30 mL, degassed

Equipment Required:

- Volatile acid distillation still: Cash still consisting of Graham condenser and boiling chamber (i.e., RD80™ Self Evacuation Volatile Acid Still, 110 or 220 V; Research & Development Glass Products & Equipment, Inc., Berkeley, CA)
- Volumetric pipette: 10 mL and 1 mL (dropper or bottle top dispenser may also be used)
- Thermometer: accurate; clean and dry with each use
- Timer
- 150-mL container

- Laboratory wipes
- Laboratory wash bottles
- Erlenmeyer flask: 250 mL
- Flask stopper
- Burette: 50-mL glass, bottle top, or auto burette

Reagents Required:

- $0.05N$ NaOH (sodium hydroxide); may substitute $0.1N$ NaOH to $0.01N$ NaOH, must alter formula to reflect normality of titer
- 1% $C_{20}H_{14}O_4$ (phenolphthalein)
- 0.3% H_2O_2 (hydrogen peroxide)
- Distilled water

Measurement Accuracy: Wine sample and titer are accurate; H_2O_2 and $C_{20}H_{14}O_4$ are approximate.

Note: Water used in boiling chamber must be void of CO_2, which can distilled over into the sample as H_2CO_3 (carbonic acid). Always have water in the boiling chamber and the heating element covered during the entire distillation process. *Never* heat an empty boiling chamber and *never* add water to a dry hot boiling chamber; turn off the heating element and allow the still to cool before adding water. Still can break or explode. Wear proper personal protection, especially safety glasses. Juice samples should be analyzed immediately after degassing.

Possible Errors: Boiling chamber water contains CO_2 not eliminated; inaccurate sample or reagent volume; water used to wash H_2O_2 into sample; wrong reagent concentrations; expired or contaminated reagents; failure to standardized noncertified NaOH titer solutions; failure to zero burette or note burette starting volume; over titration or undertitration; too vigorous mixing during titration; incorrect volume of distillate; Graham condenser water/coolant too warm (> 18°C); condenser water/coolant not turned on; boiling chamber allowed to steam too long before closing; improper pipetting or burette technique; sample allowed to uptake or create gas (SO_2, CO_2); waiting too long to analyze after degassing; dirty still; faulty still.

Procedure:

Obtain reagent and sample temperatures to ensure proper measurement temperature.

Never allow reagent volumes to become too low in their containers. Standardize when refilling reagents (unless reagent is certified).

Add distilled water to the boiling chamber fill level per the manufacturer's recommendations.

Check all glass connections and secure.

Open the stopcock to the boiling chamber. Turn on heating element and bring the water to a boil; allow the still to steam for 5–10 min to reduce the CO_2 in the water. Note the vigor of the boiling water; extreme boiling might require transforming the voltage to a lower level; contact the manufacturer for information.

Turn off heating element.

Open stopcock to sample chamber.

Pipette into the inlet funnel 10 mL of wine/juice sample and 1 mL of 0.3% H_2O_2.

Turn on water/coolant to condenser; place a clean, dry flask under the condenser with the delivery tip against the side of the glass to prevent splashing and loss of distillate.

Turn on heating element and allow the still to come to a boil. Immediately close the stopcock to the chambers, which will allow the steam to be directed to the condenser. Add distilled water to the funnel to keep it cool if performing several assays (hot funnels may volatize the sample precipitately).

Set timer to a predetermined point known to provide adequate time to allow distillation of approximately 90 mL in the flask (~ 7 min).

At the end of the preset time, watch the level of the distillate in the flask (eye level) until it reaches 100 mL; remove the flask and stopper it; turn off the heating element; place a container under the condenser to catch excess distillate. Distillate volume over 110 mL should be discarded and the process repeated using a fresh sample.

Zero the burette or note the beginning volume; immediately add several drops of $C_{20}H_{14}O_4$ to the flask and begin titration with the 0.05 NaOH. Distillate sample should be titrated while still warm.

Titrate while slowly rotating the flask to mix.

Rapidly titrate to a pink end point, which persists for 15 s then disappears; note titer volume (mL).

Calculate the VA as acetic acid g/L:

$$VA\ (g/L) = \frac{(\text{Titer volume})\ (N\ \text{of NaOH})\ (0.06)\ (1000)}{\text{Sample volume}}, \tag{1}$$

where 0.06 is the normality factor.

Using $0.05N$ NaOH with a 10-mL sample, the short formula would be

$$VA\ (g/L) = \text{Titer volume} \times 0.3. \tag{2}$$

Using $0.1N$ NaOH with a 10-mL sample, the short formula would be

$$VA\ (g/L) = \text{Titer volume} \times 0.6. \tag{3}$$

Therefore, the factor for a $0.01N$ NaOH with a 10-mL sample would be 0.06.

Discard sample and distillate.

Open inlet stopcock to the sample chamber; open sample chamber evacuation stopcock; allow sample to be aspirated and close.

Add distilled water to the sample chamber to clean; open evacuation stopcock and aspirate water; repeat. Stills should be maintained and cleaned frequently with a NaOH solution boiled in the sample chamber and thoroughly rinsed. Analyze a known acetic acid standard routinely for quality control. Water samples can be analyzed to monitor acid carryover.

9.2.18   Insoluble Solids (% Solids) Measurement

Concept: Refer to Chapter 5 (Section 5.4.2).
Method Type: Direct measurement
Sample Volume Required: 20 mL
Equipment Required:

- Benchtop centrifuge equipped with rotors to fit centrifuge tubes specified below; capable of 2500–3000 rpm
- Thermometer: accurate; clean and dry with each use
- Centrifuge tubes: 15-mL, conical bottom; glass with beaded rim; 0.1-mL graduations
- Pasture pipette
- Timer
- Tube rack

Reagents Required: None
Measurement Accuracy: Sample volume accurate
Note: Results will vary, so it is important to be consistent and follow the procedure. Volume level of solids in the centrifuge tube is not always even and, therefore, difficult to read. Read volume to the nearest mark, estimating that volume by averaging the varying heights of the solids. Each 0.1-mL division represents 1% solids, which is derived from the formula

$$\% \text{ Solids} = (0.1 \text{ mL} \div 10\text{mL}) \times 100 \tag{1}$$

or

$$\% \text{ Solids} = \text{Volume of solids} \times 10.$$

Possible Errors: Dirty centrifuge tube; sample not mixed; poor sample; improper centrifuge speed; misreading of volume; inadequate centrifuge time; sample at incorrect measurement temperature.

Procedure:

Identify each sample tube to be centrifuged. Do not use water-based inks or markers.
Obtain temperatures of all liquids to ensure proper measurement temperature.
Gently mix sample; pour sample into a dry centrifuge tube to just below the 10-mL mark.
Use the Pasteur pipette and add sample to the centrifuge tube to the 10-mL mark.

Place labeled tubes into the centrifuge; if all positions are not filled, use water-filled centrifuge tubes to balance the load. Close and lock lid.

Set the centrifuge to 2500–3000 rpm and start.

Set timer for 10 mins. *Never* open the lid of a centrifuge while it is spinning.

Carefully remove a sample tube; hold it at eye level and ascertain the level of solids; calculate the percentage.

Discard sample; clean the centrifuge tube thoroughly and dry.

9.2.19 % v/v Alcohol Measurement via Ebulliometer

Concept: The ability of alcohol to lower the boiling-point temperature of a liquid. The direct correlation of reduced boiling-point temperature to % v/v of alcohol compared to the boiling point temperature of distilled water at the same barometric pressure. Refer to Chapter 5 (Sections 5.3.4 and 5.3.5) and Figure 5.4.

Method Type: Ebulliometry

Sample Volume Required: 60 mL, clarified as needed

Equipment Required:

- Ebulliometer: alcohol burner; accurate mercury thermometer; disk scale or other table that corresponds Δ in temperature to % v/v alcohol (i.e., Dujardin & Salleron ebulliometer setup)
- Centrifuge or sample filtration for clarification (filter with glass filter or other material that will not absorb the alcohol in the sample)
- Graduated cylinders: 20 mL, 50 mL
- Ignition source (matches)
- Thermometer: accurate; clean and dry with each use
- Thermometer rack (or dry soft storage area to cool thermometer)
- Timer

Reagents Required:

- Denatured alcohol
- Distilled water
- Distilled ice water
- 1% NaOH solution

Measurement Accuracy: Wine sample, accurate; water sample approximate

Note: Appropriate safety precautions for work with hot liquids required. Barometric pressure should be monitored frequently during the day (morning, noon, and night minimally); more frequent readings might be required for heavy sample loads or unstable weather patterns. A change of 0.1575 in. of Hg (4 mm Hg) can alter the readings by 0.5% v/v (Iland et al., 2000). Sugar or solids in the sample will alter the boiling point and, therefore, the accuracy of the alcohol % v/v. Wine containing over 1.5% sugar should be diluted 1 : 1; wine containing over 2% sugar can be diluted or corrected for sugar interference (see formula in this section). High-alcohol wines might require a 1 : 1

dilution prior to testing. Maintain a clean ebulliometer to avoid errors; clean daily during heavy use. Water in the condenser, during wine analysis, must not exceed 40°C (104°F) (Iland et al., 2000). After the water zero has been obtained, analyze a wine standard with a known alcohol value for quality control and repeat; keep a log of results with corresponding barometric pressure readings. Refer to manufacturer's operation manual for information.

Possible Errors: Incorrect sample volume; contaminated wines sample from condenser water; sample too turbid; sugar level or alcohol level too high; inaccurate dilution; faulty thermometer (mercury separation); inaccurate thermometer reading; failure to read thermometer at the appropriate time; improper thermometer position in boiling chamber; inadequate heat source; dirty boiling chamber; incorrect condenser water temperature; contaminated boiling chamber (cleaning solution); failure to rinse boiling chamber with each wine sample before analysis; acute change in barometric pressure.

Procedure:

Obtain temperatures of all liquids to ensure proper measurement temperature.
Water zero point:

Locate ebulliometer in a draft-free, well-ventilated area. Close tap on ebulliometer.

Inspect alcohol burner; replace charred wick if needed; fill burner with denatured alcohol.

To clean, dry ebulliometer add 20 mL of distilled water.

Insert dry mercury thermometer into boiling chamber with the tip in the vapor area of the chamber (above the boiling liquid level).

Light burner and place in the proper position proximal to the tap.

Set timer for approximate length of time to reach boiling (usually 2–3 mins). Check the barometric pressure reading and note.

As the water is boiling, read thermometer; when reading is stable for 20–30 s, note the temperature (T_w) to the nearest 0.02°C.

Extinguish burner and remove; allow ebulliometer to cool; open tap and drain water sample; remove and place thermometer in rack to cool.

This will be the zero point (temperature) to set the disk scale to 0.0% v/v alcohol (see manufacturer's instructions).

Wine sample analysis:

Pour approximately 5 mL to 10 mL wine sample into the cooled boiling chamber; rotate instrument to rinse off sides; open tap and drain completely.

Fill reflux condenser to within a fraction of an inch from the top with distilled ice water (take care not to drip water into the chamber); close tap.

Add 50 mL wine sample to the boiling chamber.

Insert dry mercury thermometer into boiling chamber with the tip in the vapor area of the chamber (above the boiling liquid level).

Light burner and place in the proper position proximal to the tap.

Set timer for approximate length of time to reach boiling (usually 4–5 min).

As the sample is boiling, read thermometer; when reading is stable for
20–30s, note the temperature (T_s) to the nearest 0.02°C.

Extinguish burner and remove; allow ebulliometer to cool; open tap and
drain sample; remove and place thermometer in rack to cool.

If significant time has elapsed since the water zero point was obtained, consider rerunning another water sample and verify barometric pressure.

On disk scale, find the boiling-point temperature of the wine sample and
note the correlating alcohol % v/v. Correct for dilutions and sugar content;
if using a table, use the appropriate table:

$$\text{Apparent alcohol \% v/v} = \Delta T_1 - T_2 \text{ correlated to conversion}$$
$$\text{table (Fig. 9.2),} \tag{1}$$

$$\text{Alcohol \% v/v} = \text{Apparent alcohol \% v/v} - (\text{RS \% sugar} \times 0.05), \tag{2}$$

$$\text{Alcohol \% v/v} = \text{Apparent alcohol \% v/v} \times \text{Dilution factor.} \tag{3}$$

Reproducibility varies from 0.2% v/v to 0.5% v/v.

Rinse ebulliometer several times with distilled water and allow it to dry;
if other samples are to be analyzed, rinse ebulliometer with wine from
the next sample. Finish analysis by thoroughly cleaning with water washes.
Clean mercury thermometer.

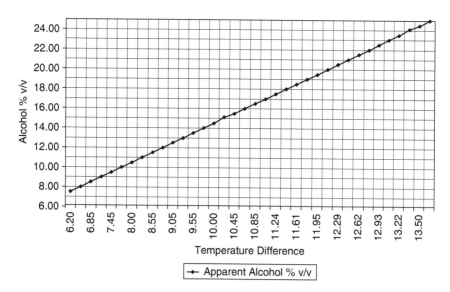

FIGURE 9.2. Apparent alcohol % v/v values at various Δ boiling points via ebulliometry.
(Data based on the Churchward table; some values have been extrapolated.)

Return burner to its appropriate storage area; return all chemicals to their proper storage area.

Clean ebulliometer regularly by adding 20 mL of a 1% NaOH solution into the boiling chamber and allow the solution to boil for a few minutes; cool; rinse thoroughly several times with distilled water.

9.2.20 Yeast and Bacteria Viability and Cell Counts

Concept: Count the number of organisms in a specified area and mathematically extrapolate a number to represent the number of cells per milliliter of sample. Employ a reducible dye that changes color when utilized by living cells as a hydrogen acceptor rather than oxygen (Zoecklein et al., 1999). Refer to Chapter 6 (Section 6.4.2) and Chapter 7 (Section 7.5.2).

Method Type: Direct count/dye reduction

Sample Volume Required: 10 mL, dilute as needed

Equipment Required:

- Microscope: $10 \times$ and $40 \times$ dry objectives; bright-field position (or phase 2 position)
- Levy–Hausser counting chamber and glass cover slip
- Pasteur pipette
- Tube: 10–15 mL
- Vortex mixer
- Counter: manual or digital
- Laboratory wash bottle
- Laboratory wipes

Reagents Required:

- PRAS (prereduced anaerobically sterilized) dilution blanks: 10 mL (peptone dilution blanks can also be used)
- 0.4% or 0.2% Methylene blue reducible dye
- 70% Isopropyl alcohol
- Distilled water

Measurement Accuracy: Approximate

Note: According to Zoecklein et al. (1999), samples with large populations must be diluted to make accurate counts, and because of the group collective nature of some organisms (primarily yeasts), the samples should be mixed by vortex to break up the groups. Counting chambers are 1 mm^2 and there are usually two counting fields. Each 1-mm^2 chamber is divided into 25 sections; each section is divided into 16 cells. Due to the uneven distribution of yeast cells in the counting chambers, all 400 cells (25×16) should be counted rather than the typical 5 sections of 16 cells, averaged, and multiplied times 25. Lower the microscope stage prior to setting the counting chamber in place and upon completion of the count to remove the chamber. Methylene blue can be added during the final dilution or direct stain. Live cells will be clear; dead cells will contain blue within the cell wall. Methylene blue is toxic to live cells, therefore counts must be made within a few minutes.

Possible Errors: Inaccurate sample volume; air bubbles in counting field; improper sample preparation; improper mixing prior to pipetting sample; dirty counting chamber and/or cover; inadequate time allowed for settling; overexposure to methylene blue; counting errors; poor microscopic illumination.

Procedure:

Mix sample thoroughly and place a small volume of sample in a 10-mL tube. Predicted low counts do not require dilution. Predicted higher counts must be diluted until cells can be counted without difficulty. Samples are diluted using the dilution blanks to preserve the viability of the microbes. Add 1 or 2 drops of methylene blue to the final dilution or add to undiluted sample before pipetting the sample onto the counting chamber. For cell count only, omit this step.

Place a clean and dry counting chamber on a flat surface. Place clean and dry cover over counting chamber.

Fill Pasteur pipette with sample; hold at a 45° angle with the tip at the edge of the glass cover; allow a few drops to fill the chamber (via capillary action); fill both fields; do not overfill.

Place chamber on the microscope stage; carefully rotate and place the 10X objective over the chamber.

Slowly raise the stage until the grid comes into clear view. Adjust the stage until the 25 sections containing 16 cells is centered; slowly rotate and place the 40× objective over the chamber. Adjust the stage until the first of the 25 sections is centered and begin counting.

Use the counter to count the number of viable, nonviable, and budding or dividing cells; note each count. Adjust the stage to the next section of 16 cells and count. Continue until all cells are taken into accounted.

Lower the stage of the microscope; remove the counting chamber and cover; wash with distilled water; rinse with a 70% isopropyl; place chamber and cover in a safe and clean area to dry. Repeated use of chamber might require the chamber and lens to be blotted dry with a wipe.

Repeat the count on the second field. Average the results of both fields. Calculate the number of cells in each category:

$$\text{Cell/mL} = [(\text{Cells counted in category}) \times 10^4] \times \text{Dilution factor}, \quad (1)$$

$$\text{Total cell count (cell/mL)} = \text{Sum of all counts or total of cells counted without dye}. \quad (2)$$

9.2.21 Cold Stability Assay for the Verification of Bitartrate Stability via Freeze Test

Concept: Refer to Chapter 7 (Section 7.9.1).
Method Type: Freeze test

Sample Volume Required: 250 mL, same temperature as cold stability tank
Equipment Required:

- Freezer: temperatures of −10°C to −20°C
- Tubes: 150 mL, glass with screw on caps
- Filtration setup: including Buchner funnel with appropriate serial filtration to the desired filtration level with the smallest-pore filter on the bottom (most often 0.45 μm) or the use of a combination filter to the desired filtration level; side-arm vacuum flask (for Buchner funnel)
- Thermometer: accurate; clean and dry with each use
- Graduated cylinder: 100 mL
- High-intensity lamp
- Timer
- Laboratory wipes

Reagents Required: None
Measurement Accuracy: Approximate
Note: It is imperative to filter the wine at the same temperature as the stabilizing tank. Wine should be filtered to the level at which it will be bottled [i.e., 0.45 μm (sterile filtration), 0.8 μm, 2.0 μm, or more coarse for a true evaluation of the postbottled wine]. Testing time varies from 4 to 24 hs depending on the quality of the freezer and the ability to maintain the subzero temperature. On average, the test time is 8–16 h. This assay has limited accuracy and other methods to determine cold stability are more acceptable.
Possible Errors: Inadequate freezer time; sample too warm; incorrect filtration; sample not observed at correct time; sample artificially warmed; ice crystals mistaken for tartrate crystals.

Procedure:

Prepare proper filtration setup prior to the sample arriving in the laboratory.

Obtain sample temperature; immediately filter into clean and dry vacuum flask or 100-mL clean and dry cylinder (or rinse with filtered wine sample and discard before filling).

Add 100 mL of filtered wine sample to two clean and dry glass tube; secure cap; identify one as sample and the other as the control.

Place the sample in the freezer compartment; set timer for the predetermined time (always use the same time frame once established to maintain consistency and repeatability of the test); the control sample will remain at room temperature for comparison.

Remove sample from freezer and allow thawing at room temperature.

Using the high-intensity light, inspect the sample for crystal formation just at the point that the sample has thawed; compare to the control sample; note the results. Bitartrate crystals will refract light.

Allow the sample to come to room temperature; inspect the sample again for crystal formation; compare to the control sample; note results.

This is a pass or fail test.

Passing for a sterile-filtered wine sample or a wine bottled under sterile filtration has no crystal formations found at either inspection, or (depending on the winemaker) crystal formations at the first inspection (thaw) but no crystal formations at room temperature.

Passing for wine samples at other levels of filtration is determined by the winemaker.

Discard sample and thoroughly clean and dry equipment.

9.2.22   Cold Stability Assay for the Verification of Bitartrate Stability via Conductivity

Concept: Refer to Chapter 7 (Section 7.9.2).
Method Type: Conductivity
Sample Volume Required: 250 mL
Equipment Required:

- Conductivity meter (calibrated): range of 100 μS (micro-Siemens) to 1.0 mS (milli-Siemens) and 1.0–10 mS; accuracy of 0.5% over the full scale (Boulton et al., 1999); with conductivity probe; and temperature probe
- Cold bath: with integrated stir plates or individual single-sample cold bath (jacketed beaker) placed on stirrer; water or glycol temperature-controlled circulator set at 0°C or lower (wine variety dictates measuring temperature)
- Stir bars
- Analytic balance: or similar precise weighing device
- Weigh boats or weigh paper
- Thermometer: accurate; clean and dry with each use
- Erlenmeyer flasks: 125 mL with stoppers
- Filtration setup: including Buchner funnel with appropriate serial filtration to the desired filtration level with the smallest-pore filter on the bottom (most often 0.45 μm) or the use of a combination filter to the desired filtration level; side-arm vacuum flask (for Buchner funnel)
- Stir plate
- Timer
- Graduated cylinder: 100 mL
- Laboratory wash bottle
- Laboratory wipes

Reagents Required:

- Cellar-grade KHTa (potassium bitartrate)
- Conductivity standard
- Distilled water

Measurement Accuracy: Accurate
Note: Usually this assay is performed on sterile-filtered wine. There are many different setups in use; the important points are to maintain a consistent sam-

ple temperature; test the sample at the stabilizing tank temperature; use an accurate meter and probes that will fit securely in the sample flask; use a sample volume appropriate for the type of probes you are using; keep sample stirring; and watch the readings.

Possible Errors: Inaccurate sample temperature; faulty meter or probes; failure to stir sample; improper timing and reading of meter; failure to add the KHTa; improper filtration; improper sample filtration temperature.

Procedure:

Turn on cold bath/circulator far in advance of the sample.

Prepare proper filtration setup prior to the sample arriving in the laboratory. Sterile-filtered bottled wine requires no filtration.

Obtain sample temperature; immediately filter into clean and dry vacuum flask or 100-mL clean and dry cylinder (or rinse with filtered wine sample and discard before filling).

For duplicate samples, add 100 mL of filtered wine sample to two clean and dry flasks. For single-sample setups, place approximately 110 mL in a flask; stopper flasks; identify the sample and its duplicate (conduct test on duplicate samples if using a multi sample setup).

Place sample(s) on stir plate; remove stopper and add stir bar; apply vacuum to degas; set timer for 5 mins.

Place sample(s) in cold bath or place a 100-mL sample in a sample-rinsed jacketed beaker (add stir bar); turn on stirring device and stir at a moderate speed.

Insert the conductivity probe and temperature probe [sample(s) should be covered to maintain temperature with openings to allow for probe insertion]; note the start time.

Allow the sample(s) to come to temperature.

Note the conductivity reading and temperature.

Very carefully add 1 g KHTa to the sample(s) and continue to stir.

Note the conductivity reading, temperature, and time after the addition (C_1).

Begin monitoring the meter readings until the sample stabilizes (C_2) or approximately 20 min depending on the setup. Sample is stable when there are three consecutive 1-min readings (Boulton et al., 1999); note conductivity reading, temperature, and time.

During the course of the test, samples that do not stabilize and continue to decrease in conductivity are considered unstable at the test temperature. Samples that do not stabilize and continue to increase in conductivity are considered very stable. Samples that reach and maintain stability are considered stable at the measuring temperature.

The change in conductivity should be minimal with acceptable ranges from 2% to 5%. Significant change would be twice the meter accuracy in the range being used (Boulton et al., 1999).

Calculation of conductivity:

$$\% \Delta \text{ Conductivity} = (C_1 - C_2) \times 100. \tag{1}$$

Turn off equipment when finished. Discard sample and thoroughly clean and dry equipment.

Carefully and thoroughly wash out jacketed beaker into a waste container and dry with wipe.

9.2.23   Heat Stability Assay for Verification of Protein Stability

Concept: Refer to Chapter 7 (Section 7.8.1).

Method Type: Heat/precipitation/nephelometry

Sample Volume Required: 200 mL (dependent on nephelometer cuvette size)

Equipment Required:

- Nephelometer: set upper range at 12 NTU; calibrate with certified standard before each reading; read manufacturer's operation manual
- Cuvettes: glass; heat resistant; clean; crystal clear no scratches; with screw caps
- Cuvette rack: heat resistant; placed in bath
- Soft polishing cloth (lint-free) or lens paper
- Filtration setup: including Buchner funnel or magnetic filter holder; side-arm vacuum flasks; appropriate serial filters to 0.45 μm with the smallest-pore filter on the bottom, or use a combination syringe filter to 0.45 μm attached to a 50-cc syringe for clear or prefiltered wine
- Timer
- High-intensity light source
- Thermometer: accurate
- Hot-water bath: set and maintain bath at 80°C (176°F); water should cover sample level in cuvette but remain well below the cap throughout the test
- Ultrasonic bath: or similar method to eliminate bubbles in sample

Reagents Required: None

Measurement Accuracy: Accurate

Note: Never fill cuvettes full; leave space for liquid expansion. Use extreme caution when working with hot water. Cuvettes should be scratch-free and lint-free. There are several similar methods that differ by water temperature and incubation time. Always check the caps of the cuvettes and tighten before removing them from the bath; caps can loosen during the incubation period. Calibrate meter before each use and place the cuvettes in the meter exactly the same each time they are read to eliminate any interference from the glass.

Possible Errors: Inaccurate water temperature; inadequate incubation time; sample read at wrong temperature; air bubbles in sample; sample not filtered; dirty or scratched cuvette; uncalibrated meter.

Procedure:

Filter wine sample.

Identify and label two cuvettes, one as sample and the other as the control.

Add filtered wine to two clean and dry cuvettes; allow space for expansion; always fill cuvettes to the same sample height for consistency in results. If there is any chance the cuvette or cap can be contaminated, rinse them both with wine filtrate before filling. Tighten cap.

Clean cuvette exterior with deionized water; use a soft cloth or lens paper to clean and dry cuvette thoroughly; do not touch glass.

Calibrate nephelometer (or confirm calibration); place sample cuvette in the measuring chamber and note cuvette position; read NTU to two places; repeat using the control cuvette; note both readings (R_1).

Observe each cuvette under a high-intensity light and note any haze or flocculation.

Verify hot-water bath temperature.

Place sample cuvette in the hot-water bath rack; ensure the water level is appropriate and does not reach the cap; cover bath

Set timer for 6 hs.

Carefully remove bath cover; very carefully tighten the cap on the cuvette; remove the cuvette and rack; allow sample to come to room temperature.

Use a soft cloth or lens paper and clean the exterior of the cuvette thoroughly. Do not agitate the samples while cleaning; do not touch the glass.

Calibrate the nephelometer (or confirm calibration); place sample in the measuring chamber exactly as in the first reading; note the NTU reading (R_2). Calculate the Δ in NTU:

$$\Delta \, NTU = R_2 - R_1. \qquad (1)$$

Acceptable differences are variable from winery to winery. Generally, a Δ NTU of 2.00 or less is acceptable.

Observe the sample and the control under high-intensity light and note any difference in haze or flocculent.

Discard sample and thoroughly clean and dry equipment. Turn off bath and meter if appropriate.

9.2.24   Carbon Dioxide (CO_2) Measurement via Carbodoseur

Concept: Refer to Chapter 8 (Section 8.4.1).
Method Type: Volumetric reduction/Carbodoseur
Sample Volume Required: 125 mL (do *not* degas)
Equipment Required:

- Carbodoseur: glass; 100 mL with hollow-tube screw-cap assembly; use appropriate table to obtain CO_2 concentration
- Laboratory wipes
- Thermometer: accurate; clean and dry with each use

Reagents Required: None
Measurement Accuracy: Accurate

Note: Most simple and widely used method for measurement of CO_2. Results are similar to those obtained via titration method (Iland et al., 2000). It is important to perform the test quickly to maintain the cold temperature.

Possible Errors: Wine sample too warm; inappropriate technique; inadequate agitation.

Procedure:

Place wine sample in refrigerator/freezer and reduce temperature to < 5°C (41°F); chill Carbodoseur.

Add cold sample to the clean, dry, and chilled Carbodoseur to the 100-mL mark (bottom of meniscus); insert glass-tube screw-cap assembly and tighten the cap.

Place a finger over the extended open end of the glass tube; vigorously shake sample for several seconds.

Hold Carbodoseur vertically over a basin; place a wipe (or small container) over the finger on the glass tube; do not touch tube top; slowly release the finger.

CO_2 will rapidly escape from the Carbodoseur; keep holding the instrument vertically until there is no more CO_2 released. Keeping the instrument vertical will lessen the amount of wine loss that occurs from the rapid release of gas.

Place a finger over the extended open end of the glass tube and repeat the agitation and release until there is no gas escaping.

Sit Carbodoseur on a flat surface; hold the bottom of the cylinder; unscrew the tube screw-cap assembly; touch the tube to the cylinder sidewall and blow the volume of wine in the tube back into the Carbodoseur.

Measure the remaining wine; note volume.

Immediately take the temperature of the wine and note.

Refer to the temperature/CO_2 table. Locate the number where the measuring temperature and the remaining volume measurement intersect; this will be the CO_2 value (in mg/L).

9.2.25   Acidulation Trial

Concept: A wine/juice sample is divided into several trial elements (100 mL each); each element receives a specific aliquot of a known acid solution (100 g/L or 0.1 g/mL); each element is evaluated and analyzed. The element having the most favorable results with the lowest volume of acid addition will indicate the rate of addition (RA, g/L). To determine the amount of acid to add to a large volume of wine/juice to obtain the same favorable results, use the formula

$$\text{Acid addition (g)} = \text{RA (g/L)} \times \text{Wine/juice volume (liters)}. \tag{1}$$

Convert g/L to lb/gal:

$$1 \text{ lb/gal} = 120 \text{ g/L} \left(\frac{454 \text{ g/lb}}{3.785 \text{ liters/gal}} \right), \tag{2}$$

and
$$\frac{1 lb}{1000 gal} = 0.12 \text{ g/L}$$

Refer to Chapter 6 (Section 6.2.2.1).
Method Type: Known addition
Sample Volume Required: 750 mL
Equipment Required:

- Graduated cylinder: 100 mL
- Erlenmeyer flasks: 125 mL
- Pipette: serological or auto pipette
- Thermometer: accurate; clean and dry with each use
- Laboratory wipes

Reagents Required:

- Acid stock solution: 100 g/L

  Tartaric acid
  Malic acid
  Citric acid

Measurement Accuracy: Accurate
Note: Titratable acidity (TA) and pH are analyzed and the elements are tasted to evaluate. It is important to note the milliequivalent weight of different acids (g/meq) (Zoecklein et al., 1999) followed by their multiplication factor:

Tartaric acid	0.075	1.00
Malic acid	0.067	0.90
Citric acid	0.064	0.85
Sulfuric acid	0.049	—

Actual changes in TA and pH by tartaric acid additions are dependent on the amount of precipitation of potassium bitartrate and the buffering capacity of the wine or juice at a given pH. Citric and malic acids are effected by the presence of mL bacteria.

In general, at wine pH, 1 g/L of tartaric acid increases the TA (expressed as tartaric acid) 1 g/L, then:

$$1 \text{ g/L tartaric acid} \uparrow \text{TA } 1.00 \text{ g/L; overall} \downarrow \text{pH } 0.1 \pm 0.02$$

$$= 1 \text{ lb/1000 gal} \uparrow \text{TA } 0.12 \text{ g/L; } 8.3 \text{ lb/1000 gal} \uparrow \text{TA } 1 \text{ g/L}$$

$$1 \text{ g/L malic acid} \uparrow \text{TA } 1.10 \text{ g/L; overall} \downarrow \text{pH } 0.08 \pm 0.02$$

$$1 \text{ g/L citric acid} \uparrow \text{TA } 1.15 \text{ g/L; overall} \downarrow \text{pH } 0.08 \pm 0.02$$

Possible Errors: Inaccurate measurement; poor pipetting technique; incorrect acid solution; failure to mix elements; identification errors.

Procedure:

Determine the range of each test and the number of trial elements; that is, the amounts of approximate additions such as 0.5 g/L, 1.0 g/L, 1.5 g/L, and 2.0 g/L. Always include a control that will have no addition.

Label each clean and dry flask, indicating the RA.

Obtain temperature of wine/juice sample and ensure it is at measurement temperature.

Add 100 mL of wine/juice to each flask.

Precisely add the appropriate predetermined aliquot of acid stock solution to the designated flask. The stock solution is 100 g/L or 0.1 g/mL; each 1.0-mL aliquot of stock solution added to the element equals 0.1 g/100 mL or 1 g/L. For example, the addition of a 2-mL aliquot of stock solution to the 100-mL of element would equate to 2 g/L, and so forth. Gently mix each element.

Perform TA and pH analyses; note results.

Pour the remainder of each element into a corresponding labeled wine glass, indicating the RA, TA, and pH; cover until tasted and evaluated.

Discard samples and thoroughly clean and dry equipment; return all chemicals to their proper storage area.

9.2.26   Deacidulation Trial: Potassium Bicarbonate $KHCO_3$ (or K_2CO_3)

Concept: A wine/juice sample is divided into several trial elements (100 mL each); each element receives a specific aliquot of a known $KHCO_3$ solution (45 g/L or 0.045 g/mL); each element is evaluated and analyzed. The element having the most favorable results with the lowest volume of $KHCO_3$ addition will indicate the rate of addition (RA, g/L). To determine the amount of $KHCO_3$ to add to a large volume of wine/juice to obtain the same favorable results, use the formula

$$KHCO_3 \text{ addition (g)} = \text{RA (g/L)} \times \text{Wine/juice volume (liters)}, \qquad (1)$$

Convert g/L to lb/gal:

$$1 \text{ lb/gal} = 120 \text{ g/L or } \left(\frac{454 \text{ g/lb}}{3.785 \text{ litres/gal}} \right), \qquad (2)$$

and
$$\frac{1 \text{ lb}}{1000 \text{ gal}} = 0.12 \text{ g/L}.$$

Refer to Chapter 6 (Section 6.2.2.2).
Method Type: Known addition
Sample Volume Required: 750 mL
Equipment Required:

- Graduated cylinder: 100 mL
- Erlenmeyer flasks: 125 mL
- Pipette: serological or auto pipette
- Thermometer: accurate; clean and dry with each use
- Laboratory wipes

Reagent Required:
KHCO$_3$ (potassium bicarbonate) solution 45 g/L; or a 31-g/L K$_2$CO$_3$ (Potassium carbonate) stock solution can be used.

Measurement Accuracy: Accurate

Note: Titratable acidity (TA) and pH are analyzed and the elements are tasted to evaluate. To reduce TA (expressed as tartaric acid), 0.9 g/L KHCO$_3$ will reduce TA by 1g/L and 0.45 g/L KHCO$_3$ will reduce TA by 0.5 g/L (Zoecklein et al., 1999). Potassium carbonate (K$_2$CO$_3$) can also be used, where 0.62 g/L K$_2$CO$_3$ equals 1 g/L reduction of TA and 0.31 g/L K$_2$CO$_3$ equals a 0.5-g/L reduction of TA.

Possible Errors: Inaccurate measurement; poor pipetting technique; incorrect stock solution concentration; failure to mix elements; identification errors.

Procedure:

Determine the range of each test and the number of trial elements; that is, the amounts of approximate additions such as 0.5 g/L, 1.0 g/L, 1.5 g/L, and 2.0 g/L. Always include a control that will have no addition.

Label each clean and dry flask, indicating the RA.

Obtain temperature of wine/juice sample and ensure it is at measurement temperature.

Add 100 mL of wine/juice to each flask.

Precisely add the appropriate predetermined aliquot of KHCO$_3$ stock solution to the designated flask. The stock solution is 45 g/L or 0.045 g/mL; each 1.0-mL aliquot of stock solution added to the element equals 0.045 g/100mL or 0.45 g/L. An addition of a 2-mL aliquot of stock solution to the 100 mL of element would equate to approximately 1 g/L. Gently mix each sample.

Perform TA and pH analyses; note results.

Pour the remainder of each element into a corresponding labeled wine glass, indicating the RA, TA, and pH; cover until tasted and evaluated.

Discard samples and thoroughly clean and dry equipment; return all chemicals to their proper storage area.

9.2.27 Gelatin Fining Trial

Concept: A wine/juice sample is divided into several trial elements (100 mL each); each element receives a specific aliquot of a known gelatin fining agent solution (1% w/v); each element is evaluated. The element having the most favorable results with the lowest volume of fining agent added will indicate the rate of addition (RA, g/L). To determine the amount of fining agent to add to a large volume of wine/juice to obtain the same favorable results, use the formula

$$\text{Fining agent addition (g)} = \left(\frac{1000}{\text{RA (mg/liter)}} \right) \times \text{Wine/juice volume (liters)}, \quad (1)$$

$$1 \text{ lb/gal} = 120 \text{ g/L or} \left(\frac{454 \text{ g/lb}}{3.785 \text{ liters/gal}} \right), \quad (2)$$

and $1000 \text{ gal } 1 \text{ lb} = 0.12 \text{ g/L}.$

Refer to Chapter 7 (Sections 7.8 and 7.8.3).
Method Type: Known addition
Sample Volume Required: 750 mL
Equipment Required:

- Graduated cylinder: 100 mL
- Erlenmeyer flasks: 125 mL
- Micropipette: manual or digital; 100 to 1000 µl range
- Pipette: manual or automatic; 1–5-mL range
- Decant or filtration setup: including Buchner funnel or magnetic filter holder; side-arm vacuum flasks; appropriate serial filters with the smallest-pore filter on the bottom ; or use a combination syringe filter attached to a 50-cc syringe for clear or prefiltered wine
- Thermometer: accurate; clean and dry with each use
- Laboratory wipes

Reagents Required:
Gelatin 1% w/v stock solution, 10 g/L
Measurement Accuracy: Accurate
Note: Fining elements will be analyzed by taste and appearance. Most wines can be decanted after incubation. Discuss the range to be tested and filtration requirements with the winemaker. Larger volumes of wine/juice sample might be desired, necessitating a more concentrate stock solution and aliquots proportionate to the new concentration.
Possible Errors: Inaccurate measurement; poor pipetting technique; incorrect concentration of stock solution; gelatin solution too old; failure to mix elements; identification errors.

Procedure:

Determine the range of each test and the number of trial elements; that is, the amounts of approximate additions such as 10 mg/L, 50 mg/L, 100 mg/L, 150 mg/L, and so forth. Always include a control that will have no addition.
Label each clean and dry flask, indicating the RA.
Obtain temperature of wine/juice sample and ensure it is at measurement temperature.
Add 100 mL of wine/juice to each flask.
Precisely add the appropriate predetermined aliquot of 1% gelatin stock solution (10 g/L) to the designated flask. Typical additions and their corresponding conversion are as follows:

0.1-mL aliquot to 100 mL = RA 10 mg/L = 0.083 lb/1000 gal = 1 g/hL
0.25-mL aliquot to 100 mL = RA 25 mg/L = 0.21 lb/1000 gal = 2.5 g/hL
0.5-mL aliquot to 100 mL = RA 50 mg/L = 0.42 lb/1000 gal = 5.0 g/hL
0.75-mL aliquot to 100 mL = RA 75 mg/L = 0.63 lb/1000 gal = 7.5 g/hL
1.0-mL aliquot to 100 mL = RA 100 mg/L = 0.83 lb/1000 gal = 10 g/hL
2.0-mL aliquot to 100 mL = RA 200 mg/L = 1.67 lb/1000 gal = 20 g/hL

2.5-mL aliquot to 100 mL = RA 250 mg/L = 2.08 lb/1000 gal = 25 g/hL
3.0-mL aliquot to 100 mL = RA 300 mg/L = 2.5 lb/1000 gal = 30 g/hL

Gently mix each element.

Allow the elements to sit overnight.

Carefully decant or filter each element into a corresponding labeled wine glass indicating the RA and filtration level (if any); cover until tasted. Evaluate each element for clarity and brightness. Determine the RA to be used.

Discard samples and thoroughly clean and dry equipment; return all chemicals to their proper storage area.

9.2.28   Isinglass Fining Trial

Concept: A wine/juice sample is divided into several trial elements (100 mL each); each element receives a specific aliquot of known isinglass fining agent solution (0.5% wt/v); each element is evaluated. The element having the most favorable results with the lowest volume of fining agent added will indicate the rate of addition (mg/L). To determine the amount of fining agent to add to a large volume of wine/juice to obtain the same favorable results, use the formula:

$$\text{Fining agent addition (g)} = \frac{1000}{\text{RA (mg/liter)}} \times \text{Wine/juice volume (liters)},$$

$$1 \text{ lb/gal} = 120 \text{ g/L or } \left(\frac{454 \text{ g/lb}}{3.785 \text{ liters/gal}} \right),$$

and
$$\frac{1 \text{ lb}}{1000 \text{ gal}} = 0.12 \text{ g/L}.$$

Refer to Chapter 7 (Sections 7.8 and 7.8.3).
Method Type: Known addition
Sample Volume Required: 750 mL
Equipment Required:

- Graduated cylinder: 100 mL
- Erlenmeyer flasks: 125 mL
- Micropipette: manual or digital; 100–1000-µL range
- Pipette: manual or automatic; 1–5 mL range
- Decant or filtration setup: including Buchner funnel or magnetic filter holder; side-arm vacuum flasks; appropriate serial filters with the smallest-pore filter on the bottom, or use a combination syringe filter attached to a 50-cc syringe for clear or prefiltered wine
- Thermometer: accurate; clean and dry with each use
- Laboratory wipes

Reagents Required:
Isinglass 0.5% w/v stock solution, 5 g/L
Measurement Accuracy: Accurate

Note: Fining elements will be analyzed by taste and appearance. Most wines can be decanted after incubation. Discuss the range to be tested and filtration requirements with the winemaker. Larger volumes of wine/juice sample might be desired, necessitating a more concentrate stock solution and aliquots proportionate to the new concentration.

Possible Errors: Inaccurate measurement; poor pipetting technique; incorrect concentration of stock solution; poor quality of isinglass; improper preparation of Isinglass; failure to mix elements; identification errors.

Procedure:

Determine the range of each test and the number of trial elements; that is, the amounts of approximate additions such as 10 mg/L, 25 mg/L, 50 mg/L, 100 mg/L, and so forth. Always include a control that will have no addition.

Label each clean and dry flask, indicating the RA.

Obtain temperature of wine/juice sample and ensure it is at measurement temperature.

Add 100 ml wine/juice to each flask.

Precisely add the appropriate predetermined aliquot of 0.5% isinglass stock solution (5 g/L) to the designated flask. Typical additions and their corresponding conversion are as follows:

0.1-mL aliquot to 100 mL = RA 5 mg/L = 0.04 lb/1000 gal = 0.5 g/hL
0.2-mL aliquot to 100 mL = RA 10 mg/L = 0.08 lb/1000 gal = 1.0 g/hL
0.5-mL aliquot to 100 mL= RA 25 mg/L = 0.21 lb/1000 gal = 2.5 g/hL
1.0-mL aliquot to 100 mL = RA 50 mg/L = 0.42 lb/1000 gal = 5.0 g/hL
1.5-mL aliquot to 100 mL = RA 75 mg/L = 0.63 lb/1000 gal = 7.5 g/hL
2.0-mL aliquot to 100 mL = RA 100 mg/L = 0.83 lb/1000 gal = 10.0 g/hL
2.5-mL aliquot to 100 mL = RA 125 mg/L = 1.04 lb/1000 gal = 12.5 g/hL
3.0-mL aliquot to 100 mL = RA 150 mg/L = 1.25 lb/1000 gal = 15.0 g/hL

Gently mix each element.

Allow the elements to sit overnight.

Carefully decant or filter each element into a corresponding labeled wine glass, indicating the RA and filtration level (if any); cover until tasted. Evaluate each element for clarity and brightness. Determine the RA to be used.

Discard samples and thoroughly clean and dry equipment; return all chemicals to their proper storage area.

9.2.29   Egg White Fining Trial

Concept: A wine/juice sample is divided into several trial elements (100 mL each); each element receives a specific aliquot of known isinglass fining agent solution (10% w/v); each element is evaluated. The element having the most favorable results with the lowest volume of fining agent added will indicate the rate of addition (mg/L). To determine the amount of fining agent to add to a large volume of wine/juice to obtain the same favorable results, use the formula

$$\text{Fining agent addition (g)} = \frac{1000}{RA\,(\text{mg/liter})} \times \text{Wine/juice volume (liters)}, \quad (1)$$

$$1\,\text{lb/gal} = 120\,\text{g/L or} \left(\frac{454\,\text{g/lb}}{3.785\,\text{liters/gal}}\right), \quad (2)$$

and $\qquad\qquad \dfrac{1\,\text{lb}}{1000\,\text{gal}} = 0.12\,\text{g/L}.$

Refer to Chapter 7 (Sections 7.8 and 7.8.2).
Method Type: Known addition
Sample Volume Required: 750 mL
Equipment Required:

- Graduated cylinder: 100 mL
- Erlenmeyer flasks: 125 mL
- Micropipette: manual or digital; 100–1000-µL range
- Decant or filtration setup: including Buchner funnel or magnetic filter holder; side-arm vacuum flasks; appropriate serial filters with the smallest-pore filter on the bottom, or use a combination syringe filter attached to a 50-cc syringe for clear or prefiltered wine
- Thermometer: accurate; clean and dry with each use
- Laboratory wipes

Reagent Required:
Egg white 10% w/v stock solution, 100 g/L
Measurement Accuracy: Accurate
Note: Fining elements will be analyzed by taste and appearance. Most wines can be decanted after incubation. Discuss the range to be tested and filtration requirements with the winemaker. Larger volumes of wine/juice sample might be desired necessitating a more concentrate stock solution and aliquots proportionate to the new concentration.
Possible Errors: Inaccurate measurement; poor pipetting technique; incorrect concentration of stock solution; nonfresh egg white solution; failure to mix elements; identification errors.

Procedure:

Determine the range of each test and the number of trial elements; that is, the amounts of approximate additions such as 100 mg/L, 200 mg/L, 300 mg/L, 400 mg/L, and so forth. Always include a control that will have no addition.
Label each clean and dry flask, indicating the RA.
Obtain temperature of wine/juice sample and ensure it is at measurement temperature.
Add 100 mL of wine/juice to each flask.
Precisely add the appropriate predetermined aliquot of 10% w/v egg white stock solution (100 g/L) to the designated flask. Typical additions and their corresponding conversion are as follows:

0.1-mL aliquot to 100 mL = RA 100 mg/L = 0.83 lb/1000 gal = 10 g/hL
0.2-mL aliquot to 100 mL = RA 200 mg/L = 1.67 lb/1000 gal = 20 g/hL

0.3-mL aliquot to 100 mL = RA 300 mg/L = 2.50 lb/1000 gal = 30 g/hL
0.4-mL aliquot to 100 mL = RA 400 mg/L = 3.33 lb/1000 gal = 40 g/hL
0.5-mL aliquot to 100 mL= RA 500 mg/L = 4.17 lb/1000 gal = 50 g/hL
0.6-mL aliquot to 100 mL = RA 600 mg/L = 5.00 lb/1000 gal = 60 g/hL
0.7-mL aliquot to 100 mL = RA 700 mg/L = 5.83 lb/1000 gal = 70 g/hL

Gently mix each element.

Allow the elements to sit overnight.

Carefully decant or filter each element into a corresponding labeled wine glass, indicating the RA and filtration level (if any); cover until tasted. Evaluate each element for clarity and brightness. Determine the RA to be used.

Discard samples and thoroughly clean and dry equipment; return all chemicals to their proper storage area.

9.2.30   Bentonite Fining Trial

Concept: A wine/juice sample is divided into several trial elements (100 mL each); each element receives a specific aliquot of known bentonite fining agent solution (5% w/v); each element is evaluated. The element having the most favorable results with the lowest volume of fining agent added will indicate the rate of addition (g/L). To determine the amount of fining agent to add to a large volume of wine/juice to obtain the same favorable results, use the formula

$$\text{Fining agent addition (g)} = \frac{1000}{RA\,(\text{mg/liter})} \times \text{Wine/juice volume (liters)}, \quad (1)$$

$$1 \text{ lb/gal} = 120 \text{ g/L or } \left(\frac{454\,\text{g/lb}}{3.785\,\text{liters/gal}}\right), \quad (2)$$

and

$$\frac{1\,\text{lb}}{1000\,\text{gal}} = 0.12 \text{ g/L}.$$

Refer to Chapter 7 (Sections 7.8 and 7.8.1).
Method Type: Known addition
Sample Volume Required: 750 mL
Equipment Required:

- Graduated cylinder: 100 mL
- Erlenmeyer flasks: 125 mL
- Micropipette: manual or digital; 100–1000-µL range
- Pipette: manual or automatic; 1–5-mL range
- Filtration setup: including Buchner funnel or magnetic filter holder; side arm vacuum flasks; appropriate serial filters to 0.45 µm with the smallest-pore filter on the bottom, or use a combination syringe filter to 0.45 µm attached to a 50-cc syringe for clear or prefiltered wine
- Stir plates, with stir bars
- Thermometer: accurate; clean and dry with each use
- Laboratory wipes

Reagent Required:
Bentonite 5% w/v stock solution, 50 g/L
Measurement Accuracy: Accurate
Note: Fining elements will be analyzed by taste, appearance, and heat stability testing. Discuss the element range to be tested with the winemaker. Larger volumes of wine/juice sample might be desired, necessitating a more concentrate stock solution and aliquots proportionate to the new concentration. Determination of RA might require two consecutive trials, the second trial being built around the first RA determination to further define the lowest addition of fining agent.
Possible Errors: Inaccurate measurement; poor pipetting technique; incorrect concentration of stock solution; improper preparation of bentonite; failure to mix elements; inadequate filtration; identification errors.

Procedure:

Determine the range of each test and the number of trial elements; that is, the amounts of approximate additions such as 0.5 g/L, 1 g/L, 1.5 g/L, 2 g/L, and so forth. Always include a control that will have no addition.
Label each clean and dry flask, indicating the RA.
Obtain temperature of wine/juice sample and ensure it is at measurement temperature.
Add 100 mL of wine/juice to each flask.
Precisely add the appropriate predetermined aliquot of 5% bentonite stock solution (50 g/L) to the designated flask. Typical additions and their corresponding conversion are as follows:

0.5-mL aliquot to 100 mL= RA 0.25 g/L = 2.08 lb/1000 gal = 25 g/hL
1.0-mL aliquot to 100 mL = RA 0.50 g/L = 4.17 lb/1000 gal = 50 g/hL
1.5-mL aliquot to 100 mL = RA 0.75 g/L = 6.25 lb/1000 gal = 75 g/hL
2.0-mL aliquot to 100 mL = RA 1.00 g/L = 8.33 lb/1000 gal = 100 g/hL
2.5-mL aliquot to 100 mL = RA 1.25 g/L = 10.42 lb/1000 gal = 125 g/hL
3.0-mL aliquot to 100 mL = RA 1.50 g/L = 12.50 lb/1000 gal = 150 g/hL

Place each element on a stir plate; insert stir bar; mix at a moderate speed for 5 min.
Allow the elements to sit undisturbed for a period of time similar to the cellar practice.
Sterile-filter elements to the 0.45-µm level.
Perform heat stability assays on each element (refer to Section 9.2.2).
Evaluate the heat stability results. Evaluate each element for clarity and brightness. Determine the RA to be used.
Discard samples and thoroughly clean and dry equipment; return all chemicals to their proper storage area.

9.2.31 Copper Sulfate ($CuSO_4$) Trial

Concept: A wine/juice sample is divided into several trial elements (100 mL each); each element receives a specific aliquot of known copper sulfate solution (394 mg/L $CuSO_4$ equates to 100 mg/L or ppm Cu^{2+}); each element is evaluated. The element having the most favorable results with the lowest volume of Cu^{2+} added will indicate the rate of addition (mg/L). To determine the amount of $CuSO_4$ to add to a large volume of wine/juice to obtain the same favorable results, use the formula

$$CuSO_4 \text{ addition (g)} = \left(\frac{RA\,(\text{mg/liter})}{1000} \right) \times \text{wine/juice}$$

$$\text{volume (liters))} \times 3.93 \text{ g} \qquad (1)$$

or

$$CuSO_4 \text{ g/gallon} = RA\,(\text{mg/L}) \times 0.01488, \qquad (2)$$

$$1 \text{ lb/gal} = 120 \text{ g/L or} \left(\frac{454\,\text{g/lb}}{3.785\,\text{liters/gal}} \right), \qquad (3)$$

and $1\text{lb}/1000 \text{ gal} = 0.12 \text{ g/L}.$

Refer to Chapter 7 (Sections 7.8 and 7.8.4).
Method Type: Known addition
Sample Volume Required: 750 mL
Equipment Required:

- Graduated cylinder: 100 mL
- Erlenmeyer flasks: 125 mL
- Micropipette: manual or digital; 100–1000-µL range
- Thermometer: accurate; clean and dry with each use
- Laboratory wipes

Reagent Required: $CuSO_4$ (copper sulfate) stock solution; 394 mg/L $CuSO_4$ contains 100 mg/L (ppm) Cu^{2+} (check purity of chemical used)
Measurement Accuracy: Accurate
Note: $CuSO_4$ elements will be analyzed by smell. Discuss the element range to be tested with the winemaker. Larger volumes of wine/juice sample might be desired, necessitating a more concentrate stock solution and aliquots proportionate to the new concentration. Regulation of Cu^{2+} in wine/juice allows 0.5-mg/L additions in the United States, with an allowable residual of 0.2 mg/L, and 0.2 mg/L in other countries. Determination of RA might require two consecutive trials, the second trial being built around the first RA determination to further define the lowest addition of $CuSO_4$. The Cu^{2+} concentration of the $CuSO_4$ is the measured RA. The amount of $CuSO_4$ required to supply the desired Cu^{2+} RA must be calculated.
Possible Errors: Inaccurate measurement; poor pipetting technique; incorrect concentration of Cu^{2+} in the $CuSO_4$ stock solution; failure to mix elements; identification errors; mathematical errors.

Procedure:

Determine the range of each test and the number of trial elements; that is, the amounts of approximate additions such as 0.1 mg/L, 0.2 mg/L, 0.3

mg/L, 0.4 mg/L, and so forth. Always include a control that will have no addition.

Label each clean and dry flask, indicating the RA.

Obtain temperature of wine/juice sample and ensure it is at measurement temperature.

Add 100 mL of wine/juice to each flask.

Precisely add the appropriate predetermined aliquot of $CuSO_4$ stock solution to the designated flask. Typical additions of Cu^{2+} and their corresponding conversion are as follows:

0.1-mL aliquot to 100 mL = RA 0.1 mg /L = 1.5 g $CuSO_4$/1000 gal = 39 mg/hL

0.2-mL aliquot to 100 mL = RA 0.2 mg/L = 3.0 g $CuSO_4$/1000 gal = 79 mg/hL

0.3-mL aliquot to 100 mL = RA 0.3 mg/L = 4.5 g $CuSO_4$/1000 gal = 118 mg/hL

0.4-mL aliquot to 100 mL = RA 0.4 mg/L = 6.0 g $CuSO_4$/1000 gal = 157 mg/hL

0.5-mL aliquot to 100 mL = RA 0.5 mg/L = 7.4 g $CuSO_4$/1000 gal = 197 mg/hL

Carefully pour each element into a corresponding labeled wine glass, indicating the RA; cover until evaluated for off odors. Determine the RA to be used.

Discard samples and thoroughly clean and dry equipment; return all chemicals to their proper storage area.

10

Reagents

10.1 Introduction

Preparation of reagents requires an understanding of the chemicals being used, excellent laboratory techniques, and adhering to all safety rules. The following are few suggestions:

- Please read Chapters 1–3 before you attempt to create reagents or use any chemical.
- Temperature is critical when measuring any liquid: sample, chemical, or water.
- Unless a container or measuring device is clean and dry, always rinse the container or measurement device with the final wine sample.
- Check and recheck chemical identification and concentration before use.
- Practice and perfect your pipetting techniques, which are paramount for accurate measurements. Know how to read a meniscus correctly. Learn to use the balance correctly.
- Delivering solutions via burette requires flushing of the delivery tube and touching off the last drop into the receiving container (read manufacturer's recommendations).
- All measurements should be exact unless otherwise stated ($\sim$ = approximately).
- Remember 1 : 4, 1 + 3, 1 × 4, 1 in 4, all mean the same thing.
- Recheck all readings, math, and notations for errors.
- Standardize reagents.
- Use proper chemical storage; note shelf life, expiration dates, chemical name and concentration, date made; and NFPA rating on all reagents made.
- Never make reagents when alone in the laboratory
- Know your chemicals; refer to MSDS sheets. Utilize fume hood and proper ventilation; always wear the appropriate personal protection.

10.2   Reagents

The majority of reagents are prepared fresh on an as-needed basis or kept for a very limited time. It is important not to purchase or make more reagents than can be used quickly. Reagents that are purchased usually contain preservatives, which prolong their shelf life and maintain the chemical quality after opening.

Reagents made in the laboratory without the addition of preservatives are subject to contamination or degradation. Some reagents are sensitive to cold or heat, some to light, and others are hydroscopic (absorb moisture from the atmosphere). Reagents prepared with water are subject to microbial contamination.

Reagents, especially standards, should be made with the purest form of chemical under the best conditions (reagent grade). The weight or volume of chemical used to make a reagent should be adjusted according to the grade of purity:

$$\text{Adjusted chemical weight} = \frac{\text{Weight of chemical}}{\% \text{ Purity}} \qquad (1)$$

or

$$\text{Normality} = \frac{\text{Weight of compound} \times \% \text{ Purity}}{\text{CFW} \times \text{Final volume}} \qquad (2)$$

where C_{FW} is the compound formula weight.

Reagents used for trials should be made from the same chemicals used in the cellar (cellar grade). Use only deionized or distilled pure water for solutions, never use tap water (unless used for cleaning solutions).

Read the MSDS information on each chemical and know the proper method of storage and handling.

Standard laboratory equipment and supplies used for reagent preparation includes the following:

- Analytical balance, calibration certified
- Weigh paper, or weigh boats
- Thermometer (calibrated and certified)
- Volumetric flasks (Class A certified and, in some cases, serialized) and closures
- Volumetric pipettes (Class A certified)
- Serological pipettes (Class A certified)
- Micropipettes: calibration certified, manual or digital
- Erlenmeyer flasks with stoppers
- Fernbach flasks with stoppers
- Griffin beakers
- Graduated cylinders (Class A certified)
- Stir plate with stir bars
- Heated stir plate with stir bars

- Hot-water bath
- Proper storage containers
- Fume hood
- Glass stirring rods
- Appropriate safety equipment

The quality of the reagents directly affects the precision of the analyses. Variances in analysis created by inaccurate reagents can multiply quickly. Analyses requiring precise concentrations of reagents might reduce the ability of some laboratories to make their own reagents, because of the lack of precision equipment. These laboratories might be able to purchase the reagents premade.

Laboratories should evaluate their capabilities and adjust their level of acceptance of analysis to those capabilities. Unacceptable ranges will require sending samples to an outside commercial laboratory that is better equipped.

10.2.1   Acetaldehyde (C_2H_4O) 10% w/v Solution

Acetaldehyde, as a liquid, is weighed cold and is a weight-to-volume measurement. The 10% w/v measurement equates to 100 g/L, 10 g/100 mL, and 0.1 g/mL (or 1 mg/mL). Note the MSDS for special storage requirements. This solution can be stored for approximately 1 week.

Solution:

- Add to a 100-mL volumetric flask, ~ 50 mL of distilled water.
- Place flask on the balance; tare the balance.
- Add quickly to the flask and weigh 10 g acetaldehyde (C_2H_4O) liquid.
- Remove flask from balance; cover immediately and invert the flask several time to mix.
- Bring flask to the 100-mL volume mark with distilled water; cover immediately and mix.
- Store in an appropriate container.
- Clean and dry equipment thoroughly; return all chemicals to their proper storage area.

10.2.2   Acetic Acid ($C_2H_4O_2$) 1% w/v Standard

A standard 10-g/L stock solution of acetic acid is made (1% w/v). From this stock solution, the individual standards are made. Glacial acetic acid, as a liquid, is weighed and is a weight-to-volume measurement. Acetic acid is very volatile and should be measured and brought to volume quickly. Acetic acid solutions do not keep well and should be made fresh. You can calculate the volume of stock solution to add to any volume to achieve the desired concentration:

$$C1 \times V1 = C2 \times V2 \quad \text{or} \quad V1 = \frac{V_2}{\left(C_1/C_2\right)} \qquad (1)$$

where $C1$ is the initial concentration, $V1$ is the volume to add, $C2$ is the desired concentration, $V2$ is the desired final volume, and $C1/C2$ is the dilution factor

Stock Solution:

- Add to 100 mL volumetric flask, ~ 50 mL of distilled water.
- Place flask on the balance; tare the balance.
- Add quickly via a funnel to the flask and weigh 1 g $C_2H_4O_2$ (glacial acetic acid liquid –99.8% or higher concentration).[*]
- Remove flask and cover; gently invert flask several times to mix.
- Bring flask to the 100-mL volume mark with distilled water; cover immediately and mix. From this stock solution, pipette the designated volume (1 mL contains 10 mg) into a 100-mL volumetric flask. Bring to volume with distilled water; cover immediately and mix.
  Typical standards are as follows:

0.1 mL stock solution → bring to 100-mL volume = 10 mg/L standard solution

0.5 mL stock solution → bring to 100-mL volume = 50 mg/L standard solution

1.0 mL stock solution → bring to 100-mL volume = 100 mg/L standard solution

1.5 mL stock solution → bring to 100-mL volume = 150 mg/L standard solution

- Store in an appropriate container.
- Clean and dry equipment thoroughly; return all chemicals to their proper storage area.

$$\text{Adjusted chemical weight} = \frac{\text{Weight of chemical}}{\% \text{ Purity}}. \qquad (1)$$

10.2.3 Ammonia (NH_3) 0.1% w/v Standard

A standard 1-g/L NH_3 stock solution is made (0.1% w/v). From this stock solution, the individual standards are made. Ammonia is derived from ammonium chloride (NH_4Cl) and makes up approximately 32% of the compound (17_{FW} NH_3 and NH_4Cl 53.45_{FW}). Ammonium chloride is a weight to volume measurement. To find the weight of NH_4Cl that will render the amount of NH3 needed, use the formula

Weight of compound required =
$$\qquad \text{Desired weight of the substance} \times \left(\frac{C_{FW}}{S_{FW}}\right), \qquad (1)$$

[*]The volume or weight of chemical used to make a reagent should be adjusted according to the grade of purity.

where C_{FW} is the compound formula weight (molar mass) and S_{FW} is the substance formula weight.

You can calculate the volume of stock solution to add to any volume to achieve the desired concentration:

$$C1 \times V1 = C2 \times V2 \quad \text{or} \quad V1 = \frac{V_2}{(C_1/C_2)}, \qquad (2)$$

where $C1$ is the initial concentration, $V1$ is the volume to add, $C2$ is the desired concentration, $V2$ is the desired final volume, and $C1/C2$ is the dilution factor.

Stock Solution:

- Add to a 1-L (1000 mL) volumetric flask, [3.181 g NH_4Cl (ammonium chloride) will equate to 1g/L NH_3], ~ 700 mL of deionized water and 0.1 mL of $1N$ HCl (hydrochloric acid).
- Cover flask and gently swirl to mix thoroughly.
- Bring flask to the 1-L volume mark with distilled water. Cover immediately and mix by inverting the flask several times.

From this stock solution, pipette the designated volume (1 ml contains 1 mg NH_3) into a 1-L volumetric flask. Bring to volume with distilled water; cover and mix well.

Typical standards are as follows:

1 mL stock solution → bring to 1 L volume = 1 mg/L standard solution

10 mL stock solution → bring to 1 L volume = 10 mg/L standard solution

50 mL stock solution → bring to 1 L volume = 50 mg/L standard solution

100 mL stock solution → bring to 1 L volume = 100 mg/L standard solution

- Store in an appropriate container.
- Clean and dry equipment thoroughly; return all chemicals to their proper storage area.

10.2.4   Bentonite 5% w/v Slurry

A 5% w/v stock solution of bentonite is made. This is a weight-to-volume measurement. From this stock solution, aliquots will be added to wine/juice samples. Obtain cellar-grade bentonite. Bentonite slurry has a shelf life of approximately 3 weeks.

Stock Solution:

- Add to a 1-L volumetric boiling flask, ~ 800 mL of deionized water.
- Place flask on heated stir plate (approximately 60°C); add stir bar.

- Add 50 g bentonite slowly to beaker using a dry funnel.
- Continue stirring until bentonite is well hydrated. Remove beaker from heat and allow the slurry to cool and swell (this will take several hours; an alternate method is to eliminate the heat and continue stirring the mixture overnight).
- Bring flask to the 1-L volume mark with deionized water; cover and mix thoroughly. Always mix before use.
- Store in an appropriate container.
- Clean and dry equipment thoroughly; return all chemicals to their proper storage area.

10.2.5   20 °Brix Sucrose ($C_{12}H_{22}O_{11}$) Standard

Brix measurement is a weight-to-weight measurement. Brix solution does not store well and should be made fresh as needed or frozen and thawed when needed.

Solution:

- Place a 250-mL Erlenmeyer flask on the balance and tare.
- Add 20 g pure sucrose to the flask and weigh.
- Add to the flask, ~ 70 mL of distilled water to a final weight of 100 g.
- Place flask on stir plate; add stir bar; mix until sucrose is dissolved.
- Store in an appropriate container.
- Clean and dry equipment thoroughly; return all chemicals to their proper storage area.

10.2.6   Copigmentation Buffer

This buffer is a weight-to-volume measurement.

Solution:

- Add 0.5 g of KHTa (potassium bitartrate, $C_4H_5KO_6$) and 24 mL of EtOH (ethanol) 95–100% v/v (190 to 200 proof) to a 200-mL volumetric flask.
- Bring flask to the 200-mL volume mark with distilled water; cover and mix thoroughly by inverting flask several times.
- Store in an appropriate container.
- Clean and dry equipment thoroughly; return all chemicals to their proper storage area.

10.2.7   Copper Sulfate ($CuSO_4$) 100 mg/L Solution

A standard 100-mg/L copper (Cu^{2+}) stock solution is made. From this stock solution, individual aliquots will be added to wine/juice. Copper is derived from copper sulfate ($CuSO_4$) and makes up approximately 25.4% of the compound (63.5_{MW} Cu^{2+} and $CuSO_4$ 249.68_{FW}). This is a weight-to-volume measurement. To find the weight of $CuSO_4$ that will provide the amount of Cu^{2+} needed, use the formula

$$C_W \text{ required} = \text{Desired } S_W \times \left(\frac{C_{FW}}{S_{MW}}\right), \tag{1}$$

where C_W is the compound weight, S_W is the substance weight, C_{FW} is the compound formula weight (molar mass), and S_{MW} is the substance molecular weight (atomic mass).

Stock Solution:

- Add 394 mg $CuSO_4$, which will equate to 100 mg/L Cu^{2+}, and ~ 800 mL of distilled water to a 1-L (1000 ml) volumetric flask.
- Place flask on stir plate; add stir bar; mix until $CuSO_4$ is dissolved.
- Bring flask to the 1-L volume mark with deionized water.
- Cover and mix by inverting flask several times.
- Store in an appropriate container. $CuSO_4$ solution has approximately a 2–3-month shelf life.
- Clean and dry equipment thoroughly; return all chemicals to their proper storage area.

10.2.8   *Isinglass 0.5% w/v Solution*

A 0.5% w/v stock solution of isinglass is made. This is a weight-to-volume measurement. From this stock solution, aliquots will be added to wine/juice samples. Obtain cellar-grade Isinglass. Isinglass should not smell fishy. This solution should be made fresh for each use. Do not heat the solution; keep as cool as possible while mixing. Drifine is one of several brands of isinglass powder, follow manufacturer's recommendations for hydration to attain a 0.5% w/v solution.

Stock Solution:

- Add 1 g of citric acid and 80 mL of deionized water to a 100-ml volumetric flask.
- Place flask on stir plate; add stir bar and mix gently for a few minutes.
- Add 0.5 g isinglass (cut into easily dissolving pieces or use powder) slowly to the flask.
- Continue stirring gently (12 h or more).
- Stop stirring and bring flask to the 1-L volume mark with deionized water.
- Continue to stir gently for a few minutes. The solution should be well blended with a gel consistency. Extensive blending time might require the addition of 200 mg/L of SO_2 to preserve the isinglass (Iland et al., 2000).
- Use immediately.
- Clean and dry equipment thoroughly; return all chemicals to their proper storage area.

10.2.9   Egg White 10% w/v Solution

A 10% w/v stock solution of egg white is made. This is a weight-to-volume measurement. From this stock solution, aliquots will be added to wine/juice samples. Fresh eggs are more effective than frozen whites. This solution should be made fresh for each use. The use of NaCl (sodium chloride) is not permitted by the Alcohol and Tobacco Tax and Trade Bureau (TTB) in the United States.

Stock Solution:

- Add 10 g of egg white (each egg contains an average of 30 g/egg white) and ~ 0.4 g of KCl (potassium chloride) carefully to a 100-mL volumetric flask.
- Place flask on stir plate; add stir bar and mix gently.
- Add ~ 80 mL of deionized water slowly to the flask.
- Continue to mix gently.
- Stop mixing and bring flask to the 100-mL volume mark with deionized water.
- Continue to mix gently until well blended.
- Use immediately.
- Clean and dry equipment thoroughly; return all chemicals to their proper storage area.

10.2.10   Ethanol (EtOH) Solutions and Standards

The preparation of EtOH solutions and standards is a volume-to-volume measurement. EtOH is available as percent purity or proof. Ethanol rated as 95% pure is listed as 190 proof and a 100% ethanol is 200 proof. Calculate the percent purity using

$$\frac{\text{EtOH proof}}{2} = \% \text{ Purity.} \tag{1}$$

The measurements are based on 100% pure EtOH, so any variation from 100% must be accounted for in the calculations of volume. The calculation is as follows:

$$\text{Final volume of EtOH (100\%) (mL)} = \frac{\text{mL of EtOH (100\%) needed}}{\% \text{ Purity of EtOH}}, \tag{2}$$

Another method to calculate the final volume of EtOH at any % purity is to use the formula

$$C1 \times V1 = C2 \times V2 \quad \text{or} \quad V1 = \frac{V2}{(C1/C2)}, \tag{3}$$

where $C1$ is the initial concentration, $V1$ is the volume to add, $C2$ is the desired concentration, $V2$ is the desired final volume, and $C1 / C2$ is the dilution factor.

   The preparation of 1-L volume solutions and 100-mL standards becomes easy and can be adapted for any use or volume (see Table 10.1). Follow this

TABLE 10.1. Preparation of ethanol standards and solutions.

Standard/solution concentration	Volume of 100% EtOH	Bring to volume with distilled water (100 mL)	Bring to volume with distilled water (1 L)
5% v/v	5 mL →	5 mL + 95 mL	50 mL + 950 mL
	50 mL →		
10% v/v	10 mL →	10 mL + 90 mL	100 mL + 900 mL
	100 mL →		
14% v/v	14 mL →	14 mL + 86 mL	140 mL + 860 mL
	140 mL →		
20% v/v	20 mL →	20 mL + 80 mL	200 mL + 800 mL
	200 mL →		
70% v/v	70 mL →	70 mL + 30 mL	700 mL + 300 mL
	700 mL →		

procedure for all solutions and standards using the designated additions to the appropriate distilled water volume. Large volumes of solutions created for use as an antimicrobial can use deionized water rather than distilled water.

Solution Example 14% Standard:

- Add 14 mL of 100% EtOH (200 proof) to a 100-mL volumetric flask.
- Bring flask to the 100-mL volume mark with distilled water; cover and mix thoroughly by inverting flask several times.
- EtOH is volatile and flammable; store in an appropriate container. High concentrations of EtOH should be stored in a flammables cabinet.
- Clean and dry equipment thoroughly; return all chemicals to their proper storage area.

10.2.11   Gelatin 1% w/v Solution

A 1% w/v stock solution of gelatin is made. This is a weight-to-volume measurement. Use cellar-grade gelatin. Do not exceed 40°C (104°F) during preparation. From this stock solution, aliquots will be added to wine/juice samples. This solution should be made fresh for each use.

Stock Solution:

- Add ~ 80 mL of deionized water and 1 g of gelatin to a 100-mL volumetric flask.
- Place flask on a heated stir plate (< 40°C); add stir bar and stir gently until well mixed.
- Stop stirring; remove stir bar carefully; bring to the 100-mL volume mark with deionized water.
- Insert the same stir bar and continue to mix gently until well blended.
- Store in an appropriate container. Use within 24 h.
- Clean and dry equipment thoroughly; return all chemicals to their proper storage area.

10.2.12 Hydrochloric Acid (HCl) Solutions

Preparation of HCl solutions is a volume-to-volume measurement ($1N$ HCl $= 1M$ HCl). Using a $10N$ HCl purchased stock solution (or more concentrate), you can calculate the volume of stock solution to add to any volume to achieve the desired concentration:

$$C1 \times V1 = C2 \times V2 \quad \text{or} \quad V1 = \frac{V2}{(C1/C2)}, \tag{1}$$

where $C1$ is the initial concentration, $V1$ is the volume to add, $C2$ is the desired concentration, $V2$ is the desired final volume, and $C1/C2$ is the dilution factor.

The preparation of solutions becomes easy and can be adapted for any use or volume:

$1N$ HCl $= 10$ mL of $10N$ HCl $\rightarrow 100$ mL of distilled water
$0.01N$ HCl $= 0.1$ mL of $10N$ HCl $\rightarrow 100$ mL of distilled water

Follow this procedure for all solutions using the designated additions to the appropriate distilled water volume. Remember to always add acid *to* water. Use extreme caution.

Solution Example 0.1N HCl:

- Add ~ 80 mL of distilled water to a 100-mL volumetric flask.
- Place flask in a cold water bath (secondary containment); cold water level should be slightly higher than water level in flask.
- Add 1 mL of $10N$ HCl stock solution slowly to the flask using a funnel.
- Bring flask to the 100-mL volume mark with distilled water.
- Cover and mix thoroughly by gently swirling the flask.
- Keep the flask in the cold bath until the reaction cools.
- Store in an appropriate container.
- Clean and dry equipment thoroughly; return all chemicals to their proper storage area.

10.2.13 Hydrogen Peroxide (H_2O_2) 0.3% v/v Solution

A 30% H_2O_2 stock solution is diluted to make a 0.3% H_2O_2 solution. This is volume-to-volume measurement. You can calculate the volume of stock solution to add to any volume to achieve the desired concentration:

$$C1 \times V1 = C2 \times V2 \quad \text{or} \quad V1 = \frac{V2}{(C1/C2)}, \tag{1}$$

where $C1$ is the initial concentration, $V1$ is the volume to add, $C2$ is the desired concentration, $V2$ is the desired final volume, and $C1/C2$ is the dilution factor.

Solution:

- Add 10 mL of 30% H_2O_2 stock solution to a 1-L volumetric flask.
- Bring flask to the 1-L volume mark with distilled water.

- Cover and mix thoroughly by inverting the flask several times.
- Store in an appropriate container. Refrigerate; use within 2 months.
- Clean and dry equipment thoroughly; return all chemicals to their proper storage area.

10.2.14   Indicator Solution

This is a weight-to-volume measurement.

Solution:

- Add the following to a 100-mL volumetric flask:

  0.1 g of $C_{15}H_{14}N_3NaO_2$ (methyl red)
  0.05 g of $C_{16}H_{18}ClN_3S$ (methylene blue)
  50 mL of 50% EtOH (ethanol) solution

- Gently swirl flask to mix and dissolve the methyl red and methylene blue.
- Bring flask to the 100-mL volume mark with 50% EtOH solution.
- Cover and mix thoroughly by gently inverting the flask several times.
- Store in an appropriate container. Volatile.
- Clean and dry equipment thoroughly; return all chemicals to their proper storage area.

10.2.15   Iodine (I_2) 0.02N Solution

The iodine solution is made from a stock solution ($0.1N$ I_2), which can be purchased ($0.1N$ $I_2 = 0.05M$ I_2). You can calculate the volume of stock solution to add to any volume to achieve the desired concentration:

$$C1 \times V1 = C2 \times V2 \quad \text{or} \quad V1 = \frac{V2}{(C1/C2)}, \qquad (1)$$

where $C1$ is the initial concentration, $V1$ is the volume to add, $C2$ is the desired concentration, $V2$ is the desired final volume, and $C1/C2$ is the dilution factor.

A $0.1N$ I_2 stock solution will be used in this formulation.

Solution:

- Add 200 mL of $0.1N$ I_2 stock solution to a 1-L volumetric flask.
- Bring flask to the 1-L volume mark with deionized water.
- Cover and mix thoroughly by gently inverting the flask several times.
- Store in an appropriate container. Extremely light sensitive.
- Clean and dry equipment thoroughly; return all chemicals to their proper storage area.

10.2.16   L-Malic Acid ($C_4H_6O_5$) 0.2 g/L Standard

A standard 0.2-g/L stock solution of L-malic acid is made. From this stock solution, the individual standards are made. Reagent-grade L-malic acid is used. This is a weight-to-volume measurement. You can calculate the volume of stock solution to add to any volume to achieve the desired concentration:

$$C1 \times V1 = C2 \times V2 \quad \text{or} \quad V1 = \frac{V2}{(C1/C2)}, \tag{1}$$

where $C1$ is the initial concentration, $V1$ is the volume to add, $C2$ is the desired concentration, $V2$ is the desired final volume, and $C1/C2$ is the dilution factor.

Stock Solution:

- Add 0.2 g of L-malic acid and ~ 80 mL of distilled water to a 100-mL volumetric flask.
- Cover flask and gently swirl to mix.
- Bring flask to the 100-mL volume mark with distilled water; mix and cover. From this stock solution, pipette the designated volume into a 100-mL volumetric flask. Bring to volume with distilled water; cover and mix.

Typical standards are as follows:

5 mL stock solution → bring to 100-mL volume = 0.1 g/L standard solution
10 mL stock solution → bring to 100-mL volume = 0.2 g/L standard solution
50 mL stock solution → bring to 100-mL volume = 1.0 g/L standard solution

- Store in an appropriate container.
- Clean and dry equipment thoroughly; return all chemicals to their proper storage area.

10.2.17   Malic Acid ($C_4H_6O_5$) 10% w/v Solution

A 10% w/v stock solution of malic acid is made. This is a weight-to-volume measurement. From this stock solution, aliquots will be added to wine/juice samples. This solution should be made from cellar-grade malic acid.

Stock Solution:

- Add 10 g of $C_4H_6O_5$ (malic acid) and ~ 80 mL of deionized water carefully to a 100-mL volumetric flask.
- Cover flask and gently swirl to mix.
- Bring flask to the 100-mL volume mark with deionized water; mix and cover.
- Store in an appropriate container; refrigerate.
- Clean and dry equipment thoroughly; return all chemicals to their proper storage area.

10.2.18   Phosphoric Acid (H_3PO_4) Solutions

Preparation of H_3PO_4 solutions is a volume-to-volume measurement ($1N$ $H_3PO_4 = 0.03M$ H_3PO_4). Using concentrated H_3PO_4 (check purity) as stock solution, you can calculate the volume of stock solution to add to any volume to achieve the desired concentration:

$$C1 \times V1 = C2 \times V2 \quad \text{or} \quad V1 = \frac{V2}{(C1/C2)}, \tag{1}$$

where $C1$ is the initial concentration, $V1$ is the volume to add, $C2$ is the desired concentration, $V2$ is the desired final volume, and $C1/C2$ is the dilution factor.

The preparation of solutions becomes easy and can be adapted for any use or volume:

5% H_3PO_4 solution = 55.6 mL of 90% pure $H_3PO_4 \rightarrow$ 1 L distilled water
10% H_3PO_4 solution = 111 mL of 90% pure $H_3PO_4 \rightarrow$ 1 L distilled water

Follow this procedure for all solutions using the designated additions to the appropriate distilled water volume. Remember to always add acid *to* water.

Solution Example 25% (1+3) H_3PO_4:

- Add 800 mL of distilled water to a 1-L volumetric flask.
- Place flask in a cold-water bath (secondary containment); cold-water level should be slightly higher than the water level in the flask.
- Add 278 mL of H_3PO_4 (90% pure) stock solution* slowly to the flask using a funnel.
- Bring flask to the 1-L volume mark with distilled water.
- Cover and mix thoroughly by gently swirling the flask.
- Keep the flask in the cold bath until the reaction cools.
- Store in an appropriate container.
- Clean and dry equipment thoroughly; return all chemicals to their proper storage area.

$$\text{Adjusted chemical weight} = \frac{\text{Weight of chemical}}{\text{\% Purity}}. \tag{1}$$

10.2.19 Potassium Bicarbonate ($KHCO_3$) 4.5% w/v Solution

A 4.5% w/v stock solution of $KHCO_3$ is made. This is a weight-to-volume measurement. From this stock solution, aliquots will be added to wine/juice samples. This solution should be made from cellar-grade $KHCO_3$.

Stock solution:

- Add 4.5 g of $KHCO_3$ and ~90 mL of deionized water carefully to a 100-mL volumetric flask.

---

*The volume or weight of chemical used to make a reagent should be adjusted according to the grade of purity.

- Place flask on a stir plate; add stir bar; mix until the $KHCO_3$ has dissolved.
- Stop stirring; carefully remove stir bar; bring flask to the 100-mL volume mark with deionized water.
- Cover flask and gently invert flask several times to mix.
- Store in an appropriate container.
- Clean and dry equipment thoroughly; return all chemicals to their proper storage area.

Potassium carbonate (K_2CO_3) made in a 31-g/L solution, or 3.1 g/100 mL, can be used .

10.2.20   Sodium Hydroxide (NaOH) Solutions

Preparation of NaOH solutions is a volume-to-volume measurement. Using a concentration of $10N$ NaOH (purchase) as a stock solution, you can calculate the volume of stock solution to add to any volume to achieve the desired concentration:

$$C1 \times V1 = C2 \times V2 \quad \text{or} \quad V1 = \frac{V2}{(C1/C2)}, \tag{1}$$

where $C1$ is the initial concentration, $V1$ is the volume to add, $C2$ is the desired concentration, $V2$ is the desired final volume and $C1/C2$ is the dilution factor.

The preparation of solutions becomes easy and can be adapted for any use or volume:

$0.01N$ NaOH solution = 1 mL of $10N$ NaOH $\rightarrow$ 1 L distilled water
$0.05N$ NaOH solution = 5 mL of $10N$ NaOH $\rightarrow$ 1 L distilled water
$0.10N$ NaOH solution = 10 mL of $10N$ NaOH $\rightarrow$ 1 L distilled water

Follow this procedure for all solutions using the designated additions to the appropriate distilled water volume.

Solution Example 1N NaOH:

- Add 100 mL of $10N$ NaOH stock solution to a 1-L volumetric flask.
- Bring flask to the 1-L volume mark with distilled water.
- Cover and mix thoroughly by gently inverting the flask several times.
- Store in an appropriate container. Discard if crystal precipitate is detected. Degraded by CO_2 in the air (forms sodium carbonate Na_2CO_3).
- Clean and dry equipment thoroughly; return all chemicals to their proper storage area.

10.2.21   Starch 1% w/v Solution

This is a weight-to-volume measurement.

Solution:

- Add 950 mL of deionized water to a 2800-mL Fernbach flask.
- Place flask on a heated stir plate (plate should be equal to or larger than flask bottom); add stir bar; begin heating the water to boiling.

- Add 10 g of starch (potato starch) and 50-mL of cold deionized water to a 250-mL beaker.
- Place starch solution on stir plate; add stir bar; mix thoroughly.
- Add the starch solution to the flask when the water is boiling.
- Continue to stir and boil solution until clear.
- Turn off heat; allow the solution to stir until cool.
- Store in an appropriate container. Refrigerate. Turbid solutions should be discarded.
- Clean and dry equipment thoroughly; return all chemicals to their proper storage area.

10.2.22   Sulfur Dioxide (SO_2) 200 mg/L Solution

A 200-mg/L (ppm) SO_2 stock solution is made. This is a weight-to-volume measurement. Sulfur dioxide is derived from potassium or sodium metabisulfite ($K_2S_2O_5$ or $Na_2S_2O_5$). Sulfur dioxide makes up approximately 58% of the compound $K_2S_2O_5$, also referred to as KMBS (128_{FW} SO_2 and 222.34_{FW} $K_2S_2O_5$). To find the weight of KMBS that will provide the amount of SO_2 needed, use the formula

$$C_W \text{ required} = \text{Desired } S_W \times \left(\frac{C_{FW}}{S_{FW}} \right), \tag{1}$$

where C_W is the compound weight, S_W is the substance weight, C_{FW} is the compound formula weight (molar mass) and S_{FW} is the substance formula weight (Extreme caution—use a fume hood).

Stock Solution:

- Add 34.74 mg $K_2S_2O_5$ (cellar grade), which will equate to 200 mg/L, to a 100-mL volumetric flask.
- Bring flask to the 100-mL volume mark with deionized water.
- Cover and mix by inverting flask several times.

Using the stock solution, you can calculate the volume of stock solution to add to any volume to achieve the desired concentration:

$$C1 \times V1 = C2 \times V2 \quad \text{or} \quad V1 = \frac{V2}{(C1/C2)}, \tag{1}$$

where $C1$ is the initial concentration, $V1$ is the volume to add, $C2$ is the desired concentration, $V2$ is the desired final volume, and $C1 / C2$ is the dilution factor.

Common stock solutions of SO_2 are as follows:

7.5 mL stock solution → bring to 100-mL volume = 15 mg/L standard solution
10 mL stock solution → bring to 100-mL volume = 20 mg/L standard solution
25 mL stock solution → bring to 100-mL volume = 50 mg/L standard solution
50 mL stock solution → bring to 100-mL volume = 100 mg/L standard solution

- Store in an appropriate container.
- Clean and dry equipment thoroughly; return all chemicals to their proper storage area.

10.2.23   Sulfuric Acid (H_2SO_4) Solutions

Preparation of H_2SO_4 solutions is a volume to volume measurement. Using concentrated H_2SO_4 as stock solution, you can calculate the volume of stock solution to add to any volume to achieve the desired concentration:

$$C1 \times V1 = C2 \times V2 \quad \text{or} \quad V1 = \frac{V2}{(C1/C2)}, \tag{1}$$

where $C1$ is the initial concentration, $V1$ is the volume to add, $C2$ is the desired concentration, $V2$ is the desired final volume and $C1/C2$ is the dilution factor.

The preparation of solutions becomes easy and can be adapted for any use or volume:

10% H_2SO_4 solution = 100 mL concentrated stock solution → 1 L distilled water

Follow this procedure for all solutions using the designated additions to the appropriate distilled water volume. Remember to always add acid *to* water. Use extreme caution. A fume hood is recommended.

Solution Example 25% (1+3) H_2SO_4:

- Add 900 mL of distilled water to a 1-L volumetric flask.
- Place flask in a very cold-water bath (secondary containment); cold-water level should be slightly higher than the water level in the flask.
- Add 25 mL of H_2SO_4 stock solution[1] to the flask, slowly pour using a funnel and/or glass stir rod.
- Bring flask to the 1-L volume mark with distilled water.
- Cover and mix thoroughly by gently swirling the flask.
- Keep the flask in the cold bath until the reaction cools. Cover loosely to vent.
- Store in an appropriate container.
- Clean and dry equipment thoroughly; return all chemicals to their proper storage area.

$$\text{Adjusted chemical weight} = \frac{\text{Weight of chemical}}{\% \text{ Purity}}. \tag{1}$$

10.2.24   Tartaric Acid ($C_4H_6O_6$) 100 g/Liter Solution

A 100-g/L stock solution of $C_4H_6O_6$ is made. This is a weight-to-volume measurement. From this stock solution, aliquots will be added to

---

*The volume or weight of chemical used to make a reagent should be adjusted according to the grade of purity.

wine/juice samples. This solution should be made from cellar-grade tartaric acid.

Stock Solution:

- Add 10 g of $C_4H_6O_6$ (tartaric acid) and ~ 80 mL of deionized water carefully to a 100-mL volumetric flask.
- Cover flask and gently swirl to mix.
- Bring flask to the 100-mL volume mark with deionized water; mix and cover.
- Store in an appropriate container; refrigerate.
- Clean and dry equipment thoroughly; return all chemicals to their proper storage area.

10.2.25   Thiosulfate (S_2O_3) Solution

This is a weight-to-volume measurement. Thiosulfate is derived from reagent-grade sodium thiosulfate ($Na_2S_2O_3$). Thiosulfate makes up approximately 71% of the compound $Na_2S_2O_3$ (112.2_{FW} S_2O_3 and 158.1_{FW} NaS_2O_3). To find the weight of $Na_2S_2O_3$ that will provide the amount of needed S_2O_3, use the formula

$$C_W \text{ required} = \text{Desired } S_W \times \left(\frac{C_{FW}}{S_{FW}} \right), \qquad (1)$$

where C_W is the compound weight, S_W is the substance weight, C_{FW} is the compound formula weight (molar mass), and S_{FW} is the substance formula weight.

Solution Example 0.02N S_2O_3:

- Add 3 g of NaS_2O_3 (95%)[1] and 0.2 mg of $NaCO_3$ (sodium bicarbonate) to a 1-L volumetric flask.
- Bring flask to the 1-L volume mark with distilled water.
- Cover and mix thoroughly by gently inverting the flask several times.
- Store in an appropriate container. Viable for several months.
- Clean and dry equipment thoroughly; return all chemicals to their proper storage area.

$$\text{Adjusted chemical weight} = \frac{\text{Weight of chemical}}{\% \text{ Purity}}. \qquad (1)$$

And

$$\text{Normality} = \frac{\text{Weight of compound} \times \% \text{ Purity}}{C_{FW} \times \text{Final volume}}$$

where C_{FW} is the compound formula weight.

10.3   Standardization

Reagents used as standards that can degrade due to exposure to other elements are tested to ensure their viability. For example, sodium hydroxide (NaOH) reacts with carbon dioxide (CO_2) in the air and forms sodium carbonate (Na_2CO_3). This reaction reduces the strength of the NaOH; therefore, the reagent in use must be standardized frequently to monitor the strength. Headspace in the reagent container and using an open burette will lead to degradation of the reagent. Always flush delivery tube with fresh reagent. Always flush delivery tube with fresh reagent.

The most common reagents in a winery laboratory that require frequent diligent standardization are NaOH (as mentioned earlier) and Iodine (I_2). Although these reagents can be purchased as prestandardized, they will deteriorate during use. Acids such as HCl, H_3PO_4, and H_2SO_4 require standardization when prepared in the laboratory, but if they are purchased prestandardized, they should remain stable if properly stored. Always restandardize when the normality of a reagent is in doubt.

Standardization of acids, bases, and oxidizing and reducing agents follows the same basic rules and use the same formulas. A standardizing agent is the opposite type of reagent. You use a base to standardize an acid, an acid to standardize a base, and buffer to standardize an oxidizing/reducing agent. The relationship of the two solutions can be shown in the formulas

$$N_1 \times V_1 = N_2 \times V_2 \tag{1}$$
$$\text{Normality of reagent } (N_1) = \left(\frac{N_2 \times V_2}{V_1} \right),$$

where N_1 is the normality of the reagent, V_1 is the volume of the reagent used, N_2 is the normality of the standardizing agent, and V_2 is the volume of the standardizing agent.

Both solutions should be close to the same normality because solutions of equal normality react in equal quantities. Standardizing a $0.1N$ acid solution with a $2N$ base will require a greater volume of $0.1N$ acid solution added to the base to reach equilibrium. This is a waste of solution and creates a larger degree of error. It is better to use a base as close to $0.1N$ as possible.

10.3.1   Standardization of Acid Solutions

• Add the following to a 125-mL Erlenmeyer flask:

10 mL of NaOH (standardizing agent, which has been standardized and is of the same normality as the acid solution) and 3–4 drops of phenolphthalein indicator solution (purchase).

• Flush and zero burette containing the acid solution or flush and note the volume in the glass burette.

- Titrate the acid quickly and steadily into the NaOH solution; mix by gently swirling the flask until a pale pink end point is reached and persists for 10 s; note the titer volume of the acid solution.
- Repeat the test; the smaller the normality, the more tests should be run; calculate the average titer volume.
- Calculate the normality from the formula in Section 10.2.

10.3.2   Standardization of Base Solutions

- Add the following to a 125 mL Erlenmeyer flask:

5 mL of HCl (standardizing agent, which has been standardized and is of the same normality as the base solution) and 2–3 drops of phenolphthalein indicator solution (purchase).

- Flush and zero burette containing the base solution or flush and note the volume in the glass burette.
- Titrate the base quickly and steadily into the HCl solution; mix by gently swirling the flask until a pale pink end point is reached and persists for 10 s; note the titer volume of base solution.
- Repeat the test. The smaller the normality, the more tests should be run; calculate the average titer volume.
- Calculate the normality from the formula in Section 10.2.

10.3.3   Standardization of 0.02N Iodine Solution

Iodine oxidizes quickly and is light sensitive.

- Add the following to to a 125-mL Erlenmeyer flask:

10 mL of $0.02N$ S_2O_3 (thiosulfate, standardizing agent) and 1–2 mL of 1% starch solution

- Flush and zero burette containing the iodine solution or flush and note the volume in the glass burette.
- Titrate the iodine solution quickly and steadily into the S_2O_3; mix by gently swirling the flask until a faint blue end point is reached and persists; note the titer volume of iodine solution.
- Repeat the test two more times; calculate the average titer volume.
- Calculate the normality from the formula in Section 10.2.

Appendix

Laboratory Chemical Safety Summaries*

Acetic Acid

Substance	Acetic acid (ethanoic acid) CAS 64-19-7
Formula	CH_3COOH
Physical Properties	Colorless liquid bp 118°C, mp 17°C Miscible in water (100 g/100 mL)
Odor	Strong, pungent, vinegar like odor detectable at 0.2–1.0 ppm
Vapor Density	2.1 (air = 1.0)
Vapor Pressure	11 mm Hg at 20°C
Flash Point	39°C
Autoignition Temperature	426°C
Toxicity Data	LD_{50} oral (rat) 3310 mg/kg LD_{50} skin (rabbit) 1060 mg/kg LC_{50} inhal (mice) 5620 ppm [1 hour(h)] PEL (OSHA) 10 ppm (25 mg/m^3) TLV-TWA (ACGIH) 10 ppm (25 mg/m^3) STEL (ACGIH) 15 ppm (37 mg/m^3)
Major Hazards	Corrosive to the skin and eyes; vapor or mist is very irritating and can be destructive to the eyes, mucous membranes, and respiratory system; ingestion causes internal irritation and severe injury.

*Reprinted with permission from *Prudent Practices in the Laboratory: Handling and Disposal of Chemicals*, © 1995 by the National Academy of Sciences; courtesy of the National Academies Press, Washington, DC.

Toxicity

The acute toxicity of acetic acid is low. The immediate toxic effects of acetic acid are the result of its corrosive action and dehydration of tissues with which it comes into contact. A 10% aqueous solution of acetic acid produced mild or no irritation on guinea pig skin. At 25–50%, severe irritation generally results. In the eye, a 4–10% solution will produce immediate pain and sometimes injury to the cornea. Acetic acid solutions of 80% or greater concentration can cause serious burns of the skin and eyes. Acetic acid is slightly toxic by inhalation; exposure to 50 ppm is extremely irritating to the eyes, nose, and throat. Acetic acid has not been found to be carcinogenic or to show reproductive or developmental toxicity in humans.

Flammability and Explosibility

Acetic acid is a combustible substance (NFPA rating = 2). Heating can release vapors that can be ignited. Vapors or gases can travel considerable distances to ignition source and "flash back." Acetic acid vapor forms explosive mixtures with air at concentrations of 4–16% (by volume). Carbon dioxide or dry-chemical extinguishers should be used for acetic acid fires.

Reactivity and Incompatibility

Contact with strong oxidizers can cause fire.

Storage and Handling

Acetic acid should be handled in the laboratory using "basic prudent practices." In particular, acetic acid should be used only in areas free of ignition sources, and quantities greater than 1 liter should be stored in tightly sealed metal containers in areas separate from oxidizers.

Accidents

In the event of skin contact, immediately wash with soap and water and remove contaminated clothing. In case of eye contact, promptly wash with copious amounts of water for 15 min (lifting upper and lower lids occasionally) and obtain medical attention. If acetic acid is ingested, obtain medical attention immediately. If large amounts of this compound are inhaled, move the person to fresh air and seek medical attention at once.

In the event of a spill, remove all ignition sources; soak up the acetic acid with a spill pillow or absorbent material; place in an appropriate container; and dispose of properly. Cleaned-up material is a RCRA Hazardous Waste. Respiratory protection might be necessary in the event of a large spill or release in a confined area.

Disposal   Excess acetic acid and waste material containing this substance should be placed in a covered metal container, clearly labeled, and handled according to your institution's waste-disposal guidelines.

Ethanol

Substance   Ethanol
(ethyl alcohol, alcohol, methylcarbinol)
CAS 64-17-5

Formula C_2H_5OH

Physical Properties   Colorless liquid
bp 78°C, mp −114°C
Miscible with water

Odor   Pleasant alcoholic odor detectable at 49–716 ppm (mean = 180 ppm)

Vapor Density   1.6 (air = 1.0)

Vapor Pressure   43 mm Hg at 20°C

Flash Point   13°C

Autoignition Temperature   363°C

Toxicity Data LD_{50} oral (rat) 7060 mg/kg
LD_{50} skin (rabbit) > 20 mL/kg
LC_{50} inhal (rat) 20,000 ppm (10 h)
PEL (OSHA) 1000 ppm (1900 mg/m^3)
TLV-TWA (ACGIH) 1000 ppm (1900 mg/m^3)

Major Hazards   Flammable liquid.

Toxicity   The acute toxicity of ethanol is very low. Ingestion of ethanol can cause temporary nervous system depression, with anesthetic effects such as dizziness, headache, confusion, and loss of consciousness; large doses (250–500 mL) can be fatal in humans. High concentrations of ethanol vapor are irritating to the eyes and upper respiratory tract. Liquid ethanol does not significantly irritate the skin, but it is a moderate eye irritant. Exposure to high concentrations of ethanol by inhalation (over 1000 ppm) can cause central nervous system (CNS) effects, including dizziness, headache, and giddiness, followed by depression, drowsiness, and fatigue. Ethanol is regarded as a substance with good warning properties. Tests in some animals indicate that ethanol could have developmental and reproductive toxicity if ingested. There is no evidence that laboratory exposure to ethanol has carcinogenic effects. To discourage deliberate ingestion, ethanol for laboratory use is often "denatured" by the addition of other chemicals; the toxicity of possible additives must also be considered when evaluation the risk of laboratory exposure to ethanol.

Flammability and Explosibility

Ethanol is a flammable liquid (NFPA rating = 3), and its vapor can travel a considerable distance to an ignition source and "flash back." Ethanol vapor forms explosive mixtures with air at concentrations of 4.3–19% (by volume). Hazardous gases produced in ethanol fires include carbon monoxide and carbon dioxide. Carbon dioxide or dry-chemical extinguishers should be used for ethanol fires.

Reactivity and Incompatibility

Contact of ethanol with strong oxidizers, peroxides, strong alkalis, and strong acids can cause fires and explosions.

Storage and Handling

Ethanol should be handled in the laboratory using "basic prudent practices" supplemented by the additional precautions for dealing with highly flammable substances. In particular, ethanol should be used only in areas free of ignition sources, and quantities greater than 1 liter should be stored in tightly sealed metal containers in areas separate from oxidizers.

Accidents

In the event of skin contact, immediately wash with soap and water and remove contaminated clothing. In case of eye contact, promptly wash with copious amounts of water for 15 min (lifting upper and lower lids occasionally) and obtain medical attention. If ethanol is ingested, obtain medical attention immediately. If large amounts of this compound are inhaled, move the person to fresh air and seek medical attention at once.

In the event of a spill, remove all ignition sources; soak up the ethanol with a spill pillow or absorbent material; place in an appropriate container; and dispose of properly. Respiratory protection might be necessary in the event of a large spill or release in a confined area.

Disposal

Excess ethanol and waste material containing this substance should be placed in an appropriate container, clearly labeled, and handled according to your institution's waste-disposal guidelines.

Hydrochloric Acid

Substance

Hydrochloric acid
(muriatic acid)
CAS 7647-01-0

Formula

Reagent-grade conc. HCl contains 37 wt% HCl in water; constant-boiling acid (an azeotrope with water) contains ~ 20% HCl

Physical Properties

Concentrated acid evolves HCl at 60°C, leading to the formation of an azeotrope of constant composition (20% HCl)

	bp 110°C, mp −24°C
	Miscible with water
Odor	Sharp, irritating odor detectable at 0.25–10 ppm
Toxicity Data	LD_{50} oral (rabbit) 900 mg/kg
	LC_{50} inhal (rat) 3124 ppm (1 h)
	PEL (OSHA) 5 ppm (7 mg/m^3; ceiling)
	TLV (ACGIH) 5 ppm (7.5 mg/m^3; ceiling)

Major Hazards   Highly corrosive; causes severe burns on eye and skin contact and upon inhalation of gas.

Toxicity   Hydrochloric acid and hydrogen chloride gas are highly corrosive substances that can cause severe burns upon contact with any body tissue. The aqueous acid and gas are strong eye irritants and lacrimators. Contact of concentrated hydrochloric acid or concentrated HCl vapor with the eyes can cause severe injury, resulting in permanent impairment of vision and possible blindness, and skin contact results in severe burns. Ingestion can cause severe burns of the mouth, throat, and gastrointestinal system and can be fatal. Inhalation of hydrogen chloride gas can cause severe irritation and injury to the upper respiratory tract and lungs, and exposure to high concentrations could cause death. HCl gas is regarded as having adequate warning properties.

Flammability and Explosibility   Noncombustible, but contact with metals could produce highly flammable hydrogen gas.

Reactivity and Incompatibility   Hydrochloric acid and hydrogen chloride react violently with any metals with the generation of highly flammable hydrogen gas, which could explode. Reaction with oxidizers such as permanganates, chlorates, chlorites, and hypochlorites can produce chlorine or bromine.

Storage and Handling   Hydrochloric acid should be handled in the laboratory using "basic prudent practices." Splash goggles and rubber gloves should be worn when handling this acid, and containers of HCl should be stored in a well-ventilated location separated from incompatible metals. Water should never be added to HCl because splattering could result; always add acid *to* water. Containers of hydrochloric acid should be stored in secondary plastic trays to avoid corrosion of metal storage shelves due to drips or spills.

Accidents

In the event of skin contact, remove contaminated clothing and immediately wash with flowing water for at least 15 min. In case of eye contact, promptly wash with copious amounts of water for 15 min while holding the eyelids open. Seek medical attention. In case of ingestion, do not induce vomiting. Give large amounts of water or milk if available and transport to medical facility. In case of inhalation, remove the person to fresh air and seek medical attention. Carefully neutralize spills of hydrochloric acid with a suitable agent such as powdered sodium bicarbonate, further dilute with absorbent material, place in an appropriate container, and dispose of properly. Dilution with water before applying the solid adsorbent can be an effective means of reducing exposure to hydrogen chloride vapor. Respiratory protection might be necessary in the event of a large spill or release in a confined area. Leaks of HCl gas are evident from the formation of dense white fumes on contact with the atmosphere. Small leaks can be detected by holding an open container of concentrated ammonium hydroxide near the site of the suspected leak; dense white fumes confirm that a leak is present. In case of accidental release of hydrogen chloride gas, such as from a leaking cylinder or associated apparatus, evacuate the area and eliminate the source of the leak if this can be done safely. Remove cylinder to a fume hood or remote area if it cannot be shut off. Full respiratory protection and protective clothing might be required to deal with a hydrogen chloride release.

Disposal   In many localities, hydrochloric acid or the residue from a spill can be disposed of down the drain after appropriate dilution and neutralization. Otherwise, hydrochloric acid and waste materials containing this substance should be placed in an appropriate container, clearly labeled, and handled according to your institution's waste-disposal guidelines.

Hydrogen Peroxide

Substance	Hydrogen peroxide (hydrogen dioxide) CAS 7722-84-1
Formula	HOOH
Physical Properties	Colorless liquid bp 150°C, mp − 0.4°C Miscible in all proportions in water
Odor	Slightly pungent, irritating odor
Vapor Density	1.15 (air = 1.0)
Vapor Pressure	1 mm Hg at 15.3°C; 5 mm Hg at 30°C
Flash Point	Noncombustible
Toxicity Data	LD_{50} oral (rat) 75 mg/kg (70%) LD_{50} skin (rabbit) 700 mg/kg (90%) LD_{50} skin (rabbit) 9200 mg/kg (70%) LC_{50} inhal (rat) > 2000 ppm (90%) PEL (OSHA) 1 ppm (1.4 mg/m³) (90%) TLV-TWA (ACGIH) 1 ppm (1.4 mg/m³) (90%)
Major Hazards	Contact with certain metals and organic compounds can lead to fires and explosions; concentrated solutions can cause severe irritation or burns of the skin, eyes, and mucous membranes.

Toxicity

Contact with aqueous concentrations of less than 50% cause skin irritation, but more concentrated solutions of H_2O_2 are corrosive to the skin. At greater than 10% concentration, H_2O_2 is corrosive to the eyes and can cause severe irreversible damage and possible blindness. Hydrogen peroxide is moderately toxic by ingestion and slightly toxic by inhalation. This substance is not considered to have adequate warning properties. Hydrogen peroxide has not been found to be carcinogenic in humans. Repeated inhalation exposures produced nasal discharge, bleached hair, and respiratory tract congestion, with some deaths occurring in rats and mice exposed to concentrations greater than 67 ppm.

Flammability and Explosibility

Hydrogen peroxide is not flammable, but concentrated solutions could undergo violent decomposition in the presence of trace impurities or upon heating.

Reactivity and Incompatibility

Contact with many organic compounds can lead to immediate fires or violent explosions. Hydrogen peroxide reacts with certain organic functional groups (ethers, acetals, etc.) to form peroxides, which could explode upon concentration. Reaction with acetone generates explosive cyclic dimeric and trimeric peroxides. Explosions can also occur on exposure of H_2O_2 to metals such as sodium, potassium, magnesium, copper, iron, and nickel.

Storage and Handling

Hydrogen peroxide should be handled in the laboratory using "basic prudent practices" supplemented by the procedures for work with reactive and explosive substances. Use extreme care when carrying out reactions with H_2O_2 because of the fire and explosion potential (immediate or delayed). The use of safety shields is advisable and is essential for experiments involving concentrated (>50%)

solutions of hydrogen peroxide. Sealed containers of hydrogen peroxide can build up dangerous pressures of oxygen because of slow decomposition.

Accidents

In the event of skin contact, immediately wash with soap and water and remove contaminated clothing. In case of eye contact, promptly wash with copious amounts of water for 15 min (lifting upper and lower lids occasionally) and obtain medical attention. If H_2O_2 is ingested, obtain medical attention immediately. If large amounts of this compound are inhaled, move the person to fresh air and seek medical attention at once.

In the event of a spill, remove all ignition sources; soak up the acetic acid with a spill pillow or absorbent material; place in an appropriate container; and dispose of properly. Respiratory protection might be necessary in the event of a large spill or release in a confined area.

Disposal

Excess H_2O_2 and waste material containing this substance should be placed in a covered metal container, clearly labeled, and handled according to your institution's waste-disposal guidelines.

Iodine

Substance	Iodine
	CAS 7553-56-2
Formula	I_2
Physical Properties	Blue-violet to black crystalline solid
	bp 185°C, mp 114°C
	Slightly soluble in water (0.03 g/100 mL at 20°C)
Odor	Sharp, characteristic odor
Vapor Density	8.8 (air = 1.0)
Vapor Pressure	0.3 mmHg at 20°C
Flash Point	Noncombustible
Toxicity Data	LD_{50} oral (rat) 14,000 mg/kg

LC_{50} inhal (rat) 80 ppm (800 mg/m^3; 1 h)

PEL (OSHA) 0.1 ppm (ceiling, 1 mg/m^3)

TLV-TWA (ACGIH) 0.1 ppm (ceiling, 1 mg/m^3)

Major Hazards

Iodine vapor is highly toxic and is a severe irritant to the eyes and respiratory tract.

Toxicity

The acute toxicity of iodine by inhalation is high. Exposure can cause severe breathing difficulties, which might be delayed in onset; headache, tightness of the chest, and congestion of the lungs might also result. In an experimental investigation, four human subjects tolerated 0.57 ppm iodine vapor for 5 min without eye irritation, but all experienced eye irritation in 2 min at 1.63 ppm. Iodine in crystalline form or in concentrated solutions is a severe skin irritant; it is not easily removed from the skin, and the lesions resemble thermal burns. Iodine is more toxic by the oral route in humans than in experimental animals; ingestion of 2–3 g of the solid could be fatal in humans. Iodine has not been found to be carcinogenic or to show reproductive or developmental toxicity in humans. Chronic absorption of iodine could cause insomnia, inflammation of the eyes and nose, bronchitis, tremor, rapid heartbeat, diarrhea, and weight loss.

Flammability and Explosibility

Iodine is noncombustible and in itself represents a negligible fire hazard when exposed to heat or flame. However, when heated, it will increase the burning rate of combustible materials.

Reactivity and Incompatibility

Iodine is stable under normal temperatures and pressures. Iodine could react violently with acetylene, ammonia, acetaldehyde, formaldehyde, acrylonitrile, powdered antimony, tetraamine copper(II) sulfate, and liquid chlorine. Iodine can form sensitive, explosive mixtures with potassium, sodium, and oxygen difluoride; ammonium hydroxide

Storage and Handling

reacts with iodine to produce nitrogen triiodide, which detonates on drying.

Iodine should be handled in the laboratory using "basic prudent practices"; in particular, safety goggles and rubber gloves should be worn when handling iodine, and operations involving large quantities should be conducted in a fume hood to prevent exposure to iodine vapor or dusts by inhalation.

Accidents

In the event of skin contact, immediately wash with soap and water and remove contaminated clothing. In case of eye contact, promptly wash with copious amounts of water for 15 min (lifting upper and lower lids occasionally) and obtain medical attention. If iodine is ingested, obtain medical attention immediately. If large amounts of this compound are inhaled, move the person to fresh air and seek medical attention at once.

In the event of a spill, sweep up solid iodine; soak up liquid spills with absorbent material; place in an appropriate container; and dispose of properly. Respiratory protection might be necessary in the event of a large spill or release in a confined area.

Disposal

Excess iodine and waste material containing this substance should be placed in an appropriate container, clearly labeled, and handled according to your institution's waste-disposal guidelines.

Mercury

Substance

Mercury
(quicksilver, hydrargyrum)
CAS 7439-97-6

Formula

Hg

Physical Properties

Silvery, mobile liquid
bp 357°C, mp − 39°C

	Very slightly soluble in water (0.002 g/100 mL at 20 °C)
Odor	Odorless
Vapor Density	6.9 (air = 1.0)
Vapor Pressure	0.0012 mm Hg at 20 °C
Flash Point	Noncombustible
Toxicity Data	LC_{LO} inhal (rat) 29 mg/m^3 (30 h)
	PEL (OSHA) 0.1 mg/m^3 (ceiling)
	TLV-TWA (ACGIH) 0.025 mg/m^3, skin

Major Hazards   Repeated or prolonged exposure to mercury vapor is highly toxic to the central nervous system.

Toxicity   The acute toxicity of mercury varies significantly with the route of exposure. Ingestion is largely without effects. Inhalation of high concentrations of mercury causes severe respiratory irritation, digestive disturbances, and marked kidney damage. There are no warning properties for exposure to mercury vapor, which is colorless, odorless, and tasteless. Toxicity caused by repeated or prolonged exposure to mercury vapor or liquid is characterized by emotional disturbances, inflammation of the mouth and gums, general fatigue, memory loss, headaches, tremors, anorexic, and weight loss. Skin absorption of mercury and mercury vapor adds to the toxic effects of vapor inhalation. At low levels, the onset of symptoms is insidious; fine tremors or the hand, eyelids, lips and tongue are often the presenting complaints. Mercury has been reported to be capable of causing sensitization dermatitis. Mercury has not been shown to be a human carcinogen or reproductive toxin.

Flammability and Explosibility   Mercury is not combustible.

Reactivity and Incompatibility   Mercury is a fairly unreactive metal that is highly resistant to corrosion. It can dissolve a number of metals, such as silver, gold, and tin, forming amalgams. Mercury can react violently with acetylene and ammonia.

Storage and Handling

Mercury should be handled in the laboratory using "basic prudent practices"; particularly, precautions should be taken to prevent spills of mercury because drops of the liquid metal can easily become lodged in floor cracks, behind cabinets, equipment, and so forth, with the result that the mercury vapor concentration in the laboratory could then exceed the safe and allowable limits. Containers of mercury should be kept tightly sealed and stored in secondary containers (such as a plastic pan or tray) in a well-ventilated area. When breakage of an instrument or apparatus containing significant quantities of Hg is possible, the equipment should be placed in a plastic tray or pan that is large enough to contain the mercury in the event of an accident. Transfers of mercury between containers should be carried out in a fume hood over a tray or pan to confine any spills.

Accidents

In the event of skin contact, immediately wash with soap and water and remove contaminated clothing. In case of eye contact, promptly wash with copious amounts of water for 15 min (lifting upper and lower lids occasionally) and obtain medical attention. If mercury is ingested, obtain medical attention immediately. If large amounts of this compound are inhaled, move the person to fresh air and seek medical attention at once.

In the event of a spill, use an appropriate spill kit; place in an appropriate container; and dispose of properly. Respiratory protection will be necessary in the event of a large spill, release in a confined area, or spill under conditions of higher than normal temperatures.

Disposal

Excess mercury should be collected for recycling, and waste material containing

mercury should be placed in an appropriate container, clearly labeled, and handled according to your institution's
waste-disposal guidelines.

Sodium Hydroxide

Substance	Sodium hydroxide
	(sodium hydroxide, caustic soda, lye, caustic)
	CAS 1310-73-2
Formula	NaOH
Physical Properties	bp 1390°C, mp 318°C
	Highly soluble in water (109 g/100 mL)
Odor	Odorless
Toxicity Data	LD_{50} oral (rat) 140–340 mg/kg
	LD_{50} skin (rabbit) 1350 mg/kg
	PEL (OSHA) 2 mg/m^3
	TLV (ACGIH) 2 mg/m^3; ceiling
Major Hazards	Extremely corrosive; causes severe burns to skin, eyes, and mucous membranes.
Toxicity	The alkali metal hydroxides are highly corrosive substances; contact of solutions, dust, or mists with the skin, eyes, and mucous membranes can lead to severe damage. Skin contact with the solid hydroxides or concentrated solutions can cause rapid tissue destruction and severe burns. In contrast to acids, hydroxides do not coagulate protein (which impede penetration), and metal hydroxide burns might not be immediately painful, whereas skin penetration occurs to produce severe and slow-healing burns. Potassium hydroxide solutions can cause severe eye damage and possible blindness. Ingestion of concentrated solutions of sodium hydroxide or potassium hydroxide can cause severe abdominal pain, as well as serious damage to the mouth, throat, esophagus, and digestive tract. Inhalation of sodium/potassium hydroxide dust or mist can cause irritation and damage to the respiratory tract, depending on the

concentration and duration of exposure. Exposure to high concentrations could result in delayed pulmonary edema. Repeated or prolonged contact could cause dermatitis. Sodium hydroxide and potassium hydroxide have not been found to be carcinogenic or to show reproductive or developmental toxicity in humans.

Flammability and Explosibility

Sodium hydroxide and potassium hydroxide are not flammable as solids or aqueous solutions.

Reactivity and Incompatibility

Concentrated sodium hydroxide and potassium hydroxide react vigorously with acids with evolution of heat, and dissolution in water is highly exothermic. Reaction with aluminum and other metals could lead to evolution of hydrogen gas. The solids in prolonged contact with chloroform, trichloroethylene, and tetrachloroethanes can produce explosive products. Many organic compounds such as propylene oxide, allyl alcohol glyoxal, acetaldehyde, acrolein, and acrylonitrile can violently polymerize on contact with concentrated base. Reaction with nitromethane and nitrophenols produces shock-sensitive explosive salts. Sodium hydroxide and potassium hydroxide as solids absorb moisture and carbon dioxide from the air to form the bicarbonates. Aqueous solutions also absorb carbon dioxide to form bicarbonate. Solutions stored in flasks with ground-glass stoppers could leak air and freeze the stoppers, preventing removal.

Storage and Handling

Sodium hydroxide and potassium hydroxide should be handled in the laboratory using "basic prudent practices"; in particular, splash goggles and impermeable gloves should be worn at all times when handling these substances to prevent eye and skin contact. Operations with metal hydroxide solutions that have the potential to create aerosols should be conducted in a fume hood to prevent exposure by inhalation. Sodium hydroxide and potassium hydroxide generate considerable heat when dissolved in water; when mixing with water, always add caustics slowly to the water and stir continuously. Never add water in limited quantities to solid hydroxides. Containers of hydroxides should be stored in a cool, dry location, separated from acids and incompatible substances.

Accidents

In case of eye contact, immediate and continuous irrigation with flowing water for at least 15 min is imperative. Prompt medical consultation is essential. In case of skin contact, immediately remove contaminated clothing, flush affected area with large amounts of water for 15 min, and obtain medical attention without delay. If sodium hydroxide or potassium hydroxide is ingested, do not induce vomiting; give large amounts of water and transport to medical facility immediately. If dusts or mists of these compounds are inhaled, move the person to fresh air and seek medical attention at once.

Disposal

In many localities, sodium/potassium hydroxide may be disposed of down the drain after appropriate dilution and neutralization. If neutralization and drain disposal is not permitted, excess hydroxide and waste material containing this substance should be placed in an appropriate container, clearly labeled, and handled according to your institution's waste-disposal guidelines.

Sulfur Dioxide

Substance

Sulfur dioxide
(sulfurous oxide, sulfur oxide, sulfurous anhydride)
CAS 7446-09-5

Formula	SO_2
Physical Properties	Colorless gas or liquid under pressure
	bp $-10.0°C$, mp $-75.5°C$
	Soluble in water (10 g/100 mL at 20 °C)
Odor	Pungent odor detectable at 0.3–5 ppm
Vapor Density	2.26 (air = 1.0)
Vapor Pressure	1779 mm Hg at 21 °C
Flash Point	Noncombustible
Toxicity Data	LC_{50} inhal (rat) 2520 ppm (6590 mg/m^3; 1 h)
	LC_{LO} inhal (human) 1000 ppm (2600 mg/m^3; 10 min)
	PEL (OSHA) 5 ppm (13 mg/m^3)
	TLV-TWA (ACGIH) 2 ppm (5.2 mg/m^3)
	STEL (ACGIH) 5 ppm (13 mg/m^3)
Major Hazards	Intensely irritating to the skin, eyes, and respiratory tract; moderate acute toxicity. The acute toxicity of sulfur dioxide is moderate. Inhalation of high concentrations could cause death as a result of respiratory paralysis and pulmonary edema. Exposure to 400–500 ppm is immediately dangerous, and 1000 ppm for 10 min is reported to have caused death in humans. Sulfur dioxide gas is a severe corrosive irritant of the eyes, mucous membranes, and skin. Its irritant properties are due to the rapidity with which it forms sulfurous acid on contact with moist membranes. When sulfur dioxide is inhaled, most of it is absorbed in the upper respiratory passages, where most of its effects then occur. Exposure to concentrations of 10–50 ppm for 5–15 min causes irritation of the eyes, nose, and throat, choking, and coughing. Some individuals are extremely sensitive to the effects of sulfur dioxide, whereas experienced workers might become adapted to its irritating properties. Sulfur dioxide is regarded as a substance with good warning properties except in the case of individuals with reactive respiratory tracts and asthmatics. Exposure of the eyes to liquid sulfur dioxide from pressurized containers can cause severe

burns, resulting in the loss of vision. Liquid SO_2 on the skin produces skin burns from the freezing effect of rapid evaporation. Sulfur dioxide has not been shown to be carcinogenic or to have reproductive or developmental effects in humans. Chronic exposure to low levels of sulfur dioxide has been shown to exacerbate pulmonary disease.

Flammability and Explosibility Sulfur dioxide is a noncombustible substance (NFPA rating = 0).

Reactivity and Incompatibility Contact with some powdered metals and with alkali metals such as sodium or potassium could cause fires and explosions. Liquid sulfur dioxide will attack some forms of plastics, rubber, and coatings.

Storage and Handling Sulfur dioxide should be handled in the laboratory using "basic prudent practices" supplemented by the procedures for work with compressed gases.

Accidents In the event of skin contact, immediately wash with water and remove contaminated clothing. In case of eye contact, promptly wash with copious amounts of water for 15 min (lifting upper and lower lids occasionally) and obtain medical attention. If large amounts of this compound are inhaled, move the person to fresh air and seek medical attention at once.

Leaks of sulfur dioxide can be detected by passing a rag dampened with aqueous NH_3 over the suspected valve or fitting. White fumes indicate escaping SO_2 gas. To respond to a release, use appropriate protective equipment and clothing. Positive-pressure air-supplied respiratory protection is required. Close cylinder valve and ventilate area. Remove cylinder to a fume hood or remote area if it cannot be shut off. If in liquid form, allow it to vaporize.

Disposal Excess SO_2 should be returned to the manufacturer if possible, according to your institution's waste-disposal guidelines.

Sulfuric Acid

Substance	Sulfuric acid
	(oil of vitriol)
	CAS 7664-93-9
Formula	H_2SO_4
Physical Properties	Clear, colorless, oily liquid
	bp 300–338°C (loses SO_3 above 300 °C), mp 11°C
	Miscible with water in all proportions
Odor	Odorless
Vapor Density	3.4 (air = 1.0)
Vapor Pressure	< 0.3 mm Hg at 25°C
Flash Point	Noncombustible
Toxicity Data	LD_{50} oral (rat) 2140 mg/kg
	LC_{50} inhal (rat) 347 mg/m³ (1 h)
	PEL (OSHA) 1 mg/m³
	TLV-TWA (ACGIH) 1 mg/m³
	STEL (ACGIH) 3 mg/m³
Major Hazards	Highly corrosive; causes severe burns on eye and skin contact and upon inhalation of sulfuric acid mist; highly reactive, reacts violently with any organic and inorganic substances.
Toxicity	Concentrated sulfuric acid is a highly corrosive liquid that can cause severe, deep burns upon skin contact. The concentrated acid destroys tissue because of its dehydrating action, whereas dilute H_2SO_4 acts as a skin irritant because of its acid character. Eye contact with concentrated H_2SO_4 causes severe burns, which can result in permanent loss of vision; contact with dilute H_2SO_4 results in more transient effects from which recovery might be complete. Sulfuric acid mist severely irritates the eyes, respiratory tract, and skin. Because of its low vapor pressure, the principal inhalation hazard from sulfuric acid involves breathing in acid mists, which could result in irritation of the upper respiratory passages and erosion of dental surfaces. Higher inhalation exposures can lead to temporary lung irritation with difficulty breathing. Ingestion of sul-

furic acid can cause severe burns to the mucous membranes of the mouth and esophagus. Animal testing with sulfuric acid did not demonstrate carcinogenic, mutagenic, embryo toxic, or reproductive effects. Chronic exposure to sulfuric acid mist could lead to bronchitis, skin lesions, conjunctivitis, and erosion of teeth.

Flammability and Explosibility   Sulfuric acid is noncombustible but can cause finely divided combustible substances to ignite. Sulfuric acid reacts with most metals, especially when dilute, to produce flammable and potentially explosive hydrogen gas.

Reactivity and Incompatibility   Concentrate sulfuric acid is stable but can react violently with water and with many organic compounds because of its action as a powerful dehydrating, oxidizing, and sulfonating agent. Ignition or explosions can occur on contact of sulfuric acid with many metals, carbides, chlorates, perchlorates, permanganates, bases, and reducing agents. Sulfuric acid reacts with a number of substances to generate highly toxic products. Examples include the reaction of H_2SO_4 with formic or oxalic acid (CO formation), with cyanide salts (HCN formation), and sodium bromide (SO_2 and Br_2 formation).

Storage and Handling   Sulfuric acid should be handled in the laboratory using "basic prudent practices" supplemented with splash goggles and rubber gloves worn when handling this acid, and containers of sulfuric acid should be stored in a well-ventilated location, separated from organic substances and other combustible materials. Containers of sulfuric acid should be stored in secondary plastic trays to avoid corrosion of metal storage shelves due to drips or spills. Water should never be added to sulfuric acid because splattering might result; always add acid *to* water.

Accidents

In the event of skin contact, immediately wash with soap and water and remove contaminated clothing. In case of eye contact, promptly wash with copious amounts of water for 15 min (lifting upper and lower lids occasionally) and obtain medical attention. If sulfuric acid is ingested, obtain medical attention immediately. If large amounts of this compound are inhaled, move the person to fresh air and seek medical attention at once.

Carefully neutralize small spills of sulfuric acid with a suitable agent such as sodium carbonate; further dilute with absorbent material; place in an appropriate container; and dispose of properly. Respiratory protection may be necessary in the event of a large spill or release in a confined area.

Disposal

Excess sulfuric acid and waste material containing this substance should be placed in a covered metal container, clearly labeled, and handled according to your institution's waste-disposal guidelines.

Bibliography

About Inc. 2004. History of Glass Wine Bottles, *Wine History*, October 12, 1999. Available from http://wine.about.com/library Newsletter/weekly/aa111299. htm (accessed August 30, 2004).

Agilent Technologies 2004. *Instrumentation Education*, Agilent Technologies, PaloAlto, CA. Available from www.chem.agilent.com/Scripts/Library.asp? (accessed March 12, 2004).

Amorim Cork 2003. *Cork History and Production*, Santa Maria de Lamas Portugal. Available from www.corkfacts.com (accessed August 30, 2004).

Anton Paar. 2000. Checking the fermentation process in wine production using the DMA 35n from Anton Paar, Application Note, Anton Paar GmbH, Graz, Austria, pp. 1–4.

Anton Paar. 2000. Alcohol measurement in wine using the Anton Paar Alcolyzer®, Lab Application Note, Anton Paar GmbH, Graz, Austria, pp. 1–6.

Armour, M. 1996. *Hazardous Laboratory Chemicals Disposal Guide,* 2nd ed., Lewis Publishers, Boca Raton, FL.

Astoria Pacific International 2002. *Astoria®Analyzer*, Product information, Astoria Pacific, Inc, Clackmas, Oregon.

Bartlett, J. 1992. *Bartlett's Familiar Quotations,* 16th ed. (J. Kaplan, ed.), Little Brown & Co, Boston, p. 297.

Berberoglu, H. 2004. *Phylloxera Vastatrix*: The most deadly vine disease, Food Reference website (March 18, 2004). Available from http://www.foodreference.com/html/artphylloxera.html (accessed November 7, 2004).

Bergey, D.H., Krieg, N.R., and Staley, J.T. 1986. *Bergey's Manual of Systemic Bacteriology, Volume 2* (P. Sneath, N. Mair, M. Sharpe, and J.G. Holt, eds), Williams & Wilkins, Baltimore, MD.

Bird, D. 2000. *Understanding Wine Technology,* DBQA Publishing, Newark, UK.

Boehringer, Mannheim. 1997. *Methods of Enzymatic Bio-analysis and Food Analysis*, Boehringer Mannheim GmbH, Mannheim, Germany, pp. 10–12, 52–56, 100–102.

Boulton, R. B., Singleton, V. L., Bisson, L. F., and Kunkee, R. E. 1999. *Principles and Practices of Winemaking*, Aspen Publishers, Gaithersburg, MD.

Budavari, S., O'Neil, M., Smith, A., Heckelman, P., and Kinneary, J. (eds.). 1996. *The Merck Index*, 12th ed., Merck Research Laboratories, Whitehouse Station, NJ.

ChemsSoft Corp. 2004. Various chemical facts, ChemFinder database, Cambridge, MA. Available from www.chemfinder.com (accessed December 3, 2003).

Cocolin, L., Manzano, M., Silvana, R., and Comi, G. 2001. Monitoring of yeast population changes during a continuous wine fermentation by molecular methods, *American Journal of Enology & Viticulture* 53(1):24–27. (2002 abstract).

Cork Quality Council. 1994. CQC Establishes standards for cork grades, *Vineyard & Winery Management* May/June; vol. 20, no. 3: 66.

Cork Supply USA 2004. Packaging, Cork Supply USA Benicia, CA: Available from www.corksupplyusa.com/packaging (accessed August 31, 2004).

Cornell University and Gordon, J. 2004. *Environmental Health & Safety, Material Safety Data Sheets Online,* Cornell University Press, Ithaca, NY. Available from www.ehs.cornell.edu (website accessed January 8, 2004).

Council for Scientific and Industrial Research. 1999. Advanced fermentation monitoring by image analysis, Council for Scientific and Industrial Research, Pretoria, South Africa, Available from www.csir.co.za/plsql (accessed June 8, 2004).

Dard, P. 1994. *Tout Savoir Sur Le Vin,* Nouvelle Edition Complétée, Comptoir Du Livre, Paris.

ETS Laboratories. 2002. Export information, ETS Laboratories, St. Helena, CA (August 5, 2003). Available from www.etslabs.com/pagetemplate/ets4.asp?pageid =213 (website accessed August 17, 2004).

Fisher Scientific 2003. various MSDS, chemical quantities, DOT, NFPA information, Fisher Scientific, Hampton, NH. Available from www.fishersci.com (accessed January 16, 2004).

Fugelsang, K. C. 1997. *Wine Microbiology,* Chapman & Hall, New York.

Fugelsang, K.C. and Zoecklein, B.W. 1993. MLF survey, *Practical Winery and Vineyard* May/June: 12–19.

Garfield, F. M. 1984. *Quality Assurance Principles for Analytical Laboratories,* AOAC, Arlington, VA.

Garfield, F. M. 1994. *Statistical Manual of the Association of Official Analytical Chemists: Statistical Techniques for Collaborative Testing,* AOAC International, Arlington, VA.

Gawel, R. 2004. Brettanomyces character in wine, Paper presented at the Australian Society of Wine Education National Convention, Hunter Valley, Australia, June 4–6, 2004, Aroma Dictionary, Recognose Pty Ltd., Kings Park, Australia. Available at www.aromadictionary.com/articles/brettanomyces_article.html (accessed August 17, 2004).

Gietz, R. D. 2003. High efficiency transformation Liac/SS-DNA/PEG method, The Trafo Page, Yeast Transformation Information article (February 5, 2003), University of Manitoba, Winnipeg, Canada. Available from www.umanitoba.ca/faculties/medicine/biochem/gietz/method.html (accessed August 11, 2004).

Hägglund, N. O. 2004. Density and specific gravity, Anton Paar, Ashland, VA (November 24, 2004) (e-mail to author).

Hess, F. C. 1955. *Chemistry Made Simple,* Doubleday, Garden City, NY.

Hoffmann, A., Sponholz, W.R., David, F., and Sandra, P. 2000. Application Note: Corkiness in Wine—Trace Analysis of 2,4,6-Trichloroanisole by Stir Bar Sorptive Extraction (SBSE) and Thermal Desorption GC/MS, Gerstel GmbH and Co. KG, Mülheim an der Ruhr, Germany.

Howe, P. 1998a. Winery laboratories and analytical quality: What does ATF require in the winery laboratory, *Vineyard & Winery Management* May/June vol. 24, no. 3: 42–48.

Howe, P. 1998b. Appropriate laboratory procedures and necessary analysis, *Vineyard & Winery Management* July/August vol. 24, no. 4: 14–16.

Howe, P. 1998c. Method validation and verification, *Vineyard & Winery Management* September/October vol. 25, no. 5: 25–28.

Howe, P. 2000a. Density and specific gravity: Pycnometry and hydrometry, *Vineyard & Winery Management* March/April vol. 26, no. 2: 19–22.

Howe, P. 2000b. Volatile acidity; the most difficult analysis, *Vineyard & Winery Management* July/August vol. 26, no. 4: 17–21.

Howe, P. 2001. References, standards, and controls, *Vineyard & Winery Management* September/October vol. 26, no. 5: 21–25.

Howe, P. 2002. Making reference material in H_2O for common winery laboratory procedures, *Vineyard & Winery Management* January/February vol. 28, no. 1: 15–19.

Iland, P., Ewart, A., Sitters, J., Markides, A., and Bruer, N. 2000. *Techniques for Chemical Analysis and Quality Monitoring During Winemaking,* Patrick Iland Wine Promotions, Campbelltown, Australia.

Industry Education Resources. 1997. *Back Safe and Sitting Safe for Office Professionals,* Future Industrial Technologies, Santa Barbara, CA.

Institut für Chemie 2004. Chemical modeling and visualization Internet Service, Freie Universität Berlin, Berlin, Germany (July 29, 2004). Available at www.chemie.fu-berlin.de/chemistry/ index_en.html#bio (accessed October 1, 2004).

Interactive Learning Paradigms, Inc. 2003. Short term exposure limit (STEL), Safety Emporium, Lexington, KY (November 8, 2003) Available from www.ilpi.com/ msds/ref/stel.html (accessed January 14, 2004).

Invasive Species Specialist Group. 2004. Homalodisca coagulate, Global Invasive Species Database, Auckland, NZ (June 10, 2004) Available from http://www.issg.org/ database/species/ecology.asp?si=240&fr=1&sts= (accessed) November 7, 2004.

Labelmaster Software, American Labelmark Company. 2003. CFR 49, Part 173 Shippers-General Requirements for Shipments and Packaging, 173.2 Hazardous materials classes and index to hazard class definitions. Available from www.myregs.com/ dotrspa (accessed February 6, 2004).

La Mar, J. 2003. *Wine 101 History,* Professional Friends of Wine Kingsburg, CA. Available from www.winepros.org/wine101/history.htm (accessed November 21, 2004).

Lamprecht, J. L. 1992. *ISO 9000 Preparing for Registration* (E. G. Schilling, W. G. Barnard, R. R. Bingham, JR, L. Rabinowitz, and T. Witt, eds.), ASQC Quality Press, Milwaukee, WI, pp. 7–38.

Lane, D.M. 2000. Rice University, Rice Virtual Lab in Statistics, HyperStat Online Glossary, Houston, TX (September 1, 2004). Available from http://www.davidm-lane.com/hyperstat (accessed October 3, 2004).

Mackison, F. W., Stricoff, R. S., and Partridge, L. J. (eds.). 1981. *Occupational Health Guideline for Chemical Hazards,* US Department of Health & Human Services, US Government Printing Office, Washington, DC.

Mallinckrodt Baker, Inc. 2000. Various MSDS, Mallinckrodit Baker, Inc., Phillipsburg, NJ (April 2, 2001). Available from www.jtbaker.com (accessed February 2, 2004).

Margalit, Y. 1996. *Winery Technology & Operations: A Handbook for Small Wineries,* The Wine Appreciation Guild, San Francisco, CA.

Margalit, Y. 1997. *Concepts in Wine Chemistry* (Crum, J., ed.), The Wine Appreciation Guild, San Francisco, CA.

National Academy of Science, *Prudent Practices in the Laboratory, Handling and Disposal of Chemicals*, Washington, DC, (1995).

NFPA (National Fire Protection Association). 1991. *The Fire Protection Guide on Hazardous Materials,* NFPA, Quincy, MA.

NFPA (National Fire Protection Association). 2001. *The Fire Protection Guide to Hazardous Materials,* 2001 ed., NFPA, Quincy, MA.

NFPA (National Fire Protection Association). 2000. *Fire Protection for Laboratories Using Chemicals*, NFPA Standard Code No. 45, NFPA, Quincy, MA.

National Research Council. 1995. *Prudent Practices in the Laboratory: Handling and Disposal of Chemicals,* National Academy Press, Washington, DC.

Olsen, E. 1994. *Brettanomyces*: Occurrence, flavor effects, and control, Chateau St. Michelle Winery, New York State Agricultural Research Station, Cornell University, Geneva, NY.

Ough, C. S., and Amerine, M. A. 1988. *Methods for Analysis of Musts and Wines,* 2nd ed., John Wiley & Sons, New York.

Oxford University Department of Chemistry. 2004. *The Physical and Theoretical Chemistry Laboratory,* Oxford University Press, Oxford, UK; Available from http://www.physchem.ox.ac.uk/MSDS/ (accessed November 21, 2004).

Patnaik, P. A. and Van Nostrand, R. 1999. *A Comprehensive Guide to the Hazardous Properties of Chemical Substances,* 2nd ed., Van Nostrand, New York.

Petrik Laboratories Inc. 2003. Glossary of organic chemistry, Petrik Laboratories Inc., Woodland, CA Available from www.petrik.com (accessed December 2, 2003).

Research and Development Glass Products and Equipment, Inc. 2004. RD80 Self evacuating volatile acid still and $RDSO_2$ apparatus, Product Information, Research and Development Glass Products and Equipment, Inc., Berkeley, CA.

Ribéreau-Gayon, P., Dubourdieu, D., Donéche, B., and Lonvaud, A., 2000a, *Handbook of Enology Volume 1: The Microbiology of Wine and Vinifications* (J. Branco Jr., transl.), John Wiley & Sons, Chichester, UK.

Ribéreau-Gayon, P., Glories, Y., Maujean, A., and Dubourdieu, D., 2000b, *Handbook of Enology Volume 2: The Chemistry of Wine Stability and Treatment* (Aquitrad Traduction, Bordeaux Translation), John Wiley & Sons, Chichester, UK.

Sigma-Aldrich. 2004. Various MSDS and chemical information, Sigma-Aldrich, St. Louis, MO. Available from www.sigmaaldrich.com (accessed January 12, 2004).

StatSoft. 2004. *Electronic Statistical Textbook, Basic Statistics and Glossary,* StatSoft, Tulsa, OK. Available from www.statsoftinc.com/textbook/stathome.html (accessed March 4, 2004).

Stricoff, R. S. and Walters, D. B. 1995. *Handbook of Laboratory Health and Safety,* 2nd ed., John Wiley & Sons, New York.

United States Commerce Department, National Institute of Standards & Technology. 2002. Chemistry WebBook database, Washington, DC.; Available from www.nist.gov/srd/chemistry.htm (accessed November 21, 2004).

United States Department of Health and Human Services. 1991. *National Toxicology Program,* US Government Printing Office, Washington, DC

United States Department of Health and Human Services Center for Disease Control, National Institute for Occupational Safety and Health. 1976. *Specification for Industrial Hygiene Laboratory Quality Program Requirements,* NIOSH, Cincinnati, OH.

United States Department of Health and Human Services Center for Disease Control. 2004. *NIOSH Manual of Analytical Methods (NMAM)*, 4th ed., 3rd Supplement 2003-154, National Institute for Occupational Safety and Health, Washington, DC. Available from www.cdc.gov/niosh/nmam/ (accessed August 22, 2004).

United States Department of Health and Human Services, Food and Drug Administration, Center for Drug Evaluation and Research, and Center for Biologics Evaluation and Research. 2000. *Analytical Procedures and Methods Validation, Guidance for Industry,* US Government Printing Office, Rockville, MD.

United States Department of Labor, Occupational Safety & Health Administration, Occupational Exposure to Hazardous Chemicals in Laboratories. 2000. OSHA Regulations (Standards-29CFR 1910.1450), Washington, DC. Available from www.osha-slc.gov/OshStd_data/1910_1450.html (accessed July 31, 2004).

United States Department of Labor, Occupational Safety & Health Administration. 2001. National Research Council Recommendations Concerning Chemical Hygiene in Laboratories (Non-Mandatory)-1910.1450 APP A, OSHA Regulations (Standards-29 CFR), Washington, D.C. Available from www.osha-slc.gov/OshStd_data/1910_1450_APP_A.html, (accessed February 15, 2003).

United States Department of Labor, Occupational Safety & Health Administration. 2004. Re: fines and penalties, (e-mail to author).

United States Department of Transportation. 2004. Re: transportation information, (e-mail to author).

United States Department of Transportation, Office of Hazardous Materials Safety. 2004. Available at http://hazmat.dot.gov/hazhome.htm (accessed November 21, 2004).

United States Department of The Treasury Alcohol and Tobacco Tax and Trade Bureau. 1998. Export information. Available from www.ttb.gov/alcohol/info/interrel.htm (accessed August 14, 2004).

United States Department of The Treasury Alcohol and Tobacco Tax and Trade Bureau. 1999. Wine label regulations, *What You Should Know About Grape Wine Labels,* TTB, Washington, DC.

United States Department of The Treasury Alcohol and Tobacco Tax and Trade Bureau. 2003. Official laboratory methods. Available at www.ttb.gov/lab (accessed April 21, 2004).

United States Department of The Treasury Alcohol and Tobacco Tax and Trade Bureau. 2004. Title 27 CFR Part 24. Available from www.access.gpo.gov/nara/cfr/index.html (accessed September 23, 2004).

United States Environmental Protection Agency. 2002. Section 261.5 Special requirements for hazardous waste generated by conditionally exempt small quantity generators, 40 CFR Ch.I (7-1-02 Edition), EPA, Washington, DC., pp. 51–53 (online).

United States Environmental Protection Agency. 2003. Substance Registry System. Available from www.epa.gov/srs (accessed April 11, 2004).

United States Environmental Protection Agency. 2004. Listed wastes and hazardous waste characteristics, EPACALLCENTER (e-mail to author).

University of Sheffield. 2003. WebElements, periodic table. Available from www.webelements.com/webelements/ properties/text/definitions/group (accessed November 21, 2004).

Vinquiry Inc. 2000. Various MSDS, product information, references, Vinquiry Inc., Santa Rosa, CA. Available at www.vinquiry.com.

Waterhouse, A.L. 1994. Winery analysis survey 1994, *American Vineyard Viticulture & Enology Lab* (online). Available at http://wineserver.ucdavis.edu/av/AV9408.html (accessed October 9, 2004).

Waterhouse, A.L. 1998. *Topics in Wine Analysis*: *Color & Tannin Measurement*, University of California–Davis, Davis, CA, pp. 1–8.

Weeks, S. 2000. Method validation: Making sure it works in your laboratory, *Practical Winery and Management* September/October, vol. xxii, no. 3, 76–77.

Weiss, K.C., Lange, L.W., and Bisson, L.F. 2001. Small-scale fining trials: Effect of method of addition on efficiency of bentonite fining, *American Journal of Enology and Viticulture* 52(3): 275–278.

Wernimont, G. T. 1985. *Use of Statistic to Develop and Evaluate Analytical Methods* (W. Spendley ed.), AOAC, Arlington, VA, pp. 162–180.

Zoecklein, B.W., Fugelsang, K.C., Gump, B.H., and Nury, F.S. 1999. *Wine Analysis and Production*, Aspen Publishers/Kluwer Academic/Plenum, New York.

Index

CPSIA information can be obtained
at www.ICGtesting.com
Printed in the USA
LVHW080030261122
733866LV00003B/50